Data Mining for Business Applications

Data Mining for Business Applications

Edited by

Longbing Cao
Philip S. Yu
Chengqi Zhang
Huaifeng Zhang

Editors
Longbing Cao
School of Software
Faculty of Engineering and
　Information Technology
University of Technology, Sydney
PO Box 123
Broadway NSW 2007, Australia
lbcao@it.uts.edu.au

Chengqi Zhang
Centre for Quantum Computation and
　Intelligent Systems
Faculty of Engineering and
　Information Technology
University of Technology, Sydney
PO Box 123
Broadway NSW 2007, Australia
chengqi@it.uts.edu.au

Philip S.Yu
Department of Computer Science
University of Illinois at Chicago
851 S. Morgan St.
Chicago, IL 60607
psyu@cs.uic.edu

Huaifeng Zhang
School of Software
Faculty of Engineering and
　Information Technology
University of Technology, Sydney
PO Box 123
Broadway NSW 2007, Australia
hfzhang@it.uts.edu.au

ISBN: 978-0-387-79419-8　　　　e-ISBN: 978-0-387-79420-4
DOI: 10.1007/978-0-387-79420-4

Library of Congress Control Number: 2008933446

© 2009 Springer Science+Business Media, LLC
All rights reserved. This work may not be translated or copied in whole or in part without the written permission of the publisher (Springer Science+Business Media, LLC, 233 Spring Street, New York, NY 10013, USA), except for brief excerpts in connection with reviews or scholarly analysis. Use in connection with any form of information storage and retrieval, electronic adaptation, computer software, or by similar or dissimilar methodology now known or hereafter developed is forbidden.
The use in this publication of trade names, trademarks, service marks, and similar terms, even if they are not identified as such, is not to be taken as an expression of opinion as to whether or not they are subject to proprietary rights.

Printed on acid-free paper

springer.com

Preface

This edited book, *Data Mining for Business Applications*, together with an upcoming monograph also by Springer, *Domain Driven Data Mining*, aims to present a full picture of the state-of-the-art research and development of *actionable knowledge discovery* (AKD) in real-world businesses and applications.

The book is triggered by ubiquitous applications of data mining and knowledge discovery (KDD for short), and the real-world challenges and complexities to the current KDD methodologies and techniques. As we have seen, and as is often addressed by panelists of SIGKDD and ICDM conferences, even though thousands of algorithms and methods have been published, very few of them have been validated in business use.

A major reason for the above situation, we believe, is the gap between academia and businesses, and the gap between academic research and real business needs. Ubiquitous challenges and complexities from the real-world complex problems can be categorized by the involvement of six types of intelligence ($6I_s$), namely *human roles and intelligence, domain knowledge and intelligence, network and web intelligence, organizational and social intelligence, in-depth data intelligence*, and most importantly, the *metasynthesis of the above intelligences*.

It is certainly not our ambition to cover everything of the $6I_s$ in this book. Rather, this edited book features the latest methodological, technical and practical progress on promoting the successful use of data mining in a collection of business domains. The book consists of two parts, one on AKD methodologies and the other on novel AKD domains in business use.

In Part I, the book reports attempts and efforts in developing domain-driven workable AKD methodologies. This includes domain-driven data mining, post-processing rules for actions, domain-driven customer analytics, roles of human intelligence in AKD, maximal pattern-based cluster, and ontology mining.

Part II selects a large number of novel KDD domains and the corresponding techniques. This involves great efforts to develop effective techniques and tools for emergent areas and domains, including mining social security data, community security data, gene sequences, mental health information, traditional Chinese medicine data, cancer related data, blog data, sentiment information, web data, procedures, moving object trajectories, land use mapping, higher education, flight scheduling, and algorithmic asset management.

The intended audience of this book will mainly consist of researchers, research students and practitioners in data mining and knowledge discovery. The book is also of interest to researchers and industrial practitioners in areas such as knowledge engineering, human-computer interaction, artificial intelligence, intelligent information processing, decision support systems, knowledge management, and AKD project management.

Readers who are interested in actionable knowledge discovery in the real world, please also refer to our monograph: *Domain Driven Data Mining*, which has been scheduled to be published by Springer in 2009. The monograph will present our research outcomes on theoretical and technical issues in real-world actionable knowledge discovery, as well as working examples in financial data mining and social security mining.

We would like to convey our appreciation to all contributors including the accepted chapters' authors, and many other participants who submitted their chapters that cannot be included in the book due to space limits. Our special thanks to Ms. Melissa Fearon and Ms. Valerie Schofield from Springer US for their kind support and great efforts in bringing the book to fruition. In addition, we also appreciate all reviewers, and Ms. Shanshan Wu's assistance in formatting the book.

Longbing Cao, Philip S. Yu, Chengqi Zhang, Huaifeng Zhang
July 2008

Contents

Part I Domain Driven KDD Methodology

1 Introduction to Domain Driven Data Mining 3
Longbing Cao
1.1 Why Domain Driven Data Mining 3
1.2 What Is Domain Driven Data Mining 5
    1.2.1 Basic Ideas 5
    1.2.2 D^3M for Actionable Knowledge Discovery 6
1.3 Open Issues and Prospects 9
1.4 Conclusions 9
References 10

2 Post-processing Data Mining Models for Actionability 11
Qiang Yang
2.1 Introduction 11
2.2 Plan Mining for Class Transformation 12
    2.2.1 Overview of Plan Mining 12
    2.2.2 Problem Formulation 14
    2.2.3 From Association Rules to State Spaces 14
    2.2.4 Algorithm for Plan Mining 17
    2.2.5 Summary 19
2.3 Extracting Actions from Decision Trees 20
    2.3.1 Overview 20
    2.3.2 Generating Actions from Decision Trees 22
    2.3.3 The Limited Resources Case 23
2.4 Learning Relational Action Models from Frequent Action Sequences 25
    2.4.1 Overview 25
    2.4.2 ARMS Algorithm: From Association Rules to Actions .. 26
    2.4.3 Summary of ARMS 28
2.5 Conclusions and Future Work 29

References ... 29

3 On Mining Maximal Pattern-Based Clusters ... 31
Jian Pei, Xiaoling Zhang, Moonjung Cho, Haixun Wang, and Philip S. Yu

- 3.1 Introduction ... 32
- 3.2 Problem Definition and Related Work ... 34
  - 3.2.1 Pattern-Based Clustering ... 34
  - 3.2.2 Maximal Pattern-Based Clustering ... 35
  - 3.2.3 Related Work ... 35
- 3.3 Algorithms *MaPle* and *MaPle+* ... 36
  - 3.3.1 An Overview of *MaPle* ... 37
  - 3.3.2 Computing and Pruning MDS's ... 38
  - 3.3.3 Progressively Refining, Depth-first Search of Maximal pClusters ... 40
  - 3.3.4 *MaPle+*: Further Improvements ... 44
- 3.4 Empirical Evaluation ... 46
  - 3.4.1 The Data Sets ... 46
  - 3.4.2 Results on Yeast Data Set ... 47
  - 3.4.3 Results on Synthetic Data Sets ... 48
- 3.5 Conclusions ... 50
- References ... 50

4 Role of Human Intelligence in Domain Driven Data Mining ... 53
Sumana Sharma and Kweku-Muata Osei-Bryson

- 4.1 Introduction ... 53
- 4.2 DDDM Tasks Requiring Human Intelligence ... 54
  - 4.2.1 Formulating Business Objectives ... 54
  - 4.2.2 Setting up Business Success Criteria ... 55
  - 4.2.3 Translating Business Objective to Data Mining Objectives ... 56
  - 4.2.4 Setting up of Data Mining Success Criteria ... 56
  - 4.2.5 Assessing Similarity Between Business Objectives of New and Past Projects ... 57
  - 4.2.6 Formulating Business, Legal and Financial Requirements ... 57
  - 4.2.7 Narrowing down Data and Creating Derived Attributes ... 58
  - 4.2.8 Estimating Cost of Data Collection, Implementation and Operating Costs ... 58
  - 4.2.9 Selection of Modeling Techniques ... 59
  - 4.2.10 Setting up Model Parameters ... 59
  - 4.2.11 Assessing Modeling Results ... 59
  - 4.2.12 Developing a Project Plan ... 60
- 4.3 Directions for Future Research ... 60
- 4.4 Summary ... 61
- References ... 61

5 Ontology Mining for Personalized Search ... 63
Yuefeng Li and Xiaohui Tao
- 5.1 Introduction ... 63
- 5.2 Related Work ... 64
- 5.3 Architecture ... 65
- 5.4 Background Definitions ... 66
  - 5.4.1 World Knowledge Ontology ... 66
  - 5.4.2 Local Instance Repository ... 67
- 5.5 Specifying Knowledge in an Ontology ... 68
- 5.6 Discovery of Useful Knowledge in LIRs ... 70
- 5.7 Experiments ... 71
  - 5.7.1 Experiment Design ... 71
  - 5.7.2 Other Experiment Settings ... 74
- 5.8 Results and Discussions ... 75
- 5.9 Conclusions ... 77
- References ... 77

Part II Novel KDD Domains & Techniques

6 Data Mining Applications in Social Security ... 81
Yanchang Zhao, Huaifeng Zhang, Longbing Cao, Hans Bohlscheid, Yuming Ou, and Chengqi Zhang
- 6.1 Introduction and Background ... 81
- 6.2 Case Study I: Discovering Debtor Demographic Patterns with Decision Tree and Association Rules ... 83
  - 6.2.1 Business Problem and Data ... 83
  - 6.2.2 Discovering Demographic Patterns of Debtors ... 83
- 6.3 Case Study II: Sequential Pattern Mining to Find Activity Sequences of Debt Occurrence ... 85
  - 6.3.1 Impact-Targeted Activity Sequences ... 86
  - 6.3.2 Experimental Results ... 87
- 6.4 Case Study III: Combining Association Rules from Heterogeneous Data Sources to Discover Repayment Patterns ... 89
  - 6.4.1 Business Problem and Data ... 89
  - 6.4.2 Mining Combined Association Rules ... 89
  - 6.4.3 Experimental Results ... 90
- 6.5 Case Study IV: Using Clustering and Analysis of Variance to Verify the Effectiveness of a New Policy ... 92
  - 6.5.1 Clustering Declarations with Contour and Clustering ... 92
  - 6.5.2 Analysis of Variance ... 94
- 6.6 Conclusions and Discussion ... 94
- References ... 95

7 Security Data Mining: A Survey Introducing Tamper-Resistance ... 97
Clifton Phua and Mafruz Ashrafi
- 7.1 Introduction ... 97
- 7.2 Security Data Mining ... 98
  - 7.2.1 Definitions ... 98
  - 7.2.2 Specific Issues ... 99
  - 7.2.3 General Issues ... 101
- 7.3 Tamper-Resistance ... 102
  - 7.3.1 Reliable Data ... 102
  - 7.3.2 Anomaly Detection Algorithms ... 104
  - 7.3.3 Privacy and Confidentiality Preserving Results ... 105
- 7.4 Conclusion ... 108
- References ... 108

8 A Domain Driven Mining Algorithm on Gene Sequence Clustering ... 111
Yun Xiong, Ming Chen, and Yangyong Zhu
- 8.1 Introduction ... 111
- 8.2 Related Work ... 112
- 8.3 The Similarity Based on Biological Domain Knowledge ... 114
- 8.4 Problem Statement ... 114
- 8.5 A Domain-Driven Gene Sequence Clustering Algorithm ... 117
- 8.6 Experiments and Performance Study ... 121
- 8.7 Conclusion and Future Work ... 124
- References ... 125

9 Domain Driven Tree Mining of Semi-structured Mental Health Information ... 127
Maja Hadzic, Fedja Hadzic, and Tharam S. Dillon
- 9.1 Introduction ... 127
- 9.2 Information Use and Management within Mental Health Domain ... 128
- 9.3 Tree Mining - General Considerations ... 130
- 9.4 Basic Tree Mining Concepts ... 131
- 9.5 Tree Mining of Medical Data ... 135
- 9.6 Illustration of the Approach ... 139
- 9.7 Conclusion and Future Work ... 139
- References ... 140

10 Text Mining for Real-time Ontology Evolution ... 143
Jackei H.K. Wong, Tharam S. Dillon, Allan K.Y. Wong, and Wilfred W.K. Lin
- 10.1 Introduction ... 144
- 10.2 Related Text Mining Work ... 145
- 10.3 Terminology and Multi-representations ... 145
- 10.4 Master Aliases Table and OCOE Data Structures ... 149
- 10.5 Experimental Results ... 152
  - 10.5.1 CAV Construction and Information Ranking ... 153

		10.5.2 Real-Time CAV Expansion Supported by Text Mining . . 154

- 10.6 Conclusion... 155
- 10.7 Acknowledgement 156
- References 156

11 Microarray Data Mining: Selecting Trustworthy Genes with Gene Feature Ranking 159
Franco A. Ubaudi, Paul J. Kennedy, Daniel R. Catchpoole, Dachuan Guo, and Simeon J. Simoff

- 11.1 Introduction 159
- 11.2 Gene Feature Ranking 161
  - 11.2.1 Use of Attributes and Data Samples in Gene Feature Ranking 162
  - 11.2.2 Gene Feature Ranking: Feature Selection Phase 1 163
  - 11.2.3 Gene Feature Ranking: Feature Selection Phase 2 163
- 11.3 Application of Gene Feature Ranking to Acute Lymphoblastic Leukemia data 164
- 11.4 Conclusion.. 166
- References 167

12 Blog Data Mining for Cyber Security Threats 169
Flora S. Tsai and Kap Luk Chan

- 12.1 Introduction 169
- 12.2 Review of Related Work 170
  - 12.2.1 Intelligence Analysis.. 171
  - 12.2.2 Information Extraction from Blogs 171
- 12.3 Probabilistic Techniques for Blog Data Mining 172
  - 12.3.1 Attributes of Blog Documents 172
  - 12.3.2 Latent Dirichlet Allocation....................................... 173
  - 12.3.3 Isometric Feature Mapping (Isomap) 174
- 12.4 Experiments and Results 175
  - 12.4.1 Data Corpus 175
  - 12.4.2 Results for Blog Topic Analysis 176
  - 12.4.3 Blog Content Visualization 178
  - 12.4.4 Blog Time Visualization 179
- 12.5 Conclusions... 180
- References 181

13 Blog Data Mining: The Predictive Power of Sentiments 183
Yang Liu, Xiaohui Yu, Xiangji Huang, and Aijun An

- 13.1 Introduction 183
- 13.2 Related Work.. 185
- 13.3 Characteristics of Online Discussions 186
  - 13.3.1 Blog Mentions 186
  - 13.3.2 Box Office Data and User Rating 187
  - 13.3.3 Discussion 187

- 13.4 S-PLSA: A Probabilistic Approach to Sentiment Mining 188
  - 13.4.1 Feature Selection 188
  - 13.4.2 Sentiment PLSA 188
- 13.5 ARSA: A Sentiment-Aware Model 189
  - 13.5.1 The Autoregressive Model 190
  - 13.5.2 Incorporating Sentiments 191
- 13.6 Experiments 192
  - 13.6.1 Experiment Settings 192
  - 13.6.2 Parameter Selection 193
- 13.7 Conclusions and Future Work 194
- References 194

14 Web Mining: Extracting Knowledge from the World Wide Web 197
Zhongzhi Shi, Huifang Ma, and Qing He
- 14.1 Overview of Web Mining Techniques 197
- 14.2 Web Content Mining 199
  - 14.2.1 Classification: Multi-hierarchy Text Classification 199
  - 14.2.2 Clustering Analysis: Clustering Algorithm Based on Swarm Intelligence and k-Means 200
  - 14.2.3 Semantic Text Analysis: Conceptual Semantic Space 202
- 14.3 Web Structure Mining: PageRank vs. HITS 203
- 14.4 Web Event Mining 204
  - 14.4.1 Preprocessing for Web Event Mining 205
  - 14.4.2 Multi-document Summarization: A Way to Demonstrate Event's Cause and Effect 206
- 14.5 Conclusions and Future Works 206
- References 207

15 DAG Mining for Code Compaction 209
T. Werth, M. Wörlein, A. Dreweke, I. Fischer, and M. Philippsen
- 15.1 Introduction 209
- 15.2 Related Work 211
- 15.3 Graph and DAG Mining Basics 211
  - 15.3.1 Graph–based versus Embedding–based Mining 212
  - 15.3.2 Embedded versus Induced Fragments 213
  - 15.3.3 DAG Mining Is *NP*–complete 213
- 15.4 Algorithmic Details of DAGMA 214
  - 15.4.1 A Canonical Form for DAG enumeration 214
  - 15.4.2 Basic Structure of the DAG Mining Algorithm 215
  - 15.4.3 Expansion Rules 216
  - 15.4.4 Application to Procedural Abstraction 219
- 15.5 Evaluation 220
- 15.6 Conclusion and Future Work 222
- References 223

16	**A Framework for Context-Aware Trajectory Data Mining** 225	
	Vania Bogorny and Monica Wachowicz	
	16.1 Introduction 225	
	16.2 Basic Concepts 227	
	16.3 A Domain-driven Framework for Trajectory Data Mining 229	
	16.4 Case Study 232	
	16.4.1 The Selected Mobile Movement-aware Outdoor Game .. 233	
	16.4.2 Transportation Application 234	
	16.5 Conclusions and Future Trends 238	
	References 239	

17 Census Data Mining for Land Use Classification 241
E. Roma Neto and D. S. Hamburger
17.1 Content Structure 241
17.2 Key Research Issues 242
17.3 Land Use and Remote Sensing 242
17.4 Census Data and Land Use Distribution 243
17.5 Census Data Warehouse and Spatial Data Mining 243
    17.5.1 Concerning about Data Quality 243
    17.5.2 Concerning about Domain Driven 244
    17.5.3 Applying Machine Learning Tools 246
17.6 Data Integration 247
    17.6.1 Area of Study and Data 247
    17.6.2 Supported Digital Image Processing 248
    17.6.3 Putting All Steps Together 248
17.7 Results and Analysis 249
References 251

18 Visual Data Mining for Developing Competitive Strategies in Higher Education 253
Gürdal Ertek
18.1 Introduction 253
18.2 Square Tiles Visualization 255
18.3 Related Work 256
18.4 Mathematical Model 257
18.5 Framework and Case Study 260
    18.5.1 General Insights and Observations 261
    18.5.2 Benchmarking 262
    18.5.3 High School Relationship Management (HSRM) 263
18.6 Future Work 264
18.7 Conclusions 264
References 265

19 Data Mining For Robust Flight Scheduling ... 267
Ira Assent, Ralph Krieger, Petra Welter, Jörg Herbers, and Thomas Seidl

- 19.1 Introduction ... 267
- 19.2 Flight Scheduling in the Presence of Delays ... 268
- 19.3 Related Work ... 270
- 19.4 Classification of Flights ... 272
  - 19.4.1 Subspaces for Locally Varying Relevance ... 272
  - 19.4.2 Integrating Subspace Information for Robust Flight Classification ... 272
- 19.5 Algorithmic Concept ... 274
  - 19.5.1 Monotonicity Properties of Relevant Attribute Subspaces ... 274
  - 19.5.2 Top-down Class Entropy Algorithm: Lossless Pruning Theorem ... 275
  - 19.5.3 Algorithm: Subspaces, Clusters, Subspace Classification ... 276
- 19.6 Evaluation of Flight Delay Classification in Practice ... 278
- 19.7 Conclusion ... 280
- References ... 280

20 Data Mining for Algorithmic Asset Management ... 283
Giovanni Montana and Francesco Parrella

- 20.1 Introduction ... 283
- 20.2 Backbone of the Asset Management System ... 285
- 20.3 Expert-based Incremental Learning ... 286
- 20.4 An Application to the iShare Index Fund ... 290
- References ... 294

Reviewer List ... 297

Index ... 299

List of Contributors

Longbing Cao
School of Software, University of Technology Sydney, Australia, e-mail: lbcao@it.uts.edu.au

Qiang Yang
Department of Computer Science and Engineering, Hong Kong University of Science and Technology, e-mail: qyang@cse.ust.hk

Jian Pei
Simon Fraser University, e-mail: jpei@cs.sfu.ca

Xiaoling Zhang
Boston University, e-mail: zhangxl@bu.edu

Moonjung Cho
Prism Health Networks, e-mail: moonjungcho@hotmail.com

Haixun Wang
IBM T.J.Watson Research Center e-mail: haixun@us.ibm.com

Philip S.Yu
University of Illinois at Chicago, e-mail: psyu@cs.uic.edu

Sumana Sharma
Virginia Commonwealth University, e-mail: sharmas5@vcu.edu

Kweku-Muata Osei-Bryson
Virginia Commonwealth University, e-mail: kmuata@isy.vcu.edu

Yuefeng Li
Information Technology, Queensland University of Technology, Australia, e-mail: y2.li@qut.edu.au

Xiaohui Tao
Information Technology, Queensland University of Technology, Australia, e-mail: x.tao@qut.edu.au

Yanchang Zhao
Faculty of Engineering and Information Technology, University of Technology, Sydney, Australia, e-mail: yczhao@it.uts.edu.au

Huaifeng Zhang
Faculty of Engineering and Information Technology, University of Technology, Sydney, Australia, e-mail: hfzhang@it.uts.edu.au

Yuming Ou
Faculty of Engineering and Information Technology, University of Technology, Sydney, Australia, e-mail: yuming@it.uts.edu.au

Chengqi Zhang
Faculty of Engineering and Information Technology, University of Technology, Sydney, Australia, e-mail: chengqi@it.uts.edu.au

Hans Bohlscheid
Data Mining Section, Business Integrity Programs Branch, Centrelink, Australia, e-mail: hans.bohlscheid@centrelink.gov.au

Clifton Phua
A*STAR, Institute of Infocomm Research, Room 04-21 (+6568748406), 21, Heng Mui Keng Terrace, Singapore 119613, e-mail: cwphua@i2r.a-star.edu.sg

Mafruz Ashrafi
A*STAR, Institute of Infocomm Research, Room 04-21 (+6568748406), 21, Heng Mui Keng Terrace, Singapore 119613, e-mail: mashrafi@i2r.a-star.edu.sg

Yun Xiong
Department of Computing and Information Technology, Fudan University, Shanghai 200433, China, e-mail: yunx@fudan.edu.cn

Ming Chen
Department of Computing and Information Technology, Fudan University, Shanghai 200433, China, e-mail: chenming@fudan.edu.cn

Yangyong Zhu
Department of Computing and Information Technology, Fudan University, Shanghai 200433, China, e-mail: yyzhu@fudan.edu.cn

Maja Hadzic
Digital Ecosystems and Business Intelligence Institute (DEBII), Curtin University of Technology, Australia, e-mail: m.hadzic@curtin.edu.au

Fedja Hadzic
Digital Ecosystems and Business Intelligence Institute (DEBII), Curtin University of Technology, Australia, e-mail: f.hadzic@curtin.edu.au

List of Contributors

Tharam S. Dillon
Digital Ecosystems and Business Intelligence Institute (DEBII), Curtin University of Technology, Australia, e-mail: t.dillon@curtin.edu.au

Jackei H.K. Wong
Department of Computing, Hong Kong Polytechnic University, Hong Kong SAR, e-mail: jwong@purapharm.com

Allan K.Y. Wong
Department of Computing, Hong Kong Polytechnic University, Hong Kong SAR, e-mail: csalwong@comp.polyu.edu.hk

Wilfred W.K. Lin
Department of Computing, Hong Kong Polytechnic University, Hong Kong SAR, e-mail: cswklin@comp.polyu.edu.hk

Franco A. Ubaudi
Faculty of IT, University of Technology, Sydney, e-mail: faubaudi@it.uts.edu.au

Paul J. Kennedy
Faculty of IT, University of Technology, Sydney, e-mail: paulk@it.uts.edu.au

Daniel R. Catchpoole
Tumour Bank, The Childrens Hospital at Westmead, e-mail: DanielC@chw.edu.au

Dachuan Guo
Tumour Bank, The Childrens Hospital at Westmead, e-mail: dachuang@chw.edu.au

Simeon J. Simoff
University of Western Sydney, e-mail: S.Simoff@uws.edu.au

Flora S. Tsai
Nanyang Technological University, Singapore, e-mail: fst1@columbia.edu

Kap Luk Chan
Nanyang Technological University, Singapore e-mail: eklchan@ntu.edu.sg

Yang Liu
Department of Computer Science and Engineering, York University, Toronto, ON, Canada M3J 1P3, e-mail: yliu@cse.yorku.ca

Xiaohui Yu
School of Information Technology, York University, Toronto, ON, Canada M3J 1P3, e-mail: xhyu@yorku.ca

Xiangji Huang
School of Information Technology, York University, Toronto, ON, Canada M3J 1P3, e-mail: jhuang@yorku.ca

Aijun An
Department of Computer Science and Engineering, York University, Toronto, ON, Canada M3J 1P3, e-mail: `ann@cse.yorku.ca`

Zhongzhi Shi
Key Laboratory of Intelligent Information Processing, Institute of Computing Technology, Chinese Academy of Sciences, No. 6 Kexueyuan Nanlu, Beijing 100080, People's Republic of China, e-mail: `shizz@ics.ict.ac.cn`

Huifang Ma
Key Laboratory of Intelligent Information Processing, Institute of Computing Technology, Chinese Academy of Sciences, No. 6 Kexueyuan Nanlu, Beijing 100080, People's Republic of China, e-mail: `mahf@ics.ict.ac.cn`

Qing He
Key Laboratory of Intelligent Information Processing, Institute of Computing Technology, Chinese Academy of Sciences, No. 6 Kexueyuan Nanlu, Beijing 100080, People's Republic of China, e-mail: `heq@ics.ict.ac.cn`

T. Werth
Programming Systems Group, Computer Science Department, University of Erlangen–Nuremberg, Germany, phone: +49 9131 85-28865, e-mail: `werth@cs.fau.de`

M. Wörlein
Programming Systems Group, Computer Science Department, University of Erlangen–Nuremberg, Germany, phone: +49 9131 85-28865, e-mail: `woerlein@cs.fau.de`

A. Dreweke
Programming Systems Group, Computer Science Department, University of Erlangen–Nuremberg, Germany, phone: +49 9131 85-28865, e-mail: `dreweke@cs.fau.de`

M. Philippsen
Programming Systems Group, Computer Science Department, University of Erlangen–Nuremberg, Germany, phone: +49 9131 85-28865, e-mail: `philippsen@cs.fau.de`

I. Fischer
Nycomed Chair for Bioinformatics and Information Mining, University of Konstanz, Germany, phone: +49 7531 88-5016, e-mail: `Ingrid.Fischer@inf.uni-konstanz.de`

Vania Bogorny
Instituto de Informatica, Universidade Federal do Rio Grande do Sul (UFRGS), Av. Bento Gonalves, 9500 - Campus do Vale - Bloco IV, Bairro Agronomia - Porto Alegre - RS -Brasil, CEP 91501-970 Caixa Postal: 15064, e-mail: `vbogorny@inf.ufrgs.br`

List of Contributors

Monica Wachowicz
ETSI Topografia, Geodesia y Cartografa, Universidad Politecnica de Madrid, KM 7,5 de la Autovia de Valencia, E-28031 Madrid - Spain, e-mail: m.wachowicz@topografia.upm.es

E.Roma Neto
Av. Eng. Euséio Stevaux, 823 - 04696-000, São Paulo, SP, Brazil, e-mail: elias.rneto@sp.senac.br

D. S. Hamburger
Av. Eng. Euséio Stevaux, 823 - 04696-000, São Paulo, SP, Brazil, e-mail: diana.hamburger@gmail.com

Gürdal Ertek
Sabancı University, Faculty of Engineering and Natural Sciences, Orhanlı, Tuzla, 34956, Istanbul, Turkey, e-mail: ertekg@sabanciuniv.edu

Ira Assent
Data Management and Exploration Group, RWTH Aachen University, Germany, phone: +492418021910, e-mail: assent@cs.rwth-aachen.de

Ralph Krieger
Data Management and Exploration Group, RWTH Aachen University, Germany, phone: +492418021910, e-mail: krieger@cs.rwth-aachen.de

Thomas Seidl
Data Management and Exploration Group, RWTH Aachen University, Germany, phone: +492418021910, e-mail: seidl@cs.rwth-aachen.de

Petra Welter
Dept. of Medical Informatics, RWTH Aachen University, Germany, e-mail: pwelter@mi.rwth-aachen.de

Jörg Herbers
INFORM GmbH, Pascalstraße 23, Aachen, Germany, e-mail: joerg.herbers@inform-ac.com

Giovanni Montana
Imperial College London, Department of Mathematics, 180 Queen's Gate, London SW7 2AZ, UK, e-mail: g.montana@imperial.ac.uk

Francesco Parrella
Imperial College London, Department of Mathematics, 180 Queen's Gate, London SW7 2AZ, UK, e-mail: f.parrella@imperial.ac.uk

Part I
Domain Driven KDD Methodology

Chapter 1
Introduction to Domain Driven Data Mining

Longbing Cao

Abstract The mainstream data mining faces critical challenges and lacks of soft power in solving real-world complex problems when deployed. Following the paradigm shift from 'data mining' to 'knowledge discovery', we believe much more thorough efforts are essential for promoting the wide acceptance and employment of knowledge discovery in real-world smart decision making. To this end, we expect a new paradigm shift from 'data-centered knowledge discovery' to 'domain-driven actionable knowledge discovery'. In the domain-driven actionable knowledge discovery, ubiquitous intelligence must be involved and meta-synthesized into the mining process, and an actionable knowledge discovery-based problem-solving system is formed as the space for data mining. This is the motivation and aim of developing *Domain Driven Data Mining* (D^3M for short). This chapter briefs the main reasons, ideas and open issues in D^3M.

1.1 Why Domain Driven Data Mining

Data mining and knowledge discovery (data mining or KDD for short) [9] has emerged to be one of the most vivacious areas in information technology in the last decade. It has boosted a major academic and industrial campaign crossing many traditional areas such as machine learning, database, statistics, as well as emergent disciplines, for example, bioinformatics. As a result, KDD has published thousands of algorithms and methods, as widely seen in regular conferences and workshops crossing international, regional and national levels.

Compared with the booming fact in academia, data mining applications in the real world has not been as active, vivacious and charming as that of academic research. This can be easily found from the extremely imbalanced numbers of pub-

---

Longbing Cao
School of Software, University of Technology Sydney, Australia, e-mail: lbcao@it.uts.edu.au

lished algorithms versus those really workable in the business environment. That is to say, there is a big gap between academic objectives and business goals, and between academic outputs and business expectations. However, this runs in the opposite direction of KDD's original intention and its nature. It is also against the value of KDD as a discipline, which generates the power of enabling smart businesses and developing business intelligence for smart decisions in production and living environment.

If we scrutinize the reasons of the existing gaps, we probably can point out many things. For instance, academic researchers do not really know the needs of business people, and are not familiar with the business environment. With many years of development of this promising scientific field, it is time and worthwhile to review the major issues blocking the step of KDD into business use widely.

While after the origin of *data mining*, researchers with strong industrial engagement realized the need from 'data mining' to 'knowledge discovery' [1, 7, 8] to deliver useful knowledge for the business decision-making. Many researchers, in particular early career researchers in KDD, are still only or mainly focusing on 'data mining', namely mining for patterns in data. The main reason for such a dominant situation, either explicitly or implicitly, is on its originally narrow focus and overemphasized by innovative algorithm-driven research (unfortunately we are not at the stage of holding as many effective algorithms as we need in the real world applications).

Knowledge discovery is further expected to migrate into *actionable knowledge discovery* (AKD). AKD targets knowledge that can be delivered in the form of business-friendly and decision-making actions, and can be taken over by business people seamlessly. However, AKD is still a big challenge to the current KDD research and development. Reasons surrounding the challenge of AKD include many critical aspects on both macro-level and micro-level.

On the macro-level, issues are related to methodological and fundamental aspects, for instance,

- An intrinsic difference existing in academic thinking and business deliverable expectation; for example, researchers usually are interested in innovative pattern types, while practitioners care about getting a problem solved;
- The paradigm of KDD, whether as a hidden pattern mining process centered by data, or an AKD-based problem-solving system; the latter emphasizes not only innovation but also impact of KDD deliverables.

The micro-level issues are more related to technical and engineering aspects, for instance,

- If KDD is an AKD-based problem-solving system, we then need to care about many issues such as system dynamics, system environment, and interaction in a system;
- If AKD is the target, we then have to cater for real-world aspects such as business processes, organizational factors, and constraints.

In scrutinizing both macro-level and micro-level of issues in AKD, we propose a new KDD methodology on top of the traditional data-centered pattern mining framework, that is *Domain Driven Data Mining* (D^3M) [2,4,5]. In the next section, we introduce the main idea of D^3M.

1.2 What Is Domain Driven Data Mining

1.2.1 Basic Ideas

The motivation of D^3M is to view KDD as AKD-based problem-solving systems through developing effective methodologies, methods and tools. The aim of D^3M is to make AKD system deliver business-friendly and decision-making rules and actions that are of solid technical significance as well. To this end, D^3M caters for the effective involvement of the following ubiquitous intelligence surrounding AKD-based problem-solving.

- *Data Intelligence*, tells stories hidden in the data about a business problem.
- *Domain Intelligence*, refers to domain resources that not only wrap a problem and its target data but also assist in the understanding and problem-solving of the problem. Domain intelligence consists of qualitative and quantitative intelligence. Both types of intelligence are instantiated in terms of aspects such as domain knowledge, background information, constraints, organization factors and business process, as well as environment intelligence, business expectation and interestingness.
- *Network Intelligence*, refers to both web intelligence and broad-based network intelligence such as distributed information and resources, linkages, searching, and structured information from textual data.
- *Human Intelligence*, refers to (1) explicit or direct involvement of humans such as empirical knowledge, belief, intention and expectation, run-time supervision, evaluating, and expert group; (2) implicit or indirect involvement of human intelligence such as imaginary thinking, emotional intelligence, inspiration, brainstorm, and reasoning inputs.
- *Social Intelligence*, consists of interpersonal intelligence, emotional intelligence, social cognition, consensus construction, group decision, as well as organizational factors, business process, workflow, project management and delivery, social network intelligence, collective interaction, business rules, law, trust and so on.
- *Intelligence Metasynthesis*, the above ubiquitous intelligence has to be combined for the problem-solving. The methodology for combining such intelligence is called *metasynthesis* [10, 11], which provides a human-centered and human-machine-cooperated problem-solving process by involving, synthesizing and using ubiquitous intelligence surrounding AKD as need for problem-solving.

1.2.2 D^3M for Actionable Knowledge Discovery

Real-world data mining is a complex problem-solving system. From the view of systems and microeconomy, the endogenous character of actionable knowledge discovery (AKD) determines that it is an optimization problem with certain objectives in a particular environment. We present a formal definition of AKD in this section. We first define several notions as follows.

Let *DB* be a database collected from business problems (Ψ), $X = \{\mathbf{x}_1, \mathbf{x}_2, \cdots, \mathbf{x}_L\}$ be the set of items in the *DB*, where $\mathbf{x}_l$ ($l = 1, \ldots, L$) be an itemset, and the number of attributes (v) in *DB* be S. Suppose $E = \{e_1, e_2, \cdots, e_K\}$ denotes the environment set, where e_k represents a particular environment setting for AKD. Further, let $M = \{m_1, m_2, \cdots, m_N\}$ be the data mining method set, where m_n ($n = 1, \ldots, N$) is a method. For the method m_n, suppose its identified pattern set $P^{m_n} = \{p_1^{m_n}, p_2^{m_n}, \cdots, p_U^{m_n}\}$ includes all patterns discovered in *DB*, where $p_u^{m_n}$ ($u = 1, \ldots, U$) denotes a pattern discovered by the method m_n.

In the real world, data mining is a problem-solving process from business problems (Ψ, with problem status τ) to problem-solving solutions (Φ):

$$\Psi \to \Phi \tag{1.1}$$

From the modeling perspective, such a problem-solving process is a state transformation process from source data $DB(\Psi \to DB)$ to resulting pattern set $P(\Phi \to P)$.

$$\Psi \to \Phi :: DB(v_1, \ldots, v_S) \to P(f_1, \ldots, f_Q) \tag{1.2}$$

where v_s ($s = 1, \ldots, S$) are attributes in the source data *DB*, while f_q ($q = 1, \ldots, Q$) are features used for mining the pattern set *P*.

Definition 1.1. (Actionable Patterns)
Let $\widetilde{P} = \{\tilde{p}_1, \tilde{p}_2, \cdots, \tilde{p}_Z\}$ be an *Actionable Pattern Set* mined by method m_n for the given problem Ψ (its data set is *DB*), in which each pattern $\tilde{p}_z$ is *actionable* for the problem-solving if it satisfies the following conditions:

1.a. $t_i(\tilde{p}_z) \geq t_{i,0}$; indicating the pattern $\tilde{p}_z$ satisfying technical interestingness t_i with threshold $t_{i,0}$;
1.b. $b_i(\tilde{p}_z) \geq b_{i,0}$; indicating the pattern $\tilde{p}_z$ satisfying business interestingness b_i with threshold $b_{i,0}$;
1.c. $R: \tau_1 \xrightarrow{A, m_n(\tilde{p}_z)} \tau_2$; the pattern can support business problem-solving (R) by taking action A, and correspondingly transform the problem status from initially nonoptimal state τ_1 to greatly improved state τ_2.

Therefore, the discovery of actionable knowledge (AKD) on data set *DB* is an iterative optimization process toward the actionable pattern set $\widetilde{P}$.

$$AKD: DB \xrightarrow{e, \tau, m_1} P_1 \xrightarrow{e, \tau, m_2} P_2 \cdots \xrightarrow{e, \tau, m_n} \widetilde{P} \tag{1.3}$$

1 Introduction to Domain Driven Data Mining

Definition 1.2. (Actionable Knowledge Discovery)
The *Actionable Knowledge Discovery* (AKD) is the procedure to find the *Actionable Pattern Set* $\tilde{P}$ through employing all valid methods M. Its mathematical description is as follows:

$$AKD^{m_i \in M} \longrightarrow O_{p \in P} Int(p), \tag{1.4}$$

where $P = P^{m_1} \cup P^{m_2}, \cdots, \cup P^{m_n}$, $Int(.)$ is the evaluation function, $O(.)$ is the optimization function to extract those $\tilde{p} \in \tilde{P}$ where $Int(\tilde{p})$ can beat a given benchmark.

For a pattern p, $Int(p)$ can be further measured in terms of *technical interestingness* ($t_i(p)$) and *business interestingness* ($b_i(p)$) [3].

$$Int(p) = I(t_i(p), b_i(p)) \tag{1.5}$$

where $I(.)$ is the function for aggregating the contributions of all particular aspects of interestingness.

Further, $Int(p)$ can be described in terms of *objective* (o) and *subjective* (s) factors from both *technical* (t) and *business* (b) perspectives.

$$Int(p) = I(t_o(), t_s(), b_o(), b_s()) \tag{1.6}$$

where $t_o()$ is objective technical interestingness, $t_s()$ is subjective technical interestingness, $b_o()$ is objective business interestingness, and $b_s()$ is subjective business interestingness.

We say p is truly *actionable* (i.e., $\tilde{p}$) both to academia and business if it satisfies the following condition:

$$Int(p) = t_o(\mathbf{x}, \tilde{p}) \wedge t_s(\mathbf{x}, \tilde{p}) \wedge b_o(\mathbf{x}, \tilde{p}) \wedge b_s(\mathbf{x}, \tilde{p}) \tag{1.7}$$

where $I \to `\wedge'$ indicates the 'aggregation' of the interestingness.

In general, $t_o()$, $t_s()$, $b_o()$ and $b_s()$ of practical applications can be regarded as independent of each other. With their normalization (expressed by ^), we can get the following:

$$\begin{aligned} Int(p) &\to \hat{I}(\hat{t}_o(), \hat{t}_s(), \hat{b}_o(), \hat{b}_s()) \\ &= \alpha \hat{t}_o() + \beta \hat{t}_s() + \gamma \hat{b}_o() + \delta \hat{b}_s() \end{aligned} \tag{1.8}$$

So, the AKD optimization problem can be expressed as follows:

$$\begin{aligned} AKD^{e,\tau,m \in M} &\longrightarrow O_{p \in P}(Int(p)) \\ &\to O(\alpha \hat{t}_o()) + O(\beta \hat{t}_s()) + \\ &\quad O(\gamma \hat{b}_o()) + O(\delta \hat{b}_s()) \end{aligned} \tag{1.9}$$

Definition 1.3. (Actionability of a Pattern)
The *actionability* of a pattern p is measured by $act(p)$:

$$act(p) = O_{p \in P}(Int(p))$$
$$\to O(\alpha \hat{t_o}(p)) + O(\beta \hat{t_s}(p)) +$$
$$O(\gamma \hat{b_o}(p)) + O(\delta \hat{b_s}(p))$$
$$\to t_o^{act} + t_s^{act} + b_o^{act} + b_s^{act}$$
$$\to t_i^{act} + b_i^{act} \tag{1.10}$$

where $t_o^{act}, t_s^{act}, b_o^{act}$ and b_s^{act} measure the respective actionable performance in terms of each interestingness element.

Due to the inconsistency often existing at different aspects, we often find the identified patterns only fitting in one of the following sub-sets:

$$Int(p) \to \{\{t_i^{act}, b_i^{act}\}, \{\neg t_i^{act}, b_i^{act}\},$$
$$\{t_i^{act}, \neg b_i^{act}\}, \{\neg t_i^{act}, \neg b_i^{act}\}\} \tag{1.11}$$

where '¬' indicates the corresponding element is not satisfactory.

Ideally, we look for actionable patterns p that can satisfy the following:

IF

$$\forall p \in \widetilde{P}, \exists \mathbf{x} : t_o(\mathbf{x}, p) \wedge t_s(\mathbf{x}, p) \wedge b_o(\mathbf{x}, p)$$
$$\wedge b_s(\mathbf{x}, p) \to act(p) \tag{1.12}$$

THEN:

$$p \to \widetilde{p}. \tag{1.13}$$

However, in real-world mining, as we know, it is very challenging to find the most actionable patterns that are associated with both 'optimal' t_i^{act} and b_i^{act}. Quite often a pattern with significant $t_i()$ is associated with unconfident $b_i()$. Contrarily, it is not rare that patterns with low $t_i()$ are associated with confident $b_i()$. Clearly, AKD targets patterns confirming the relationship $\{t_i^{act}, b_i^{act}\}$.

Therefore, it is necessary to deal with such possible conflict and uncertainty amongst respective interestingness elements. However, it is a kind of artwork and needs to involve domain knowledge and domain experts to tune thresholds and balance difference between $t_i()$ and $b_i()$. Another issue is to develop techniques to balance and combine all types of interestingness metrics to generate uniform, balanced and interpretable mechanisms for measuring knowledge deliverability and extracting and selecting resulting patterns. A reasonable way is to balance both sides toward an acceptable tradeoff. To this end, we need to develop interestingness aggregation methods, namely the $I-function$ (or '∧') to aggregate all elements of interestingness. In fact, each of the interestingness categories may be instantiated into more than one metric. There could be several methods of doing the aggregation, for instance, empirical methods such as business expert-based voting, or more quantitative methods such as multi-objective optimization methods.

1.3 Open Issues and Prospects

To effectively synthesize the above ubiquitous intelligence in AKD-based problem-solving systems, many research issues need to be studied or revisited.

- Typical research issues and techniques in *Data Intelligence* include mining in-depth data patterns, and mining structured knowledge in unstructured data.
- Typical research issues and techniques in *Domain Intelligence* consist of representation, modeling and involvement of domain knowledge, constraints, organizational factors, and business interestingness.
- Typical research issues and techniques in *Network Intelligence* include information retrieval, text mining, web mining, semantic web, ontological engineering techniques, and web knowledge management.
- Typical research issues and techniques in *Human Intelligence* include human-machine interaction, representation and involvement of empirical and implicit knowledge.
- Typical research issues and techniques in *Social Intelligence* include collective intelligence, social network analysis, and social cognition interaction.
- Typical issues in *intelligence metasynthesis* consist of building metasynthetic interaction (m-interaction) as working mechanism, and metasynthetic space (m-space) as an AKD-based problem-solving system [6].

Typical issues in actionable knowledge discovery through m-spaces consist of

- Mechanisms for acquiring and representing unstructured and ill-structured, uncertain knowledge such as empirical knowledge stored in domain experts' brains, such as unstructured knowledge representation and brain informatics;
- Mechanisms for acquiring and representing expert thinking such as imaginary thinking and creative thinking in group heuristic discussions;
- Mechanisms for acquiring and representing group/collective interaction behavior and impact emergence, such as behavior informatics and analytics;
- Mechanisms for modeling learning-of-learning, i.e., learning other participants' behavior which is the result of self-learning or ex-learning, such as learning evolution and intelligence emergence.

1.4 Conclusions

The mainstream data mining research features its dominating focus on the innovation of algorithms and tools yet caring little for their workable capability in the real world. Consequently, data mining applications face significant problem of the workability of deployed algorithms, tools and resulting deliverables. To fundamentally change such situations, and empower the workable capability and performance of advanced data mining in real-world production and economy, there is an urgent need to develop next-generation data mining methodologies and techniques that target the paradigm shift from data-centered hidden pattern mining to domain-driven actionable knowledge discovery. Its goal is to build KDD as an AKD-based problem-solving system.

Based on our experience in conducting large-scale data analysis for several domains, for instance, finance data mining and social security mining, we have proposed the *Domain Driven Data Mining* (D^3M for short) methodology. D^3M emphasizes the development of methodologies, techniques and tools for *actionable knowledge discovery*. It involves relevantly ubiquitous intelligence surrounding the business problem-solving, such as human intelligence, domain intelligence, network intelligence and organizational/social intelligence, and the meta-synthesis of such ubiquitous intelligence into a human-computer-cooperated closed problem-solving system.

Our current work includes an attempt on theoretical studies and working case studies on a set of typically open issues in D^3M. The results will come into a monograph named *Domain Driven Data Mining*, which will be published by Springer in 2009.

Acknowledgements This work is sponsored in part by Australian Research Council Grants (DP0773412, LP0775041, DP0667060).

References

1. Ankerst, M.: Report on the SIGKDD-2002 Panel the Perfect Pata Mining Tool: Interactive or Automated? ACM SIGKDD Explorations Newsletter, 4(2):110-111, 2002.
2. Cao, L., Yu, P., Zhang, C., Zhao, Y., Williams, G.: DDDM2007: Domain Driven Data Mining, ACM SIGKDD Explorations Newsletter, 9(2): 84-86, 2007.
3. Cao, L., Zhang, C.: Knowledge Actionability: Satisfying Technical and Business Interestingness, International Journal of Business Intelligence and Data Mining, 2(4): 496-514, 2007.
4. Cao, L., Zhang, C.: The Evolution of KDD: Towards Domain-Driven Data Mining, International Journal of Pattern Recognition and Artificial Intelligence, 21(4): 677-692, 2007.
5. Cao, L.: Domain-Driven Actionable Knowledge Discovery, IEEE Intelligent Systems, 22(4): 78-89, 2007.
6. Cao, L., Dai, R., Zhou, M.: Metasynthesis, M-Space and M-Interaction for Open Complex Giant Systems, technical report, 2008.
7. Fayyad, U., Shapiro, G., Smyth, P.: From Data Mining to Knowledge Discovery in Databases, AI Magazine, 37-54, 1996.
8. Fayyad, U., Shapiro, G., Uthurusamy, R.: Summary from the KDD-03 Panel - Data mining: The Next 10 Years, ACM SIGKDD Explorations Newsletter, 5(2): 191-196, 2003.
9. Han, J., Kamber, M.: Data Mining: Concepts and Techniques, 2nd edition, Morgan Kaufmann, 2006.
10. Qian, X.S., Yu, J.Y., Dai, R.W.: A New Scientific Field–Open Complex Giant Systems and the Methodology, Chinese Journal of Nature, 13(1) 3-10, 1990.
11. Qian, X.S. (Tsien H.S.): Revisiting issues on open complex giant systems, Pattern Recognition and Artificial Intelligence, 4(1): 5-8, 1991.

Chapter 2
Post-processing Data Mining Models for Actionability

Qiang Yang

Abstract Data mining and machine learning algorithms are, in the most part, aimed at generating statistical models for decision making. These models are typically mathematical formulas or classification results on the test data. However, many of the output models do not themselves correspond to actions that can be executed. In this paper, we consider how to take the output of data mining algorithms as input, and produce collections of high-quality actions to perform in order to bring out the desired world states. This article gives an overview on two of our approaches in this actionable data mining framework, including an algorithm that extracts actions from decision trees and a system that generates high-utility association rules and an algorithm that can learn relational action models from frequent item sets for automatic planning. These two problems and solutions highlight our novel computational framework for actionable data mining.

2.1 Introduction

In data mining and machine learning areas, much research has been done on constructing statistical models from the underlying data. These models include Bayesian probability models, decision trees, logistic and linear regression models, kernel machines and support vector machines as well as clusters and association rules, to name a few [1,11]. Most of these techniques are what we refer to as predictive pattern-based models, in that they summarize the distributions of the training data in one way or another. Thus, they typically stop short of achieving the final objectives of data mining by maximizing utility when tested on the test data. The real action work is waiting to be done by humans, who read the patterns, interpret them and decide which ones to select to put into actions.

Qiang Yang
Department of Computer Science and Engineering, Hong Kong University of Science and Technology, e-mail: qyang@cse.ust.hk

In short, the predictive pattern-based models are aimed for human consumption, similar to what the World Wide Web (WWW) was originally designed for. However, similar to the movement from Web pages to XML pages, we also wish to see knowledge in the form of machine-executable patterns, which constitutes truly actionable knowledge.

In this paper, we consider how to take the output of data mining algorithms as input and produce collections of high-quality actions to perform in order to bring out the desired world states. We argue that the data mining methods should not stop when a model is produced, but rather give collections of actions that can be executed either automatically or semi-automatically, to effect the final outcome of the system. The effect of the generated actions can be evaluated using the test data in a cross-validation manner. We argue that only in this way can a data mining system be truly considered as *actionable*.

In this paper, we consider three approaches that we have adopted in post-processing data mining models for generation actionable knowledge . We first consider in the next section how to postprocess association rules into action sets for direct marketing [14]. Then, we give an overview of a novel approach that extracts actions from decision trees in order to allow each test instance to fall in a desirable state (a detailed description is in [16]). We then describe an algorithm that can learn relational action models from frequent item sets for automatic planning [15].

2.2 Plan Mining for Class Transformation

2.2.1 Overview of Plan Mining

In this section, we first consider the following challenging problem: how to convert customers from a less desirable class to a highly desirable class. In this section, we give an overview of our approach in building an actionable plan from association mining results. More detailed algorithms and test results can be found in [14].

We start with a motivating example. A financial company might be interested in transforming some of the valuable customers from reluctant to active customers through a series of marketing actions. The objective is find an unconditional sequence of actions, a plan, to transform as many from a group of individuals as possible to a more desirable status. This problem is what we call the class-transformation problem. In this section, we describe a planning algorithm for the class-transformation problem that finds a sequence of actions that will transform an initial *undesirable* customer group (e.g., brand-hopping low spenders) into a *desirable* customer group (e.g., brand-loyal big spenders).

We consider a state as a group of customers with similar properties. We apply machine learning algorithms that take as input a database of individual customer profiles and their responses to past marketing actions and produce the customer groups and the state space information including initial state and the next states after action executions. We have a set of actions with state-transition probabilities. At each state, we can identify whether we have arrived at a *desired class* through a classifier.

Suppose that a company is interested in marketing to a large group of customers in a financial market to promote a special loan sign-up. We start with a customer-loan database with historical customer information on past loan-marketing results in Table 2.1. Suppose that we are interested in building a 3-step plan to market to the selected group of customers in the new customer list. There are many candidate plans to consider in order to transform as many customers as possible from non-sign-up status to a sign-up one. The sign-up status corresponds to a positive class that we would like to move the customers to, and the non-signup status corresponds to the initial state of our customers. Our plan will choose not only low-cost actions, but also highly successful actions from the past experience. For example, a candidate plan might be:

Step 1: Offer to reduce interest rate;
Step 2: Send flyer;
Step 3: Follow up with a home phone call.

Table 2.1 An example of *Customer* table

Customer	Interest Rate	Flyer	Salary	Signup
John	5%	Y	110K	Y
Mary	4%	N	30K	Y
...	...	...	...	...
Steve	8%	N	80K	N

This example introduces a number of interesting aspects for the problem at hand. We consider the input data source, which consists of customer information and their desirability class labels. In this database of customers, not all people should be considered as candidates for the class transformation, because for some people it is too costly or nearly impossible to convert them to the more desirable states. Our output plan is assumed to be an *unconditional* sequence of actions rather than conditional plans. When these actions are executed in sequence, no intermediate state information is needed. This makes the *group marketing problem* fundamentally different from the direct marketing problem. In the former, the aim is to find a single sequence of actions with maximal chance of success without inserting if-branches in the plan. In contrast, for direct marketing problems, the aim is to find conditional plans such that a best decision is taken depending on the customers' intermediate state. These are best suited for techniques such as the Markov Decision Processes (MDP) [5, 10, 13].

2.2.2 Problem Formulation

To formulate the problem as a data mining problem, we first consider how to build a state space from a given set of customer records and a set of plan traces in the past. We have two datasets as input. As in any machine learning and data mining schemes, the input customer records consist of a set of attributes for each customer, along with a class attribute that describes the customer status. A second source of input is the previous plans recorded in a database. We also have the costs of actions. As an example, after a customer receives a promotional mail, the customer's response to the marketing action is obtained and recorded. As a result of the mailing, the action count for the customer in this marketing campaign is incremented by one, and the customer may have decided to respond by filling out a general information form and mailing it back to the bank. Table 2.2 shows an example of *plan trace table*.

Table 2.2 A set of plan traces as input

Plan #	State0	Action0	State1	Action1	State2
$Plan_1$	S_0	A_0	S_1	A_1	S_5
$Plan_2$	S_0	A_0	S_1	A_2	S_5
$Plan_3$	S_0	A_0	S_1	A_2	S_6
$Plan_4$	S_0	A_0	S_1	A_2	S_7
$Plan_5$	S_0	A_0	S_2	A_1	S_6
$Plan_6$	S_0	A_0	S_2	A_1	S_8
$Plan_7$	S_0	A_1	S_3		
$Plan_8$	S_0	A_1	S_4		

2.2.3 From Association Rules to State Spaces

From the customer records, a can be constructed by piecing together the association rule mining [1]. Each state node corresponds to a state in planning, on which a classification model can be built to classify a customer falling onto this state into either a positive (+) or a negative (-) class based on the training data. Between two states in this state space, an edge is defined as a state-action sequence which allows a probabilistic mapping from a state to a set of states. A cost is associated with each action.

To enable planning in this state space, we apply sequential association rule mining [1] to the plan traces. Each rule is of the form: $S_1, a_1, a_2, \ldots, \rightarrow S_n$, where each a_i is an action, S_1 and S_n are the initial and end states for this sequence of actions. All actions in this rule start from S_1 and follow the order in the given sequence to result in S_n. By only keeping the sequential rules that have high enough support, we can get segments or paths that we can piece together to form a search space. In particular, in this space, we can gather the following information:

- $f_s(r_i) = s_j$ maps a customer record r_i to a state s_j. This function is known as the customer-state mapping function. In our work, this function is obtained by applying odd-log ratio analysis [8] to perform a feature selection in the customer database. Other methods such as Chi-squared methods or PCA can also be applied.
- $p(+|s)$ is the classification function that is represented as a probability function. This function returns the conditional probability that state s is in a desirable class. We call this function the state-classification function;
- $p(s_k|s_i, a_j)$ returns the transition probability that, after executing an action a_j in state s_i, one ends up in state s_k.

Once the customer records have been converted to states and the state transitions, we are now ready to consider the notion of a plan. To clarify matters, we describe the state space as an AND/OR graph. In this graph, there are two types of node. A *state node* represents a state. From each state node, an action links the state node to an *outcome node*, which represents the outcome of performing the action from the state. An outcome node then splits into multiple state nodes according to the probability distribution given by the $p(s_k|s_i, a_j)$ function. This AND/OR graph unwraps the original state space, where each state is an OR node and the actions that can be performed on the node form the OR branches. Each outcome node is an AND node, where the different arcs connecting the outcome node to the state nodes are the AND edges. Figure 2.1 is an example AND/OR graph. An example plan in this space is shown in Figure 2.2.

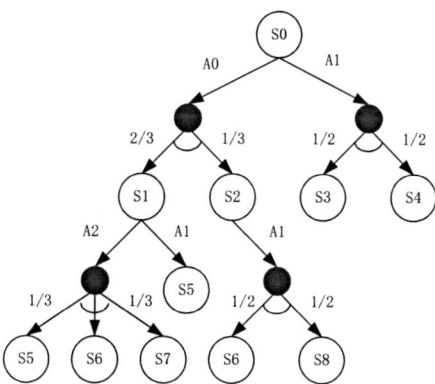

Fig. 2.1 An example of AND/OR graph

We define the utility $U(s, P)$ of the plan $P = a_1 a_2 \ldots a_n$ from an initial state s as follows. Let P' be the subplan of P after taking out the first action a_1; that is, $P = a_1 P'$. Let S be a set of states. Then the utility of the plan P is defined recursively

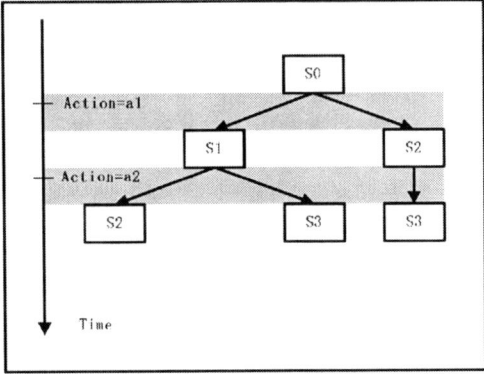

Fig. 2.2 An example of a plan

$$U(s,P) = (\sum_{s' \in S} p(s'|s,a_1) * U(s',P')) - cost(a_1) \qquad (2.1)$$

where s' is the next state resulting from executing a_1 in state s. The plan from the leaf node s is empty and has a utility

$$U(s,\{\}) = p(+|s) * R(s) \qquad (2.2)$$

$p(+|s)$ is the probability of leaf node s being in the desired class, $R(s)$ is a reward (a real value) for a customer to be in state s.

Using Equations 2.1 and 2.2, we can evaluate the utility of a plan P under an initial state $U(s_0,P)$.

Let $next(s,a)$ be the set of states resulting from executing action a in state s. Let $P(s,a,s')$ be the probability of landing in s' after executing a in state s. Let $R(s,a)$ be the immediate reward of executing a in state s. Finally, let $U(s,a)$ be the utility of the optimal plan whose initial state is s and whose first action is a. Then

$$U(s,a) = R(s,a) + \gamma \max_{a'}\{\Sigma_{s' \in next(s,a)} U(s',a') P(s,a,s')\} \qquad (2.3)$$

This equation provides the foundation for the class-transformation planning solution: in order to increase the utility of plans, we need to reduce costs (-R(s,a)) and increase the utility of the expected utility of future plans. In our algorithm below, we achieve this by minimizing the cost of the plans while at the same time, increase the expected probability for the terminal states to be in the positive class.

2.2.4 Algorithm for Plan Mining

We build an AND-OR space using the retained sequences that are both beginning and ending with states and have high enough frequency. Once the frequent sequences are found, we piece together the segments of paths corresponding to the sequences to build an abstract AND-OR graph in which we will search for plans. If $\langle s1, a1, s2 \rangle$ and $\langle s2, a3, s3 \rangle$ are two segments found by the string-mining algorithm, then $\langle s1, a1, s2, a2, s3 \rangle$ is a new path in the AND-OR graph.

We use a utility function to denote how "good" a plan is. Let s_0 be an initial state and P be a plan. Let be a function that sums up the cost of each action in the plan. Let $U(s, P)$ be a heuristic function estimating how promising the plan is for transferring customers initially belonging to state s. We use this function to perform a best-first search in the space of plans until the termination conditions are met. The termination conditions are determined by the probability or the length constraints in the problem domain.

The overall algorithm follows the following steps.

Step 1. Association Rule Mining.

Significant state-action sequences in the state space can be discovered through a association-rule mining algorithm. We start by defining a minimum-support threshold for finding the frequent state-action sequences. *Support* represents the number of occurrences of a state-action sequence from the plan database. Let $count(seq)$ be the number of times sequence "seq" appears in the database for all customers. Then the support for sequence "seq" is defined as

$$sup(seq) = count(seq),$$

Then, association-rule mining algorithms based on moving windows will generate a set of state-action subsequences whose supports are no less than a user-defined minimum support value. For connection purpose, we only retained substrings both beginning and ending with states, in the form of $\langle s_i, a_j, s_{i+1}, ..., s_n \rangle$.

Step 2: Construct an AND-OR space.

Our first task is to piece together the segments of paths corresponding to the sequences to build an abstract AND/OR graph in which we will search for plans. Suppose that $\langle s_0, a_1, s_2 \rangle$ and $\langle s_2, a_3, s_4 \rangle$ are two segments from the plan trace database. Then $\langle s_0, a_1, s_2, a_3, s_4 \rangle$ is a new path in the AND/OR graph. Suppose that we wish to find a plan starting from a state s_0, we consider all action sequences in the AND/OR graph that start from s_0 satisfying the length or probability constraints.

Step 3. Define a heuristic function

We use a function $U(s,P) = g(P) + h(s,P)$ to estimate how "good" a plan is. Let s be an initial state and P be a plan. Let $g(P)$ be a function that sums up the cost of each action in the plan. Let $h(s,P)$ be a heuristic function estimating how promising the plan is for transferring customers initially belonging to state s. In A* search, this function can be designed by users in different specific applications. In our work, we estimate $h(s,P)$ in the following manner. We start from an initial state and follow a plan that leads to several terminal states $s_i, s_{i+1},..., s_{i+j}$. For each of these terminal states, we estimate the state-classification probability $p(+|s_i)$. Each state has a probability of $1 - p(+|s_i)$ to belong to a negative class. The state requires at least one further action to proceed to transfer the $1 - p(+|s_i)$ percent who remain negative, the cost of which is at least the minimum of the costs of all actions in the action set. We compute a heuristic estimation for all terminal states where the plan leads. For an intermediate state leading to several states, an expected estimation is calculated from the heuristic estimation of its successive states weighted by the transition probability $p(s_k|s_i,a_j)$. The process starts from terminal states and propagates back to the root, until reaching the initial state. Finally, we obtain the estimation of $h(s,P)$ for the initial state s under the plan P.

Based on the above heuristic estimation methods, we can express the heuristic function as follows.

$$h(s,P) = \Sigma_a P(s,a,s') h(s',P') \text{ for non terminal states} \quad (2.4)$$
$$(1 - P(+|s))cost(a_m) \text{ for terminal states}$$

where P' is the subplan after the action a such that $P = aP'$. In the MPlan algorithm, we next perform a best-first search based on the cost function in the space of plans until the termination condition is met.

Step 4. Search Plans using *MPlan*

In the AND/OR graph, we carry out a procedure *MPlan* search to perform a best-first search for plans. We maintain a priority queue Q by starting with a single-action plan. Plans are sorted in the priority queue in terms of the evaluation function $U(s,P)$.

In each iteration of the algorithm, we select the plan with the minimum value of $U(s,P)$ from the queue. We then estimate how promising the plan is. That is, we compute the expected state-classification probability $E(+|s_0,P)$ from back to front in a similar way as with $h(s,P)$ calculation, starting with the $p(+|s_i)$ of all terminal states the plan leads to and propagating back to front, weighted by the transition probability $p(s_k|s_i,a_j)$. We compute $E(+|s_0,P)$, the expected value of the state-classification probability of all terminal states. If this expected value exceeds a predefined threshold $Success\_Threshold$ p_θ, i.e. the probability constraint, we consider the plan to be good enough whereupon the search process terminates. Otherwise, one more action is appended to this plan and the new plans are inserted into the priority queue. $E(+|s_0,P)$ is the expected state-classification probability estimating how "effective" a plan is at transferring customers from state s_i. Let $P = a_j P'$. The $E()$ value can be defined in the following recursive way:

$$E(+|s_i,P) = \sum p(s_k|s_i,a_j) * E(+|s_k,P'), \text{if } s_i \text{ is a non-terminal state} \quad (2.5)$$
$$E(+|s_i,\{\}) = p(+|s_i), \text{if } s_i \text{ is a terminal state}$$

We search for plans from all given initial states that corresponds to negative-class customers. We find a plan for each initial state. It is possible that in some AND/OR graphs, we cannot find a plan whose $E(+|s_0,P)$ exceeds the *Success_Threshold*, either because the AND/OR graph is over simplified or because the success threshold is too high. To avoid search indefinitely, we define a parameter *maxlength* which defines the maximum length of a plan, i.e. applying the length constraint. We will discard a candidate plan which is longer than the *maxlength* and $E(+|s_0)$ value less than the *Success_Threshold*.

2.2.5 Summary

We have evaluated the MPlan algorithm using several datasets, and compared to a variety of algorithms. One evaluation was done with the IBM Synthetic Generator (http://www.almaden.ibm.com/software/quest/Resources
/datasets/syndata.html) to generate a *Customer* data set with two classes (positive and negative) and nine attributes. The attributes include both numerical values and discrete values. In this data set, the positive class has 30,000 records representing successful customers and the negative class corresponds to 70,000 representing unsuccessful customers. Those 70,000 negative records are treated as starting points for plan trace generation. For the plan traces, the 70,000 negative-class records are treated as an initially failed customer. A trace is then generated for the customer, transforming the customer through intermediate states to a final state. We defined four types of action, each of which has a cost and associated impact on attribute transitions. The total utility of plans is TU, which is $TU = \sum_{s \in S} U(s,P_s)$, where P_s is the plan found starting from a state s, and S is the set of all initial states in the test data set.400 states serve as the initial states. The total utility is calculated on these states in the test data set.

For comparison, we implemented the *QPlan* algorithm in [12] which uses Q-learning to get an optimal policy and then extracts the unconditional plans from the state space. This algorithm is known as QPlan. Q-learning is carried out in the way called batch reinforcement learning [10], because we are processing a very large amount of data accumulated from past transaction history. The traces consisting of sequences of states and actions in plan database are training data for Q-learning. Q-learning tries to estimate the value function $Q(s,a)$ by value iteration. The major computational complexity of *QPlan* is on Q-learning, which is carried out once before the extraction phase starts.

Figure 2.3 shows the relative utility of different algorithms versus plan lengths. *OptPlan* has the maximal utility by exhaustive search; thus its plan's utility is at 100%. *MPlan* comes next, with about 80% of the optimal solution. *QPlan* have less than 70% of the optimal solution.

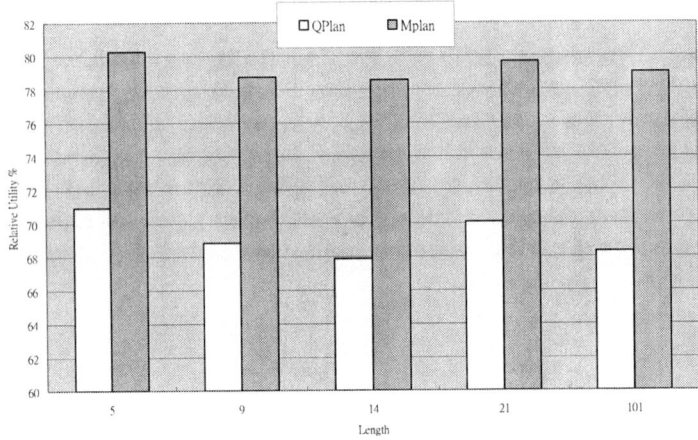

Fig. 2.3 Relative utility plan lengths

In this section, we explored data mining for planning . Our approach combines both classification and planning in order to build an state space in which high utility plans are obtained. The solution plans transform groups of customers from a set of initial states to positive class states.

2.3 Extracting Actions from Decision Trees

2.3.1 Overview

In the section above, we have considered how to construct a state space from association rules. From the state space we can then build a plan. In this section, we consider how to build a decision tree first, from which we can extract actions to improving the current standing of individuals (a more detailed description can be found in [16]). Such examples often occur in customer relationship management (CRM) industry, which is experiencing more and more competitions in recent years. The battle is over their most valuable customers. An increasing number of customers are switching from one service provider to another. This phenomenon is called customer "attrition" , which is a major problem for these companies to stay profitable.

It would thus be beneficial if we could convert a valuable customer from a likely attrition state to a loyal state. To this end, we exploit decision tree algorithms.

Decision-tree learning algorithms, such as ID3 or C4.5 [11], are among the most popular predictive methods for classification. In CRM applications, a decision tree can be built from a set of examples (customers) described by a set of features including customer personal information (such as name, sex, birthday, etc.), financial information (such as yearly income), family information (such as life style, number of children), and so on. We assume that a decision tree has already been generated.

To generate actions from a decision tree, our first step is to consider how to extract actions when there is no restriction on the number of actions to produce. In the training data, some values under the class attribute are more desirable than others. For example, in the banking application, the loyal status of a customer "stay" is more desirable than "not stay". For each of the test data instance, which is a customer under our consideration, we wish to decide what sequences of actions to perform in order to transform this customer from "not stay" to "stay" classes. This set of actions can be extracted from the decision trees.

We first consider the case of unlimited resources where the case serves to introduce our computational problem in an intuitive manner. Once we build a decision tree we can consider how to "move" a customer into other leaves with higher probabilities of being in the desired status. The probability gain can then be converted into an expected gross profit. However, moving a customer from one leaf to another means some attribute values of the customer must be changed. This change, in which an attribute A's value is transformed from v_1 to v_2, corresponds to an action. These actions incur costs. The cost of all changeable attributes are defined in a cost matrix by a domain expert. The `leaf-node search` algorithm searches all leaves in the tree so that for every leaf node, a best destination leaf node is found to move the customer to. The collection of moves are required to maximize the net profit, which equals the gross profit minus the cost of the corresponding actions.

For continuous attributes, such as interest rates that can be varied within a certain range, the numerical ranges can be discretized first using a number of techniques for feature transformation. For example, the entropy based discretization method can be used when the class values are known [7]. Then, we can build a cost matrix for each attribute using the discretized ranges as the index values.

Based on a domain-specific cost matrix for actions, we define the net profit of an action to be as follows.

$$P_{Net} = P_E \times P_{gain} - \sum_i COST_i \quad (2.6)$$

where P_{Net} denotes the net profit, P_E denotes the total profit of the customer in the desired status, P_{gain} denotes the probability gain, and $COST_i$ denotes the cost of each action involved.

2.3.2 Generating Actions from Decision Trees

The overall process of the algorithm can be briefly described in the following four steps:

1. Import customer data with data collection, data cleaning, data pre-processing, and so on.
2. Build customer profiles using an improved decision-tree learning algorithm [11] from the training data. In this case, a decision tree is built from the training data to predict if a customer is in the desired status or not. One improvement in the decision tree building is to use the area under the curve (AUC) of the ROC curve [4] to evaluate probability estimation (instead of the accuracy). Another improvement is to use Laplace Correction to avoid extreme probability values.
3. Search for optimal actions for each customer. This is a critical step in which actions are generated. We consider this step in detail below.
4. Produce reports for domain experts to review the actions and selectively deploy the actions.

The following `leaf-node search` algorithm for searching the best actions is the simplest of a series of algorithms that we have designed. It assumes that there is an unlimited number of actions that can be taken to convert a test instance to a specified class:

Algorithm `leaf-node search`

1. For each customer x, do
2.   Let S be the source leaf node in which x falls into;
3.   Let D be a destination leaf node for x the maximum net profit P_{Net};
4.   Output (S, D, P_{Net});

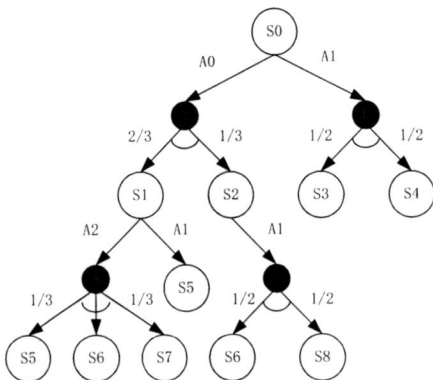

Fig. 2.4 An example of action generation from a decision tree

2 Post-processing Data Mining Models for Actionability

To illustrate, consider an example shown in Figure 2.4, which represents an overly simplified, hypothetical decision tree as the customer profile of loyal customers built from a bank. The tree has five leaf nodes (A, B, C, D, and E), each with a probability of customers' being loyal. The probability of attritors is simply 1 minus this probability. Consider a customer Jack who's record states that the Service = Low (service level is low), Sex = M (male), and Rate=L (mortgage rate is low). The customer is classified by the decision tree. It can be seen that Jack falls into the leaf node B, which predicts that Jack will have only 20% chance of being loyal (or Jack will have 80% chance to churn in the future). The algorithm will now search through all other leaves (A, C, D, E) in the decision tree to see if Jack can be "replaced" into a best leaf with the highest net profit.

Consider leaf A. It does have a higher probability of being loyal (90%), but the cost of action would be very high (Jack should be changed to female), so the net profit is a negative infinity. Now consider leaf node C. It has a lower probability of being loyal, so the net profit must be negative, and we can safely skip it.

Notice that in the above example, the actions suggested for a customer-status change imply only correlations rather than causality between customer features and status.

2.3.3 The Limited Resources Case

Our previous case considered each leaf node of the decision tree to be a separate customer group. For each such customer group, we were free to design actions to act on it in order to increase the net profit. However, in practice, a company may be limited in its resources. For example, a mutual fund company may have a limited number k (say three) of account managers, each manager can take care of only one customer group. Thus, when such limitations exist, it is a difficult problem to *optimally* merge all leave nodes into k segments, such that each segment can be assigned to an account manager. To each segment, the responsible manager can several apply actions to increase the overall profit.

This limited-resource problem can be formulated as a precise computational problem. Consider a decision tree DT with a number of source leaf nodes that correspond to customer segments to be converted and a number of candidate destination leaf nodes, which correspond to the segments we wish customers to fall in.

A solution is a set of k targetted nodes $\{G_i, i = 1, 2, \ldots, k\}$, where each node corresponds to a 'goal' that consists of a set of source leaf nodes S_{ij} and one designation leaf node D_i, denoted as: $(\{S_{ij}, j = 1, 2, \ldots, |G_i|\} \rightarrow D_i)$, where S_{ij} and D_i are leaf nodes from the decision tree DT. The goal node is meant to transform customers that belong to the source nodes S to the destination node D via a number of attribute-value changing actions. Our aim is to find a solution with the maximal net profit.

In order to change the classification result of a customer x from S to D, one may need to apply more than one attribute-value changing action. An action A is defined as a change to an attribute value for an attribute $Attr$. Suppose that for a customer x, the attribute $Attr$ has an original value u. To change its value to v, an action is needed. This action A is denoted as $A = \{Attr, u \rightarrow v\}$.

To achieve a goal of changing a customer x from a leaf node S to a destination node D, a set of actions that contains more than one action may be needed. Specifically, consider the path between the root node and D in the tree DT. Let $\{(Attr_i = v_i), i = 1, 2, \ldots, N_D\}$ be set of attribute-values along this path. For x, let the corresponding attribute-values be $\{(Attr_i = u_i), i = 1, 2, \ldots N_D\}$. Then, the actions of the form can be generated: $ASet = \{(Attr_i, u_i \rightarrow v_i), i = 1, 2, \ldots, N_D\}$, where we remove all null actions where u_i is identical to v_i (thus no change in value is needed for an $Attr_i$). This action set $ASet$ can be used for achieving the goal $S \rightarrow D$.

The net profit of converting one customer x from a leaf node S to a destination node D is defined as follows. Consider a set of actions $ASet$ for achieving the goal $S \rightarrow D$. For each action $Attr_i, u \rightarrow v$ in $ASet$, there is a cost as defined in the cost matrix: $C(Attr_i, u, v)$. Let the sum of the cost for all of $ASet$ be $C_{\text{total}, S \rightarrow D}(x)$.

The BSP problem is to find best k groups of source leaf nodes $\{Group_i, i = 1, 2, \ldots, k\}$ and their corresponding goals and associated action sets to maximize the total net profit for a given test dataset C_{test}.

The BSP problem is essentially a maximum coverage problem [9], which aims at finding k sets such that the total weight of elements covered is maximized, where the weight of each element is the same for all the sets. A special case of the BSP problem is equivalent to the maximum coverage problem with unit costs. Thus, we know that the BSP problem is NP-Complete. Our aim will then be to find approximation solutions to the BSP problem.

To solve the BSP problem, one needs to examine every combination of k action sets, the computational complexity is $O(n^k)$, which is exponential in the value of k. To avoid the exponential worst-case complexity, we have also developed a greedy algorithm which can reduce the computational cost and guarantee the quality of the solution at the same time.

Initially, our greedy search based algorithm Greedy-BSP starts with an empty result set $C = \emptyset$. The algorithm then compares all the column sums that corresponds to converting all leaf nodes S_1 to S_4 to each destination leaf node D_i in turn. It found that $ASet_2 = (\rightarrow D_2)$ has the current maximum profit of 3 units. Thus, the resultant action set C is assigned to $\{ASet_2\}$.

Next, Greedy-BSP considers how to expand the customer groups by one. To do this, it considers which additional column will increase the total net profit to a highest value, if we can include one more column. In [16], we present a large number of experiments to show that the greedy search algorithm performs close to the optimal result.

2.4 Learning Relational Action Models from Frequent Action Sequences

2.4.1 Overview

Above we have considered how to postprocess traditional models that are obtained from data mining in order to generate actions. In this section, we will give an overview on how take a data mining model and postprocess it into a action model that can be executed for plan generation. These actions can be used by robots, software agents and process management software for many advanced applications. A more detailed discussion can be found in [15].

To understand how actions are used, we can recall that automatic planning systems can take formal definitions of actions, an initial state and a goal state description as input, and produce plans for execution. In the past, the task of building action models has been done manually. In the past, various approaches have been explored to learn action models from examples. In this section, we describe our approach in automatically acquiring action models from recorded user plans. Our system is known as ARMS , which stands for *Action-Relation Modelling System* ; a more detailed description is given in [15]. The input to the ARMS system is a collection of observed traces. Our algorithm applies frequent itemset mining algorithm to these traces to find out the collection of frequent action-sets. These actions sets are then taken as the input to another modeling system known as weighted MAX-SAT, which can generate relational actions.

Consider an example input and output of our algorithm in the Depot problem domain from an AI Planning competition [2, 3]. As part of the input, we are given relations such as (clear ?x:surface) to denote that ?x is clear on top and that ?x is of type "surface", relation (at ?x:locatable ?y:place) to denote that a locatable object ?x is located at a place ?y . We are also given a set of plan examples consisting of action names along with their parameter list, such as drive(?x:truck ?y:place ?z:place), and then lift(?x:hoist ?y:crate ?z:surface ?p:place). We call the pair consisting of an action name and the associated parameter list an *action signature*; an example of an action signature is drive(?x:truck ?y:place ?z:place). Our objective is to learn an *action model* for each action signature, such that the relations in the preconditions and postconditions are fully specified.

A complete description of the example is shown in Table 2.3, which lists the actions to be learned, and Table 2.4, which displays the training examples. From the examples in Table 2.4, we wish to learn the preconditions, add and delete lists of all actions. Once an action is given with the three lists, we say that it has a complete action model. Our goal is to learn an action model for every action in a problem domain in order to "explain" all training examples successfully. An example output

from our learning algorithms for the load(?x ?y ?z ?p) action signature is:
action load(?x:hoist ?y:crate ?z:truck ?p:place)
pre: (at ?x ?p), (at ?z ?p), (*lifting* ?x ?y)
del: (lifting ?x ?y)
add: (at ?y ?p), (in ?y ?z), (available ?x), (clear ?y)

Table 2.3 Input Domain Description for Depot Planning Domain

domain	Depot
types	place locatable - object depot distributor - place truck hoist surface - locatable pallet crate - surface
relations	(at ?x:locatable ?y:place) (on ?x:crate ?y:surface) (in ?x:crate ?y:truck) (lifting ?x:hoist ?y:crate) (available ?x:hoist) (clear ?x:surface)
actions	drive(?x:truck ?y:place ?z:place) lift(?x:hoist ?y:crate ?z:surface ?p:place) drop(?x:hoist ?y:crate ?z:surface ?p:place) load(?x:hoist ?y:crate ?z:truck ?p:place) unload(?x:hoist ?y:crate ?z:truck ?p:place)

As part of the input, we need sequences of example plans that have been executed in the past, as shown in Table 2.4. Our job is to formally describe actions such as lift such that automatic planners can use them to generate plans. These training plan examples can be obtained through monitoring devices such as sensors and cameras, or through a sequence of recorded commands through a computer system such as UNIX domains. These action models can then be revised using interactive systems such as GIPO.

2.4.2 ARMS Algorithm: From Association Rules to Actions

To build action models, ARMS proceeds in two phases. *Phase one* of the algorithm applies association rule mining algorithms to find the *frequent action sets* from plans that share a common set of parameters. In addition, ARMS finds some frequent relation-action pairs with the help of the initial state and the goal state. These relation-action pairs give us an initial guess on the preconditions, add lists and delete lists of actions in this subset. These action subsets and pairs are used to obtain a set of constraints that must hold in order to make the plans correct.

In *phase two*, ARMS takes the frequent item sets as input, and transforms them into constraints in the form of a weighted MAX-SAT representation [6]. It then solves it using a weighted MAX-SAT solver and produces action models as a result.

2 Post-processing Data Mining Models for Actionability

Table 2.4 Three plan traces as part of the training examples

	Plan1	Plan2	Plan3
Initial	I_1	I_2	I_3
Step1	lift(h1 c0 p1 ds0), drive(t0 dp0 ds0)	lift(h1 c1 c0 ds0)	lift(h2 c1 c0 ds0)
State		(lifting h1 c1)	
Step2	load(h1 c0 t0 ds0)	load(h1 c1 t0 ds0)	load(h2 c1 t1 ds0)
Step3	drive(t0 ds0 dp0)	lift(h1 c0 p1 ds0)	lift(h2 c0 p2 ds0), drive(t1 ds0 dp1)
State	(available h1)		
Step4	unload(h0 c0 t0 dp0)	load(h1 c0 t0 ds0)	unload(h1 c1 t1 dp1), load(h2 c0 t0 ds0)
State		(lifting h0 c0)	
Step5	drop (h0 c0 p0 dp0)	drive(t0 ds0 dp0)	drop(h1 c1 p1 dp1), drive(t0 ds0 dp0)
Step6		unload(h0 c1 t0 dp0)	unload(h0 c0 t0 dp0)
Step7		drop(h0 c1 p0 dp0)	drop(h0 c0 p0 dp0)
Step8		unload(h0 c0 t0 dp0)	
Step9		drop(h0 c0 c1 dp0)	
Goal	(on c0 p0)	(on c1 p0) (on c0 c1)	(on c0 p0) (on c1 p1)

I_1 : (at p0 dp0), (clear p0), (available h0), (at h0 dp0), (at t0 dp0), (at p1 ds0), (clear c0), (on c0 p1), (available h1), (at h1 ds0)

I_2 : (at p0 dp0), (clear p0), (available h0), (at h0 dp0), (at t0 ds0), (at p1 ds0), (clear c1), (on c1 c0), (on c0 p1), (available h1), (at h1 ds0)

I_3 : (at p0 dp0), (clear p0), (available h0), (at h0 dp0), (at p1 dp1), (clear p1), (available h1), (at h1 dp1), (at p2 ds0), (clear c1), (on c1 c0), (on c0 p2), (available h2), (at h2 ds0), (at t0 ds0), (at t1 ds0)

The process iterates until all actions are modeled. While the action models that ARMS learns are deterministic in nature, in the future we will extend this framework to learning probabilistic action models to handle uncertainty. Additional constraints are added to allow partial observations to be made between actions, prove the formal properties of the system. In [15], ARMS was tested successfully on all STRIPS planning domains from a recent AI Planning Competition based on training action sequences.

The algorithm starts by initializing the plans by replacing the actual parameters of the actions by variables of the same types. This ensures that we learn action models for the schemata rather than for the individual instantiated actions. Subsequently, the algorithm iteratively builds a weighted MAX-SAT representation and solves it. In each iteration, a few more actions are explained and are removed from the incomplete action set Λ. The learned action models in the middle of the program help reduce the number of clauses in the SAT problem. ARMS terminates when all action schemata in the example plans are learned.

Below, we explain the major steps of the algorithm in detail.

Step 1: Initialize Plans and Variables

A plan example consists of a sequence of action instances. We *convert* all such plans by substituting all occurrences of an instantiated object in every action instance with the variables of the same type. If the object has multiple types, we generate a clause to represent each possible type for the object. For example, if an object o has two types *Block* and *Table*, the clause becomes: $\{(?o = Block)\ or\ (?o = Table)\}$. We then extract from the example plans all sets of actions that are *connected* to each other; two actions $a1$ and $a2$ are said to be *connected* if their parameter-type list has non-empty intersection. The parameter mapping $\{?x1 = ?x2, \ldots\}$ is called a connector.

Step 2: Build Action and Plan Constraints

A weighted MAX-SAT problem consists of a set of clauses representing their conjunction, where each clause is associated with a weight value representing the priority in satisfying the constraint. Given a weighted MAX-SAT problem, a weighted MAX-SAT solver finds a solution by maximizing the sum of the weight values associated with the satisfied clauses.

In the ARMS system, we have four kinds of constraints to satisfy, representing three types of clauses. They are action, information and plan and relation constraints.

Action constraints are imposed on individual actions. These constraints are derived from the general axioms of correct action representations. A relation r is said to be *relevant* to an action a if they are the same parameter type. Let pre_i, add_i and del_i represent a_i's precondition list, add-list and delete list.

Step 3: Build and Solve a Weighted MAX-SAT Problem

In solving a weighted MAX-SAT problem in Step 3, each clause is associated with a weight value between zero and one. The higher the weight, the higher the priority in satisfying the clause. ARMS assigns weights to the three types of constraints in the weighted MAX-SAT problem described above. For example, every action constraint receives a constant weight $W_A(a)$ for an action a. The weight for action constraints is set to be higher than the weight of information constraints.

2.4.3 Summary of ARMS

In this section, we have considered how to obtain action models from a set of plan examples. Our method is to first apply association rule mining algorithm on the plan traces to obtain the frequent action sequences. We then convert these frequent action sequences into constraints that are fed into a MAXSAT solver. The solution can then be converted to action models. These action models can be used by automatic planners to generate new plans.

2.5 Conclusions and Future Work

Most data mining algorithms and tools produce only statistical models in their outputs. In this paper, we present a new framework to take these results as input and produce a set of actions or action models that can bring about the desired changes. We have shown how to use the result of association rule mining to build a state space graph, based on which we then performed automatic planning for generating marketing plans. From decision trees, we have explored how to extract action sets to maximize the utility of the end states. For association rule mining, we have considered how to construct constraints in a weighted MAX-SAT representation in order to determine the relational representation of action models.

In our future work, we will research on other methods for actionable data mining, to generate collections of useful actions that a decision maker can apply in order to generated the needed changes.

Acknowledgement

We thank the support of Hong Kong RGC 621307.

References

1. R. Agrawal and R. Srikant. Fast algorithms for mining association rules. In *Proceedings of 20th International Conference on Very Large Data Bases(VLDB'94)*, pages 487–499. Morgan Kaufmann, September 1994.
2. Maria Fox and Derek Long. PDDL2.1: An extension to pddl for expressing temporal planning domains. *Journal of Artificial Intelligence Research*, 20:61–124, 2003.
3. Malik Ghallab, Adele Howe, Craig Knoblock, Drew McDermott, Ashwin Ram, Manuela Veloso, Dan Weld, and David Wilkins. PDDL—the planning domain definition language, 1998.
4. Jin Huang and Charles X. Ling. Using auc and accuracy in evaluating learning algorithms. *IEEE Trans. Knowl. Data Eng*, 17(3):299–310, 2005.
5. L. Kaelbling, M. Littman, and A. Moore. Reinforcement learning: A survey. *Journal of Artificial Intelligence Research*, 4:237–285, 1996.
6. Henry Kautz and Bart Selman. Pushing the envelope: Planning, propositional logic, and stochastic search. In *Proceedings of the Thirteenth National Conference on Artificial Intelligence (AAAI 1996)*, pages 1194–1201, Portland, Oregon USA, 1996.
7. Ron Kohavi and Mehran Sahami. Error-based and entropy-based discretization of continuous features. In *Proceedings of the Second International Conference on Knowledge Discovery and Data Mining*, pages 114–119, Portland, Oregon USA, 1996.

8. D. Mladenic and M. Grobelnik. Feature selection for unbalanced class distribution and naive bayes. In *Proceedings of ICML 1999.*, 1999.
9. M.R.Garey and D.S. Johnson. *Computers and Intractability: A guide to the Theory of NPCompleteness.* 1979.
10. E. Pednault, N. Abe, and B. Zadrozny. Sequential cost-sensitive decision making with reinforcement learning. In *Proceedings of the Eighth International Conference on Knowledge Discovery and Data Mining (KDD'02)*, 2002.
11. J.Ross Quinlan. *C4.5 Programs for machine learning.* Morgan Kaufmann, 1993.
12. R. Sun and C. Sessions. Learning plans without a priori knowledge. *Adaptive Behavior*, 8(3/4):225–253, 2001.
13. R. Sutton and A. Barto. *Reinforcement Learning: An Introduction.* MIT Press, Cambridge, MA, 1998.
14. Qiang Yang and Hong Cheng. Planning for marketing campaigns. In *International Conference on Automated Planning and Scheduling (ICAPS 2003)*, pages 174–184, 2003.
15. Qiang Yang, Kangheng Wu, and Yunfei Jiang. Learning action models from plan examples using weighted max-sat. *Artif. Intell.*, 171(2-3):107–143, 2007.
16. Qiang Yang, Jie Yin, Charles Ling, and Rong Pan. Extracting actionable knowledge from decision trees. *IEEE Trans. on Knowl. and Data Eng.*, 19(1):43–56, 2007.

Chapter 3
On Mining Maximal Pattern-Based Clusters

Jian Pei, Xiaoling Zhang, Moonjung Cho, Haixun Wang, and Philip S. Yu

Abstract Pattern-based clustering is important in many applications, such as DNA micro-array data analysis in bio-informatics, as well as automatic recommendation systems and target marketing systems in e-business. However, pattern-based clustering in large databases is still challenging. On the one hand, there can be a huge number of clusters and many of them can be redundant and thus make the pattern-based clustering ineffective. On the other hand, the previous proposed methods may not be efficient or scalable in mining large databases.

In this paper, we study the problem of *maximal pattern-based clustering*. The major idea is that the redundant clusters are avoided completely by mining only the *maximal pattern-based clusters*. We show that maximal pattern-based clusters are skylines of all pattern-based clusters. Two efficient algorithms, *MaPle* and *MaPle+* (*MaPle* is for Maximal Pattern-based Clustering) are developed. The algorithms conduct a depth-first, progressively refining search and prune unpromising branches smartly. *MaPle+* integrates several interesting heuristics further. Our extensive performance study on both synthetic data sets and real data sets shows that maximal pattern-based clustering is effective – it reduces the number of clusters substantially. Moreover, *MaPle* and *MaPle+* are more efficient and scalable than the previously proposed pattern-based clustering methods in mining large databases, and *MaPle+* often performs better than *MaPle*.

Jian Pei
Simon Fraser University, e-mail: `jpei@cs.sfu.ca`

Xiaoling Zhang
Boston University, e-mail: `zhangxl@bu.edu`

Moonjung Cho
Prism Health Networks, e-mail: `moonjungcho@hotmail.com`

Haixun Wang
IBM T.J.Watson Research Center e-mail: `haixun@us.ibm.com`

Philip S.Yu
University of Illinois at Chicago, e-mail: `psyu@cs.uic.edu`

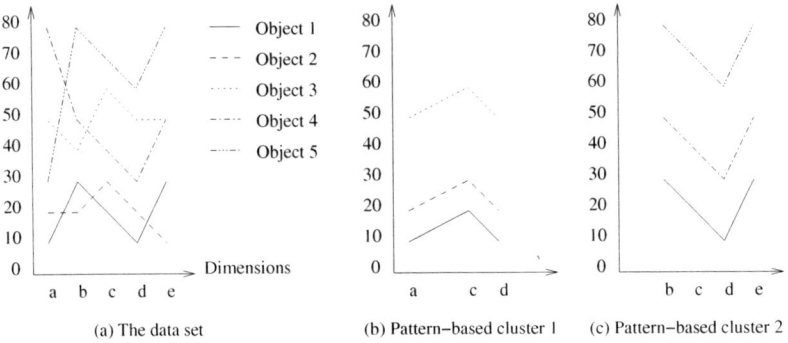

Fig. 3.1 A set of objects as a motivating example.

3.1 Introduction

Clustering large databases is a challenging data mining task with many important applications. Most of the previously proposed methods are based on similarity measures defined globally on a (sub)set of attributes/dimensions. However, in some applications, it is hard or even infeasible to define a good similarity measure on a global subset of attributes to serve the clustering.

To appreciate the problem, let us consider clustering the 5 objects in Figure 3.1(a). There are 5 dimensions, namely a, b, c, d and e. No patterns among the 5 objects are visibly explicit. However, as elaborated in Figure 3.1(b) and Figure 3.1(c), respectively, objects 1, 2 and 3 follow the same pattern in dimensions a, c and d, while objects 1, 4 and 5 share another pattern in dimensions b, c, d and e. If we use the patterns as features, they form two *pattern-based clusters*.

As indicated by some recent studies, such as [14, 15, 18, 22, 25], pattern-based clustering is useful in many applications. In general, given a set of data objects, a subset of objects form a pattern-based clusters if these objects follow a similar pattern in a subset of dimensions. Comparing to the conventional clustering, pattern-based clustering has two distinct features. First, pattern-based clustering does not require a globally defined similarity measure. Instead, it specifies quality constraints on clusters. Different clusters can follow different patterns on different subsets of dimensions. Second, the clusters are not necessary to be exclusive. That is, an object can appear in more than one cluster.

The flexibility of pattern-based clustering may provide interesting and important insights in some applications where conventional clustering methods may meet difficulties. For example, in DNA micro-array data analysis, the gene expression data is organized as matrices, where rows represent genes and columns represent samples/conditions. The value in each cell records the expression level of the particular gene under the particular condition. The matrices often contain thousands of genes and tens of conditions. It is important to identify subsets of genes whose expression levels change coherently under a subset of conditions. Such information is critical in revealing the significant connections in gene regulatory networks. As another example, in the applications of automatic recommendation and target marketing, it is essential to identify sets of customers/clients with similar behavior/interest.

In [22], the pattern-based clustering problem is proposed and a mining algorithm is developed. However, some important problems remain not thoroughly explored. In particular, we address the following two fundamental issues and make the corresponding contributions in this paper.

- *What is the effective representation of pattern-based clusters?* As can be imagined, there can exist many pattern-based clusters in a large database. Given a pattern-based cluster C, any non-empty subset of the objects in the cluster is trivially a pattern-based cluster on any non-empty subset of the dimensions. Mining and analyzing a huge number of pattern-based clusters may become the bottleneck of effective analysis. *Can we devise a succinct representation of the pattern-based clusters?*

  Our contributions. In this paper, we propose the mining of *maximal pattern-based clusters*. The idea is to report only those non-redundant pattern-based clusters, and skip their trivial sub-clusters. We show that, by mining maximal pattern-based clusters, the number of clusters can be reduced substantially. Moreover, many unfruitful searches for sub-clusters can be pruned and thus the mining efficiency can be improved dramatically as well.

- *How to mine the maximal pattern-based clusters efficiently?* Our experimental results indicate that the algorithm *p-Clustering* developed in [22] may not be satisfactorily efficient or scalable in large databases. The major bottleneck is that it has to search many possible combinations of objects and dimensions.

  Our contributions. In this paper, we develop two novel mining algorithms, *MaPle* and *MaPle+* (*MaPle* is for M̲aximal P̲attern-based Cl̲ustering). They conduct a depth-first, progressively refining search to mine maximal pattern-based clusters. We propose techniques to guarantee the completeness of the search and also prune unpromising search branches whenever it is possible. *MaPle+* also integrates several interesting heuristics further.

  An extensive performance study on both synthetic data sets and real data sets is reported. The results show that *MaPle* and *MaPle+* are significantly more efficient and more scalable in mining large databases than method *p-Clustering* in [22]. In many cases, *MaPle+* performs better than *MaPle*.

The remainder of the paper is organized as follows. In Section 3.2, we define the problem of mining maximal pattern-based clusters, review related work, compare pattern-based clustering and traditional partition-based clustering, and discuss the complexity. Particularly, we exemplify the idea of method *p-Clustering* [22]. In Section 3.3, we develop algorithms *MaPle* and *MaPle+*. An extensive performance study is reported in Section 3.4. The paper is concluded in Section 3.5.

3.2 Problem Definition and Related Work

In this section, we propose the problem of maximal pattern-based clustering and review related work. In particular, *p-Clustering*, a pattern-based clustering method developed in [22], will be examined in detail.

3.2.1 Pattern-Based Clustering

Given a set of objects, where each object is described by a set of attributes, a pattern-based cluster (R, D) is a subset of objects R that exhibits a coherent pattern on a subset of attributes D. To formulate the problem, it is essential to describe how coherent a subset of objects R are on a subset of attributes D. A measure *pScore* proposed in [22] can serve this purpose.

Definition 3.1 (pScore [22]). Let $DB = \{r_1, \ldots, r_n\}$ be a database with n *objects*. Each object has m *attributes* $A = \{a_1, \ldots, a_m\}$. We assume that each attribute is in the domain of real numbers. The value of object r_j on attribute a_i is denoted as $r_j.a_i$. For any objects $r_x, r_y \in DB$ and any attributes $a_u, a_v \in A$, the *pScore* is defined as

$$pScore\left(\begin{bmatrix} r_x.a_u & r_x.a_v \\ r_y.a_u & r_y.a_v \end{bmatrix}\right) = |(r_x.a_u - r_y.a_u) - (r_x.a_v - r_y.a_v)|.$$ ∎

Pattern-based clusters can be defined as follows.

Definition 3.2 (Pattern-based cluster [22]). Let $R \subseteq DB$ be a subset of objects in the database and $D \subseteq A$ be a subset of attributes. (R, D) is said a δ-*pCluster* (*pCluster* is for pattern-based cluster) if for any objects $r_x, r_y \in R$ and any attributes $a_u, a_v \in D$,

$$pScore\left(\begin{bmatrix} r_x.a_u & r_x.a_v \\ r_y.a_u & r_y.a_v \end{bmatrix}\right) \leq \delta,$$

where $\delta \geq 0$. ∎

Given a database of objects, pattern-based clustering is to find the pattern-based clusters from the database. In a large database with many attributes, there can be many coincident, statistically insignificant pattern-based clusters, which consist of very few objects or on very few attributes. A cluster may be considered *statistically insignificant* if it contains a small number of objects, or a small number of attributes. Thus, in addition to the quality requirement on the pattern-based clusters using an upper bound on *pScore*, a user may want to impose constraints on the minimum number of objects and the minimum number of attributes in a pattern-based cluster.

In general, given (1) a cluster threshold δ, (2) an *attribute threshold* min_a (i.e., the minimum number of attributes), and (3) an *object threshold* min_o (i.e., the minimum number of objects), the task of *mining δ-pClusters* is to find the complete set of δ-pClusters (R, D) such that $(|R| \geq min_o)$ and $(|D| \geq min_a)$. A δ-pCluster satisfying the above requirement is called *significant*.

3.2.2 Maximal Pattern-Based Clustering

Although the attribute threshold and the object threshold are used to filter out insignificant pClusters, there still can be some "*redundant*" significant pClusters. For example, consider the objects in Figure 3.1. Let $\delta = 5$, $min_a = 3$ and $min_o = 3$. Then, we have 6 significant pClusters: $C_1 = (\{1,2,3\},\{a,c,d\})$, $C_2 = (\{1,4,5\},\{b,c,d\})$, $C_3 = (\{1,4,5\},\{b,c,e\})$, $C_4 = (\{1,4,5\},\{b,d,e\})$, $C_5 = (\{1,4,5\},\{c,d,e\})$, and $C_6 = (\{1,4,5\},\{b,c,d,e\})$. Among them, C_2, C_3, C_4 and C_5 are subsumed by C_6, i.e., the objects and attributes in the four clusters, C_2-C_5, are subsets of the ones in C_6.

In general, a pCluster $C_1 = (R_1, D_1)$ is called a *sub-cluster* of $C_2 = (R_2, D_2)$ provided $(R_1 \subseteq R_2) \wedge (D_1 \subseteq D_2) \wedge (|R_1| \geq 2) \wedge (|D_1| \geq 2)$. C_1 is called a *proper sub-cluster* of C_2 if either $R_1 \subset R_2$ or $D_1 \subset D_2$. Pattern-based clusters have the following property.

Property 3.1 (Monotonicity). Let $C = (R, D)$ be a δ-pCluster. Then, every sub-cluster (R', D') is a δ-pCluster. ∎

Clearly, mining the redundant sub-clusters is tedious and ineffective for analysis. Therefore, it is natural to mine only the "maximal clusters", i.e., the pClusters that are not sub-cluster of any other pClusters.

Definition 3.3 (maximal pCluster). A δ-pCluster C is said *maximal* (or called a δ-MPC for short) if there exists no any other δ-pCluster C' such that C is a proper sub-cluster of C'. ∎

Problem Statement (mining maximal δ-pClusters). Given (1) a cluster threshold δ, (2) an attribute threshold min_a, and (3) an object threshold min_o, the task of mining maximal δ-pClusters is to find the complete set of maximal δ-pClusters with respect to min_a and min_o. ∎

3.2.3 Related Work

The study of pattern-based clustering is related to the previous work on subspace clustering and frequent itemset mining.

The meaning of clustering in high dimensional data sets is often unreliable [7]. Some recent studies (e.g. [2–4, 8]) focus on mining clusters embedded in some subspaces. For example, CLIQUE [4] is a density and grid based method. It divides the data into hyper-rectangular cells and use the dense cells to construct subspace clusters.

Subspace clustering can be used to semantically compress data. An interesting study in [13] employs a randomized algorithm to find fascicles, the subsets of data that share similar values in some attributes. While their method is effective for compression, it does not guarantee the completeness of mining the clusters.

In some applications, global similarity-based clustering may not be effective. Still, strong correlations may exist among a set of objects even if they are far away from each other as measured by distance functions (such as Euclidean) used frequently in traditional clustering algorithms. Many scientific projects collect data in the form of Figure 3.1(a), and it is essential to identify clusters of objects that manifest coherent patterns. A variety of applications, including DNA microarray analysis, E-commerce collaborative filtering, will benefit from fast algorithms that can capture such patterns.

In [9], Cheng and Church propose the biclustering model, which captures the coherence of genes and coditions in a sub-matrix of a DNA micro-array. Yang et al. [23] develop a move-based algorithm to find biclusters more efficiently.

Recently, some variations of pattern-based clustering have been proposed. For example, in [18], the notion of OP-clustering is developed. The idea is that, for an object, the list of dimensions sorted in the value ascending order can be used as its signature. Then, a set of objects can be put into a cluster if they share the a part of their signature. OP-clustering can be viewed as a (very) loose pattern-based clustering. That is, every pCluster is an OP-cluster, but not vice versa.

On the other hand, a transaction database can be modelled as a binary matrix, where columns and rows stand for items and transactions, respectively. A cell $r_{i,j}$ is set to 1 if item j is contained in transaction i. Then, the problem of mining frequent itemsets [5] is to find subsets of rows and columns such that the sub-matrix is all 1's, and the number of rows is more than a given support threshold. If a minimum length constraint min_a is imposed to find only frequent itemsets of no less than min_a items, then it becomes a problem of mining 0-pClusters on binary data. Moreover, a maximal pattern-based cluster in the transaction binary matrix is a closed itemset [19]. Interestingly, a maximal pattern-based cluster in this context can also be viewed as a formal concept, and the sets of objects and attributes are exactly the extent and intent of the concept, respectively [11].

Although there are many efficient methods for frequent itemset mining, such as [1, 6, 10, 12, 16, 17, 24], they cannot be extended to handle the general pattern-based clustering problem since they can only handle the binary data.

3.3 Algorithms *MaPle* and *MaPle+*

In this section, we develop two novel pattern-based clustering algorithms, *MaPle* (for Maximal Pattern-based Clustering) and *MaPle+*. An early version of *MaPle* is preliminarily reported in [20]. *MaPle+* integrates some interesting heuristics on the top of *MaPle*.

We first overview the intuitions and the major technical features of *MaPle*, and then present the details.

3.3.1 An Overview of MaPle

Essentially, *MaPle* enumerates all the maximal pClusters systematically. It guarantees the completeness of the search, i.e., every maximal pCluster will be found. At the same time, *MaPle* also guarantees that the search is not redundant, i.e., each combination of attributes and objects will be tested at most once.

The general idea of the search in *MaPle* is as follows. *MaPle* enumerates every combination of attributes systematically according to an order of attributes. For example, suppose that there are four attributes, a_1, a_2, a_3 and a_4 in the database, and the alphabetical order, i.e., a_1-a_2-a_3-a_4, is adopted. Let attribute threshold $min_a = 2$. For each subset of attributes, we can list the attributes alphabetically. Then, we can enumerate the subsets of two or more attributes according to the dictionary order, i.e., a_1a_2, $a_1a_2a_3$, $a_1a_2a_3a_4$, $a_1a_2a_4$, a_1a_3, $a_1a_3a_4$, a_1a_4, a_2a_3, $a_2a_3a_4$, a_2a_4, a_3a_4.

For each subset of attributes D, *MaPle* finds the maximal subsets of objects R such that (R,D) is a δ-pCluster. If (R,D) is not a sub-cluster of another pCluster (R,D') such that $D \subset D'$, then (R,D) is a maximal δ-pCluster.

There can be a huge number of combinations of attributes. *MaPle* prunes many combinations unpromising for δ-pClusters. Following Property 3.1, for a subset of attributes D, if there exists no subset of objects R such that (R,D) is a significant pCluster, then we do not need to search any superset of D. On the other hand, when searching under a subset of attributes D, *MaPle* only checks those subsets of objects R such that (R,D') is a pCluster for every $D' \subset D$. Clearly, only subsets $R' \subseteq R$ may achieve δ-pCluster (R',D). Such pruning techniques are applied recursively. Thus, *MaPle* progressively refines the search step by step.

Moreover, *MaPle* also prunes searches that are unpromising to find maximal pClusters. It detects the attributes and objects that can be used to assemble a larger pCluster from the current pCluster. If *MaPle* finds that the current subsets of attributes and objects as well as all possible attributes and objects together turn out to be a sub-cluster of a pCluster having been found before, then the recursive searches rooted at the current node are pruned, since it cannot lead to a maximal pCluster.

Why does MaPle *enumerate attributes first and then objects later, but not in the reverse way?* In real databases, the number of objects is often much larger than the number of attributes. In other words, the number of combinations of objects is often dramatically larger than the number of combinations of attributes. In the pruning using maximal pClusters discussed above, if the attribute-first-object-later approach is adopted, once a set of attributes and its descendants are pruned, all searches of related subsets of objects are pruned as well. Heuristically, the attribute-first-object-later search may bring a better chance to prune a more bushy search subtree.[1] Symmetrically, for data sets that the number of objects is far smaller than the number of attributes, a symmetrical object-first-attribute-later search can be applied.

Essentially, we rely on MDS's to determine whether a subset of objects and a subset of attributes together form a pCluster. Therefore, as a preparation of the min-

---

[1] However, there is no theoretical guarantee that the attribute-first-object-later search is optimal. There exist counter examples that object-first-attribute-later search wins.

Object	a_1	a_2	a_3	a_4	a_5
o_1	5	6	7	7	1
o_2	4	4	5	6	10
o_3	5	5	6	1	30
o_4	7	7	15	2	60
o_5	2	0	6	8	10
o_6	3	4	5	5	1

(a) The database

Objects	Attribute-pair
$\{o_1,o_2,o_3,o_4,o_6\}$	$\{a_1,a_2\}$
$\{o_1,o_2,o_3,o_6\}$	$\{a_1,a_3\}$
$\{o_1,o_2,o_6\}$	$\{a_1,a_4\}$
$\{o_1,o_2,o_3,o_6\}$	$\{a_2,a_3\}$
$\{o_1,o_2,o_6\}$	$\{a_2,a_4\}$
$\{o_1,o_2,o_6\}$	$\{a_3,a_4\}$

(b) The attribute-pair MDS's

Fig. 3.2 The database and attribute-pair MDS's in our running example.

ing, we compute all non-redundant MDS's and store them as a database before we conduct the progressively refining, depth-first search.

Comparing to *p-Clustering*, *MaPle* has several advantages. First, in the third step of *p-Clustering*, for each node in the prefix tree, combinations of the object registered at the node will be explored to find pClusters. This can be expensive if there are many objects at a node. In *MaPle*, the information of pClusters is inherited from the "parent node" in the depth-first search and the possible combinations of objects can be reduced substantially. Moreover, once a subset of attributes D is determined hopeless for pClusters, the searches of any superset of D will be pruned.

Second, *MaPle* prunes non-maximal pClusters. Many unpromising searches can be pruned in their early stages.

Last, new pruning techniques are adopted in the computing and pruning of MDS's. They also speed up the mining.

In the remainder of the section, we will explain the two steps of *MaPle* in detail.

3.3.2 Computing and Pruning MDS's

Given a database DB and a cluster threshold δ. A δ-pCluster $C_1 = (\{o_1,o_2\},D)$ is called an *object-pair MDS* if there exists no δ-pCluster $C_1' = (\{o_1,o_2\},D')$ such that $D \subset D'$. On the other hand, a δ-pCluster $C_2(R,\{a_1,a_2\})$ is called an *attribute-pair MDS* if there exists no δ-pCluster $C_2' = (R',\{a_1,a_2\})$ such that $R \subset R'$.

MaPle computes all attribute-pair MDS's as *p-Clustering* does. For the correctness and the analysis of the algorithm, please refer to [22].

Example 3.1 (Running example – finding attribute-pair MDS's). Let us consider mining maximal pattern-based clusters in a database DB as shown in Figure 3.2(a). The database has 6 objects, namely $o_1, \ldots, o_6$, while each object has 5 attributes, namely $a_1, \ldots, a_5$.

Suppose $min_a = 3$, $min_o = 3$ and $\delta = 1$. For each pair of attributes, we calculate the attribute pair MDS's. The attribute-pair MDS's returned are shown in Figure 3.2(b). ∎

3 On Mining Maximal Pattern-Based Clusters

Generally, a pair of objects may have more than one object-pair MDS. Symmetrically, a pair of attributes may have more than one attribute-pair MDS.

We can also generate all the object-pair MDS's similarly. However, if we utilize the information on the number of occurrences of objects and attributes in the attribute-pair MDS's, the calculation of object-pair MDS's can be speeded up.

Lemma 3.1 (Pruning MDS's). *Given a database DB and a cluster threshold δ, object threshold min_o and attribute threshold min_a.*

1. *An attribute a cannot appear in any significant δ-pCluster if a appears in less than $\frac{min_o \cdot (min_o - 1)}{2}$ object-pair MDS's, or if a appears in less than $(min_a - 1)$ attribute-pair MDS's;*
2. *Symmetrically, an object o cannot appear in any significant δ-pCluster if o appears in less than $\frac{min_a \cdot (min_a - 1)}{2}$ attribute-pair MDS's, or if o appears in less than $(min_o - 1)$ object-pair MDS's.* ∎

Example 3.2 (Pruning using Lemma 3.1). Let us check the attribute-pair MDS's in Figure 3.2(b). Object o_5 does not appear in any attribute-pair MDS, and object o_4 appears in only 1 attribute-pair MDS. According to Lemma 3.1, o_4 and o_5 cannot appear in any significant δ-pCluster. Therefore, we do not need to check any object-pairs containing o_4 or o_5.

There are 6 objects in the database. Without this pruning, we have to check $\frac{6 \times 5}{2} = 15$ pairs of objects. With this pruning, only four objects, o_1, o_2, o_3 and o_6 survive. Thus, we only need to check $\frac{4 \times 3}{2} = 6$ pairs of objects. A 60% of the original searches is pruned.

Moreover, since attribute a_5 does not appear in any attribute-pair MDS, it cannot appear in any significant δ-pCluster. The attribute can be pruned. That is, when generating the object-pair MDS, we do not need to consider attribute a_4.

In summary, after the pruning, only attributes a_1, a_2, a_3 and a_4, and objects o_1, o_2, o_3 and o_6 survive. We use these attributes and objects to generate object-pair MDS's. The result is shown in Figure 3.3(a). In method *p-Clustering*, it uses all attributes and objects to generate object-pair MDS's. The result is shown in Figure 3.3(b). As can be seen, not only the computation cost in *MaPle* is less, the number of object-pair MDS's in *MaPle* is also one less than that in method *p-Clustering*. ∎

Once we get the initial object-pair MDS's and attribute-pair MDS's, we can conduct a mutual pruning between the object-pair MDS's and the attribute-pair MDS's, as method *p-Clustering* does. Furthermore, Lemma 3.1 can be applied in each round to get extra pruning. The pruning algorithm is shown in Figure 3.4.

Object-pair	Attributes
$\{o_1, o_2\}$	$\{a_1, a_2, a_3, a_4\}$
$\{o_1, o_3\}$	$\{a_1, a_2, a_3\}$
$\{o_1, o_6\}$	$\{a_1, a_2, a_3, a_4\}$
$\{o_2, o_3\}$	$\{a_1, a_2, a_3\}$
$\{o_2, o_6\}$	$\{a_1, a_2, a_3, a_4\}$
$\{o_3, o_6\}$	$\{a_1, a_2, a_3\}$

(a) Object-pair MDS's in *MaPle*.

Object-pair	Attributes
$\{o_1, o_2\}$	$\{a_1, a_2, a_3, a_4\}$
$\{o_1, o_6\}$	$\{a_1, a_2, a_3, a_4\}$
$\{o_2, o_3\}$	$\{a_1, a_2, a_3\}$
$\{o_1, o_3\}$	$\{a_1, a_2, a_3\}$
$\{o_2, o_6\}$	$\{a_1, a_2, a_3, a_4\}$
$\{o_3, o_4\}$	$\{a_1, a_2, a_4\}$
$\{o_3, o_6\}$	$\{a_1, a_2, a_3\}$

(b) Object-pair MDS's in method *p-Clustering*

Fig. 3.3 Pruning using Lemma 3.1.

```
(1) REPEAT
(2)     count the number of occurrences of objects and attributes in the attribute-pair MDS's;
(3)     apply Lemma 3.1 to prune objects and attributes;
(4)     remove object-pair MDS's containing less than min_a attributes;
(5)     count the number of occurrences of objects and attributes in the object-pair MDS's;
(6)     apply Lemma 3.1 to prune objects and attributes;
(7)     remove attribute-pair MDS's containing less than min_o objects;
(8) UNTIL no pruning can take place
```

Fig. 3.4 The algorithm of pruning MDS's.

3.3.3 Progressively Refining, Depth-first Search of Maximal pClusters

The algorithm of the progressively refining, depth-first search of maximal pClusters is shown in Figure 3.5. We will explain the algorithm step by step in this subsection.

3.3.3.1 Dividing Search Space

By a list of attributes, we can enumerate all combinations of attributes systematically. The idea is shown in the following example.

Example 3.3 (Enumeration of combinations of attributes). In our running example, there are four attributes survived from the pruning: a_1, a_2, a_3 and a_4. We list the attributes in any subset of attributes in the order of a_1-a_2-a_3-a_4. Since $min_a = 3$, every maximal δ-pCluster should have at least 3 attributes. We divide the complete set of maximal pClusters into 3 exclusive subsets according to the first two attributes in the pClusters: (1) the ones having attributes a_1 and a_2, (2) the ones having attributes a_1 and a_3 but not a_2, and (3) the ones having attributes a_2 and a_3 but not a_1. ∎

Since a pCluster has at least 2 attributes, *MaPle* first partitions the complete set of maximal pClusters into exclusive subsets according to the first two attributes,

3 On Mining Maximal Pattern-Based Clusters   41

```
(1)     let n be the number of attributes; make up an attribute list AL = a₁-····-aₙ;
(2)     FOR i = 1 TO n − min₀ + 1 DO //Theorem 3.1, item 1
(3)        FOR j = i + 1 TO n − min₀ + 2 DO
(4)           find attribute-pair MDS's (R,{aᵢ,aⱼ}); //Section 3.3.3.2
(5)           FOR EACH lcoal maximal pCluster (R,{aᵢ,aⱼ}) DO
(6)              call search(R,{aᵢ,aⱼ});
(7)           END FOR EACH
(8)        END FOR
(9)     END FOR
(10)
(11)    FUNCTION search(R,D); // (R,D) is a attribute-maximal pCluster.
(12)       compute PD, the set of possible attributes; //Optimization 1 in Section 3.3.3.3
(13)       apply optimizations in Section 3.3.3.3 to prune, if possible;
(14)       FOR EACH attribute a ∈ PD DO //Theorem 3.1, item 2
(15)          find attribute-maximal pClusters (R′,D∪{a}); //Section 3.3.3.2
(16)          FOR EACH attribute-maximal pCluster (R′,D∪{a}) DO
(17)             call search(R′,D∪{a});
(18)          END FOR EACH
(19)          IF (R′,D∪{a}) isn't a subcluster of some maximal pCluster having been found
(20)          THEN output (R′,D∪{a});
(21)       END FOR EACH
(22)       IF (R,D) is not a subcluster of some maximal pCluster having been found
(23)       THEN output (R,D);
(24)    END FUNCTION
```

Fig. 3.5 The algorithm of projection-based search.

and searches the subsets one by one in the depth-first manner. For each subset, *MaPle* further divides the pClusters in the subset into smaller exclusive sub-subsets according to the third attributes in the pClusters, and search the sub-subsets. Such a process proceeds recursively until all the maximal pClusters are found. This is implemented by line (1)-(3) and (14) in Figure 3.5. The correctness of the search is justified by the following theorem.

Theorem 3.1 (Completeness and non-redundancy of *MaPle*). *Given an attribute-list* $AL : a_1\text{-}\cdots\text{-}a_m$, *where m is the number of attributes in the database. Let* min_a *be the attribute threshold.*

1. *All attributes in each pCluster are listed in the order of AL. Then, the complete set of maximal δ-pClusters can be divided into $\frac{(m-min_a+2)(m-min_a+1)}{2}$ exclusive subsets according to the first two attributes in the pClusters.*
2. *The subset of maximal pClusters whose first 2 attributes are a_i and a_j can be further divided into $(m - min_a + 3 - j)$ subsets: the kth ($1 \leq k \leq (m - j - min_a - 1)$) subset contains pClusters whose first 3 attributes are a_i, a_j and a_{j+k}.* ∎

3.3.3.2 Finding Attribute-maximal pClusters

Now, the problem becomes how to find the maximal δ-pClusters on the subsets of attributes. For each subset of attributes D, we will find the maximal subsets of objects R such that (R, D) is a pCluster. Such a pCluster is a maximal pCluster if it is not a sub-cluster of some others.

Given a set of attributes D such that $(|D| \geq 2)$. A pCluster (R, D) is called a *attribute-maximal δ-pCluster* if there exists no any δ-pCluster (R', D) such that $R \subset R'$. In other words, a attribute-maximal pCluster is maximal in the sense that no more objects can be included so that the objects are still coherent on the same subset of attributes. For example, in the database shown in Figure 3.2(a), $(\{o_1, o_2, o_3, o_6\}, \{a_1, a_2\})$ is a attribute-maximal pCluster for subset of attributes $\{a_1, a_2\}$.

Clearly, a maximal pCluster must be a attribute-maximal pCluster, but not vice versa. In other words, if a pCluster is not a attribute-maximal pCluster, it cannot be a maximal pCluster.

Given a subset of attributes D, how can we find all attribute-maximal pClusters efficiently? We answer this question in two cases.

If D has only two attributes, then the attribute-maximal pClusters are the attribute-pair MDS's for D. Since the MDS's are computed and stored before the search, they can be retrieved immediately.

Now, let us consider the case where $|D| \geq 3$. Suppose $D = \{a_{i_1}, \ldots, a_{i_k}\}$ where the attributes in D are listed in the order of attribute-list AL. Intuitively, (R, D) is a pCluster if R is shared by attribute-pair MDS's for any two attributes from D. (R, D) is an attribute-maximal pCluster if R is a maximal set of objects.

One subtle point here is that, in general, there can be more than one attribute-pair MDS for a pair of attributes a_1 and a_2. Thus, there can be more than one attribute-maximal pCluster on a subset of attributes D. Technically, (R, D) is an attribute-maximal pCluster if $R = \bigcap_{\{a_u, a_v\} \subset D} R_{uv}$ where $(R_{uv}, \{a_u, a_v\})$ is an attribute-pair MDS. Recall that *MaPle* searches the combinations of attributes in the depth-first manner, all attribute-maximal pClusters for subset of attributes $D - \{a\}$ is found before we search for D, where a is the last attribute in D according to the attribute list. Therefore, we only need to find the subset of objects in a attribute-maximal pCluster of $D - \{a\}$ that are shared by attribute-pair MDS's of a_{i_j}, a_{i_k} ($j < k$).

3.3.3.3 Pruning and Optimizations

Several optimizations can be used to prune the search so that the mining can be more efficient. The first two optimizations are recursive applications of Lemma 3.1.

Optimization 1: Only *possible attributes* should be considered to get larger pClusters.

Suppose that (R, D) is a attribute-maximal pCluster. *For every attribute a such that a is behind all attributes in D in the attribute-list, can we always find a significant pCluster $(R', D \cup \{a\})$ such that $R' \subseteq R$?*

3 On Mining Maximal Pattern-Based Clusters   43

If $(R', D \cup \{a\})$ is significant, i.e., has at least min_o objects, then a must appear in at least $\frac{min_o(min_o-1)}{2}$ object-pair MDS's $(\{o_i, o_j\}, D_{ij})$ such that $\{o_i, o_j\} \subseteq R'$. In other words, for an attribute a that appears in less than $\frac{min_o(min_o-1)}{2}$ object-pair MDS's of objects in R, there exists no attribute-maximal pCluster with respect to $D \cup \{a\}$.

Based on the above observation, an attribute a is called a *possible attribute* with respect to attribute-maximal pCluster (R, D) if a appears in $\frac{min_o(min_o-1)}{2}$ object-pair MDS's $(\{o_i, o_j\}, D_{ij})$ such that $\{o_i, o_j\} \subseteq R$. In line (12) of Figure 3.5, we compute the possible attributes and only those attributes are used to extend the set of attributes in pClusters.

Optimization 2: Pruning local maxiaml pClusters having insufficient possible attributes.

Suppose that (R, D) is a attribute-maximal pCluster. Let PD be the set of possible attributes with respect to (R, D). Clearly, if $|D \cup PD| < min_a$, then it is impossible to find any maximal pCluster of a subset of R. Thus, such a attribute-maximal pCluster should be discarded and all the recursive search can be pruned.

Optimization 3: Extracting common attributes from possible attribute set directly.

Suppose that (R, D) is a attribute-maximal pCluster with respect to D, and D' is the corresponding set of possible attributes. If there exists an attribute $a \in D'$ such that for every pair of objects $\{o_i, o_j\}$, $\{a\} \cup D$ appears in an object pair MDS of $\{o_i, o_j\}$, then we immediately know that $(R, D \cup \{a\})$ must be a attribute-maximal pCluster with respect to $D \cup \{a\}$. Such an attribute is called a *common attribute* and should be extracted directly.

Example 3.4 (Extracting common attributes). In our running example, $(\{o_1, o_2, o_3, o_6\}, \{a_1, a_2\})$ is a attribute-maximal pCluster with respect to $\{a_1, a_2\}$. Interestingly, as shown in Figure 3.3(a), for every object pair $\{o_i, o_j\} \subset \{o_1, o_2, o_3, o_6\}$, the object-pair MDS contains attribute a_3. Therefore, we immediately know that $(\{o_1, o_2, o_3, o_6\}, \{a_1, a_2, a_3\})$ is a attribute-maximal pCluster. ∎

Optimization 4: Prune non-maximal pClusters.

Our goal is to find maximal pClusters. If we can find that the recursive search on a attribute-maximal pCluster cannot lead to a maximal pCluster, the recursive search thus can be pruned. The earlier we detect the impossibility, the more search efforts can be saved.

We can use the *dominant attributes* to detect the impossibility. We illustrate the idea in the following example.

Example 3.5 (Using dominant attributes to detect non-maximal pClusters). Again, let us consider our running example. Let us try to find the maximal pClusters whose first two attributes are a_1 and a_3. Following the above discussion, we identify a attribute-maximal pCluster $(\{o_1, o_2, o_3, o_6\}, \{a_1, a_3\})$.

One interesting observation can be made from the object-pair MDS's on objects in $\{o_1, o_2, o_3, o_6\}$ (Figure 3.3(a)): attribute a_2 appears in every object pair. We called

a_2 a *dominant attribute*. That means $\{o_1, o_2, o_3, o_6\}$ also coherent on attribute a_2. In other words, we cannot have a maximal pCluster whose first two attributes are a_1 and a_3, since a_2 must also be in the same maximal pCluster. Thus, the search of maximal pClusters whose first two attributes are a_1 and a_3 can be pruned. ∎

The idea in Example 3.5 can be generalized. Suppose (R,D) is a attribute-maximal pCluster. If there exists an attribute a such that a is before the last attribute in D according to the attribute-list, and $\{a\} \cup D$ appears in an object-pair MDS $(\{o_i, o_j\}, D_{ij})$ for every $(\{o_i, o_j\} \subseteq R)$, then the search from (R,D) can be pruned, since there cannot be a maximal pCluster having attribute set D but no a. Attribute a is called a *dominant attribute* with respect to (R,D).

3.3.4 MaPle+: *Further Improvements*

MaPle+ is an enhanced version of *MaPle*. In addition to the techniques discussed above, the following two ideas are used in *MaPle+* to speed up the mining.

3.3.4.1 Block-based Pruning of Attribute-pair MDS's

In *MaPle* (please see Section 3.3.2), an MDS can be pruned if it cannot be used to form larger pClusters. The pruning is based on comparing an MDS with the other MDS's.

Since there can be a large number of MDS's, the pruning may not be efficient. Instead, we can adopt a block-based pruning as follows.

For an attribute a, all attribute-pair MDS's that a is an attribute form the *a-block*. We consider the blocks of attributes in the attribute-list order.

For the first attribute a_1, the a_1-block is formed. Then, for an object o, if o appears in any significant pCluster that has attribute a_1, o must appear in at least $(min_a - 1)$ different attribute-pair MDS's in the a_1-block. In other words, we can remove an object o from the a_1-block MDS's if its count in the a_1-block is less than $(min_a - 1)$. After removing the objects, the attribute-pair MDS's in the block that do not have at least $(min_o - 1)$ objects can also be removed.

Moreover, according to Lemma 3.1, if there are less than $(min_a - 1)$ MDS's in the resulted a_1-block, then a_1 cannot appear in any significant pCluster, and thus all the MDS's in the block can be removed.

The blocks can be considered one by one. Such a block-based pruning is more effective. In Section 3.3.2, we prune an object from attribute-pair MDS's if it appears in less than $\frac{min_a \cdot (min_a - 1)}{2}$ different attribute-pair MDS's (Lemma 3.1). In the block-based pruning, we consider pruning an object with respect to every possible attribute. It can be shown that any object pruned by Lemma 3.1 must also be pruned in some block, but not vice versa, as shown in the following example.

Attribute-pairs	objects
$\{a_1,a_2\}$	$\{o_1,o_2,o_4\}$
$\{a_1,a_3\}$	$\{o_2,o_3,o_4\}$
$\{a_1,a_4\}$	$\{o_2,o_4,o_5\}$
$\{a_2,a_3\}$	$\{o_1,o_2,o_3\}$
$\{a_2,a_4\}$	$\{o_1,o_3,o_4\}$
$\{a_2,a_5\}$	$\{o_2,o_3,o_5\}$

Fig. 3.6 The attribute-pair MDS's in Example 3.6.

Example 3.6 (Block-based pruning of attribute-pair MDS's). Suppose we have the attribute-pair MDS's as shown in Figure 3.6, and $min_o = min_a = 3$.

In the a_1-block, which contains the first three attribute-pair MDS's in the table, objects o_1, o_3 and o_5 can be pruned. Moreover, all attribute-pair MDS's in the a_1-block can be removed.

However, in *MaPle*, since o_1 appears 3 times in all the attribute-pair MDS's, it cannot be pruned by Lemma 3.1, and thus attribute-pair MDS $(\{a_1,a_2\},\{o_1,o_2,o_4\})$ cannot be pruned, either. ∎

The block-based pruning is also more efficient. To use Lemma 3.1 to prune in *MaPle*, we have to check both the attribute-pair MDS's and the object-pair MDS's mutually. However, in the block-based pruning, we only have to look at the attribute-pair MDS's in the current block.

3.3.4.2 Computing Attribute-pair MDS's Only

In many data sets, the number of objects and the number of attributes are different dramatically. For example, in the microarray data sets, there are often many genes (thousands or even tens of thousands), but very few samples (up to one hundred). In such cases, a significant part of the runtime in both *p-Clustering* and *MaPle* is to compute the object-pair MDS's.

Clearly, computing object-pair MDS's for a large set of objects is very costly. For example, for a data set of 10,000 objects, we have to consider $10000 \times 9999 \div 2 = 49,995,000$ object pairs!

Instead of computing those object-pair MDS's, we develop a technique to compute only the attribute-pair MDS's. The idea is that we can compute the attribute-maximal pClusters on-the-fly without materializing the object-pair MDS's.

Example 3.7 (Computing attribute-pair MDS's only). Consider the attribute-pair MDS's in Figure 3.2(b) again. We can compute the attribute-maximal pCluster for attribute set $\{a_1,a_2,a_3\}$ using the attribute-pair MDS's only.

We observe that an object pair o_u and o_v are in an attribute-maximal pCluster of $\{a_1,a_2,a_3\}$ if and only if there exist three attribute-pair MDS's for $\{a_1,a_2\}$, $\{a_1,a_3\}$, and $\{a_2,a_3\}$, respectively, such that $\{o_u,o_v\}$ are in the object sets of all those three attribute-pair MDS's. Thus, the intersection of the three object sets in those three attribute-pair MDS's is the set of objects in the attribute-maximal pCluster.

In this example, $\{a_1,a_2\}$, $\{a_1,a_3\}$, and $\{a_2,a_3\}$ have only one attribute-pair MDS, respectively. The intersection of their object sets are $\{o_1,o_2,o_3,o_6\}$. Therefore, the attribute-maximal pCluster is $(\{o_1,o_2,o_3,o_6\},\{a_1,a_2,a_3\})$. ∎

When the number of objects is large, computing the attribute-maximal pClusters directly from attribute-pair MDS's and smaller attribute-maximal pClusters can avoid the costly materialization of object-pair MDS's. The computation can be conducted level-by-level from smaller attribute sets to their supersets.

Generally, if a set of attributes D has multiple attribute-maximal pClusters, then its superset D' may also have multiple attribute-maximal pClusters. For example, suppose $\{a_1,a_2\}$ has attribute-pair MDS's $(R_1,\{a_1,a_2\})$ and $(R_2,\{a_1,a_2\})$, and $(R_3,\{a_1,a_3\})$ and $(R_4,\{a_2,a_3\})$ are attribute-pair MDS's for $\{a_1,a_3\}$ and $\{a_1,a_3\}$, respectively. Then, $(R_1 \cap R_3 \cap R_4, \{a_1,a_2,a_3\})$ and $(R_2 \cap R_3 \cap R_4, \{a_1,a_2,a_3\})$ should be checked. If the corresponding object set has at least min_o objects, then the pCluster is an attribute-maximal pCluster. We also should check whether $(R_1 \cap R_3 \cap R_4) = (R_2 \cap R_3 \cap R_4)$. If so, we only need to keep one attribute-maximal pCluster for $\{a_1,a_2,a_3\}$.

To compute the intersections efficiently, the sets of objects can be represented as bitmaps. Thus, the intersection operations can be implemented using the bitmap AND operations.

3.4 Empirical Evaluation

We test *MaPle*, *MaPle+* and *p-Clustering* extensively on both synthetic and real life data sets. In this section, we report the results.

MaPle and *MaPle+* are implemented using C/C++. We obtained the executable of the improved version of *p-Clustering* from the authors of [22]. Please note that the authors of *p-Clustering* improved their algorithm dramatically after their publication in SIGMOD'02. The authors of *p-Clustering* also revised the program so that only maximal pClusters are detected and reported. Thus, the output of the two methods are comparable directly. All the experiments are conducted on a PC with a P4 1.2 GHz CPU and 384 M main memory running a Microsoft Windows XP operating system.

3.4.1 The Data Sets

The algorithms are tested against both synthetic and real life data sets. Synthetic data sets are generated by a synthetic data generator reported in [22]. The data generator takes the following parameters to generate data sets: (1) the number of objects;

δ	min_a	min_o	# of max-pClusters	# of pClusters
0	9	30	5	5520
0	7	50	11	N/A
0	5	30	9370	N/A

Fig. 3.7 Number of pClusters on Yeast raw data set.

(2) the number of attributes; (3) the average number of rows of the embedded pClusters; (4) the average number of columns; and (5) the number of pClusters embedded in the data sets. The synthetic data generator can generate only perfect pClusters, i.e., $\delta = 0$.

We also report the results on a real data set, the Yeast microarray data set [21]. This data set contains the expression levels of 2,884 genes under 17 conditions. The data set is preprocessed as described in [22].

3.4.2 Results on Yeast Data Set

The first issue we want to examine is whether there exist significant pClusters in real data sets. We test on the Yeast data set. The results are shown in Figure 3.7. From the results, we can obtain the following interesting observations.

- There are significant pClusters existing in real data. For example, we can find pure pCluster (i.e., $\delta = 0$) containing more than 30 genes and 9 attributes in Yeast data set. That shows the effectiveness and utilization of mining maximal pClusters in the real data sets.
- While the number of maximal pClusters is often small, the number of all pClusters can be huge, since there are many different combinations of objects and attributes as sub-clusters to the maximal pClusters. This shows the effectiveness of the notation of maximal pClusters.
- Among the three cases shown in Figure 3.7, *p-Clustering* can only finish in the first case. In the other two cases, it cannot finish and outputs a huge number of pClusters that overflow the hard disk. In Contrast, *MaPle* and *MaPle+* can finish and output a small number of pClusters, which cover all the pClusters found by *p-Clustering*.

To test the efficiency of mining the Yeast data set with respect to the tolerance of noise, we fix the thresholds of $min_a = 6$ and $min_o = 60$, and vary the δ from 0 to 4. The results are shown in Figure 3.8.

As shown, both *p-Clustering* and *MaPle+* are scalable on the real data set with respect to δ. When δ is small, *MaPle* is fast. However, it scales poorly with respect to δ. The reason is that, as the value of δ increases, a subset of attribute has more and more attribute-maximal pClusters on average. Similarly, there are more and more object-pair MDS's. Managing a large number of MDS's and conducting iteratively

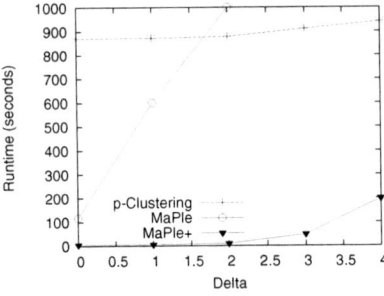
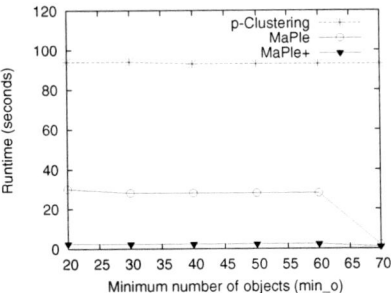

Fig. 3.8 Runtime vs. δ on the Yeast data set, $min_a = 6$ and $min_o = 60$.

Fig. 3.9 Runtime vs. minimum number of objects in pClusters.

pruning still can be costly. The block-based pruning technique and the technique of computing attribute-maximal pClusters from attribute-pair MDS's, as described in Section 3.3.4, helps *MaPle+* to reduce the cost effectively. Thus, *MaPle+* is substantially faster than *p-Clustering* and *MaPle*.

3.4.3 Results on Synthetic Data Sets

We test the scalability of the algorithms on the three parameters, the minimum number of objects min_o, the minimum number of attributes min_a in pClusters, and δ. In Figure 3.9, the runtime of the algorithms versus min_o is shown. The data set has 6000 objects and 30 attributes.

As can be seen, all the three algorithms are in general insensitive to parameter min_o, but *MaPle+* is much faster than *p-Clustering* and *MaPle*. The major reason that the algorithms are insensitive is that the number of pClusters in the synthetic data set does not changes dramatically as min_o decreases and thus the overhead of the search does not increase substantially. Please note that we do observe the slight increases of runtime in all the three algorithms as min_o goes down.

One interesting observation here is that, when $min_o > 60$, the runtime of *MaPle* decreases significantly. The runtime of *MaPle+* also decreases from 2.4 seconds to 1 second. That is because there is no pCluster in such a setting. *MaPle+* and *MaPle* can detect this in an early stage and thus can stop early.

We observe the similar trends on the runtime versus parameter min_a. That is, both algorithms are insensitive to the minimum number of attributes in pClusters, but *MaPle* is faster than *p-Clustering*. The reasoning similar to that on min_o holds here.

We also test the scalability of the algorithms on δ. The result is shown in Figure 3.10. As shown, both *MaPle+* and *pClustering* are scalable with respect to the value of δ, while *MaPle* is efficient when the δ is small. When the δ value becomes large, the performance of *MaPle* becomes poor. The reason is as analyzed before:

3 On Mining Maximal Pattern-Based Clusters

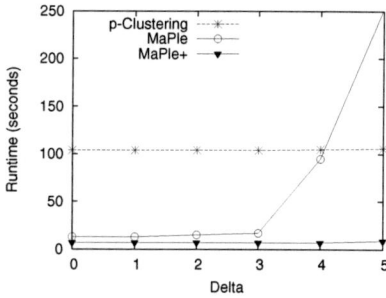

Fig. 3.10 Runtime with respect to δ.

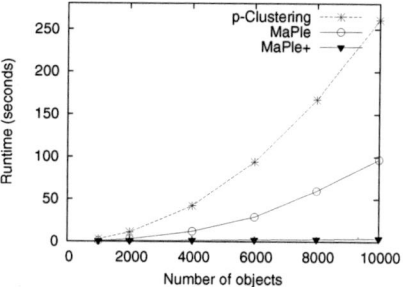

Fig. 3.11 Scalability with respect to the number of objects in the data sets.

when the value of δ increases, some attribute pairs may have multiple MDS's and some object pairs may have multiple MDS's. *MaPle* has to check many combinations. *MaPle+* uses the block-based pruning technique to reduce the cost substantially. Among the three algorithms, *MaPle+* is clearly the best.

We test the scalability of the three algorithms on the number of objects in the data sets. The result is shown in Figure 3.11. The data set contains 30 attributes, where there are 30 embedded clusters. We fix $min_a = 5$ and set $min_o = n_{obj} \cdot 1\%$, where n_{obj} is the number of objects in the data set. $\delta = 1$.

The result in Figure 3.11 clearly shows that *MaPle* performs substantially better than *p-Clustering* in mining large data sets. *MaPle+* is up to two orders of magnitudes faster than *p-Clustering* and *MaPle*. The reason is that both *p-Clustering* and *MaPle* use object-pair MDS's in the mining. When there are 10000 objects in the database, there are $\frac{10000 \times 9999}{2} = 49995000$ object-pairs. Managing a large database of object-pair MDS's is costly. *MaPle+* only uses attribute-pair MDS's in the mining. In this example, there are only $\frac{30 \times 29}{2} = 435$ attribute pairs. Thus, *MaPle+* does not suffer from the problem.

To further understand the difference, Figure 3.12 shows the numbers of local maximal pClusters searched by *MaPle* and *MaPle+*. As can be seen, *MaPle+* searches substantially less than *MaPle*. That partially explains the difference of performance of the two algorithms.

We also test the scalability of the three algorithms on the number of attributes. The result is shown in Figure 3.13. In this test, the number of objects is fixed to 3,000 and there are 30 embedded pClusters. We set $min_o = 30$ and $min_a = n_{attr} \cdot 20\%$, where n_{attr} is the number of attributes in the data set.

The curves show that all the three algorithms are approximately linearly scalable with respect to number of attributes, and *MaPle+* performs consistently better than *p-Clustering* and *MaPle*.

In summary, from the tests on synthetic data sets, we can see that *MaPle+* outperforms both *p-Clustering* and *MaPle* clearly. *MaPle+* is efficient and scalable in mining large data sets.

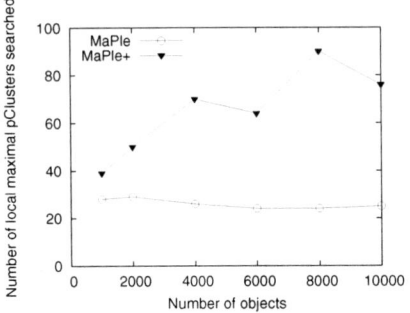

Fig. 3.12 Number of local maximal pClusters searched by *MaPle* and *MaPle+*.

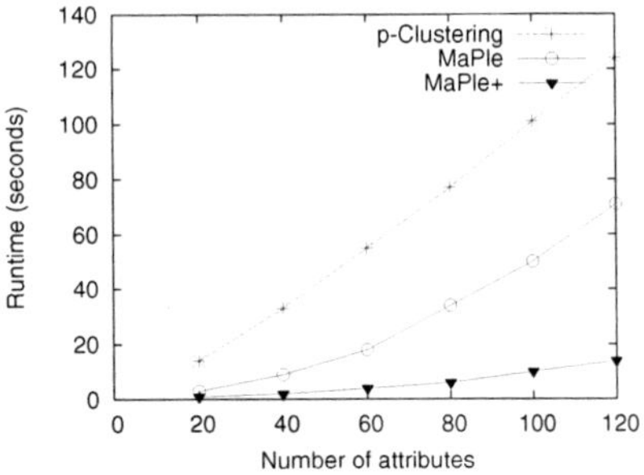

Fig. 3.13 Scalability with respect to the number of attributes in the data sets.

3.5 Conclusions

As indicated by previous studies, pattern-based clustering is a practical data mining task with many applications. However, efficiently and effectively mining pattern-based clusters is still challenging. In this paper, we propose the mining of maximal pattern-based clusters, which are non-redundant pattern-based clusters. By removing the redundancy, the effectiveness of the mining can be improved substantially.

Moreover, we develop *MaPle* and *MaPle+*, two efficient and scalable algorithms for mining maximal pattern-based clusters in large databases. We test the algorithms on both real life data sets and synthetic data sets. The results show that *MaPle+* clearly outperforms the best method previously proposed.

References

1. Ramesh C. Agarwal, Charu C. Aggarwal, and V. V. V. Prasad. A tree projection algorithm for generation of frequent item sets. *Journal of Parallel and Distributed Computing*, 61(3):350–371, 2001.
2. C.C. Aggarwal, J.L. Wolf, P.S. Yu, C. Procopiuc, and J.S. Park. Fast algorithms for projected clustering. In *Proc. 1999 ACM-SIGMOD Int. Conf. Management of Data (SIGMOD'99)*, pages 61–72, Philadelphia, PA, June 1999.
3. C.C. Aggarwal and P.S. Yu. Finding generalized projected clusters in high dimensional spaces. In *Proc. 2000 ACM-SIGMOD Int. Conf. Management of Data (SIGMOD'00)*, pages 70–81, Dallas, TX, May 2000.
4. R. Agrawal, J. Gehrke, D. Gunopulos, and P. Raghavan. Automatic subspace clustering of high dimensional data for data mining applications. In *Proc. 1998 ACM-SIGMOD Int. Conf. Management of Data (SIGMOD'98)*, pages 94–105, Seattle, WA, June 1998.
5. R. Agrawal, T. Imielinski, and A. Swami. Mining association rules between sets of items in large databases. In *Proc. 1993 ACM-SIGMOD Int. Conf. Management of Data (SIGMOD'93)*, pages 207–216, Washington, DC, May 1993.
6. R. Agrawal and R. Srikant. Fast algorithms for mining association rules. In *Proc. 1994 Int. Conf. Very Large Data Bases (VLDB'94)*, pages 487–499, Santiago, Chile, Sept. 1994.

7. K. S. Beyer, J. Goldstein, R. Ramakrishnan, and U. Shaft. When is "nearest neighbor" meaningful? In C. Beeri and P. Buneman, editors, *Proceedings of the 7th International Conference on Database Theory (ICDT'99)*, pages 217–235, Berlin, Germany, January 1999.
8. C. H. Cheng, A. W-C. Fu, and Y. Zhang. Entropy-based subspace clustering for mining numerical data. In *Proc. 1999 Int. Conf. Knowledge Discovery and Data Mining (KDD'99)*, pages 84–93, San Diego, CA, Aug. 1999.
9. Yizong Cheng and George M. Church. Biclustering of expression data. In *Proc. of the 8th International Conference on Intelligent System for Molecular Biology*, pages 93–103, 2000.
10. Mohammad El-Hajj and Osmar R. Zaïane. Inverted matrix: efficient discovery of frequent items in large datasets in the context of interactive mining. In *KDD '03: Proceedings of the ninth ACM SIGKDD international conference on Knowledge discovery and data mining*, pages 109–118. ACM Press, 2003.
11. B. Ganter and R. Wille. *Formal Concept Analysis – Mathematical Foundations*. Springer, 1996.
12. J. Han, J. Pei, and Y. Yin. Mining frequent patterns without candidate generation. In *Proc. 2000 ACM-SIGMOD Int. Conf. Management of Data (SIGMOD'00)*, pages 1–12, Dallas, TX, May 2000.
13. H. V. Jagadish, J. Madar, and R. Ng. Semantic compression and pattern extraction with fascicles. In *Proc. 1999 Int. Conf. Very Large Data Bases (VLDB'99)*, pages 186–197, Edinburgh, UK, Sept. 1999.
14. D. Jiang, J. Pei, M. Ramanathan, C. Tang, and A. Zhang. Mining coherent gene clusters from gene-sample-time microarray data. In *Proceedings of the tenth ACM SIGKDD international conference on Knowledge discovery and data mining (KDD'04)*, pages 430–439. ACM Press, 2004.
15. Daxin Jiang, Jian Pei, and Aidong Zhang. DHC: A density-based hierarchical clustering method for gene expression data. In *The Third IEEE Symposium on Bioinformatics and Bioengineering (BIBE'03)*, Washington D.C., March 2003.
16. Guimei Liu, Hongjun Lu, Wenwu Lou, and Jeffrey Xu Yu. On computing, storing and querying frequent patterns. In *KDD '03: Proceedings of the ninth ACM SIGKDD international conference on Knowledge discovery and data mining*, pages 607–612. ACM Press, 2003.
17. J. Liu, Y. Pan, K. Wang, and J. Han. Mining frequent item sets by opportunistic projection. In *Proc. 2002 ACM SIGKDD Int. Conf. on Knowledge Discovery and Data Mining (KDD'02)*, pages 229–238, Edmonton, Alberta, Canada, July 2002.
18. J. Liu and W. Wang. Op-cluster: Clustering by tendency in high dimensional space. In *Proceedings of the Third IEEE International Conference on Data Mining (ICDM'03)*, Melbourne, Florida, Nov. 2003. IEEE.
19. N. Pasquier, Y. Bastide, R. Taouil, and L. Lakhal. Discovering frequent closed itemsets for association rules. In *Proc. 7th Int. Conf. Database Theory (ICDT'99)*, pages 398–416, Jerusalem, Israel, Jan. 1999.
20. J. Pei, X. Zhang, M. Cho, H. Wang, and P. S. Yu. Maple: A fast algorithm for maximal pattern-based clustering. In *Proceedings of the Third IEEE International Conference on Data Mining (ICDM'03)*, Melbourne, Florida, Nov. 2003. IEEE.
21. S. Tavazoie, J. Hughes, M. Campbell, R. Cho, and G. Church. Yeast micro data set. In *http://arep.med.harvard.edu/biclustering/yeast.matrix*, 2000.
22. H. Wang, W. Wang, J. Yang, and P.S. Yu. Clustering by pattern similarity in large data sets. In *Proc. 2002 ACM-SIGMOD Int. Conf. on Management of Data (SIGMOD'02)*, Madison, WI, June 2002.
23. Jiong Yang, Wei Wang, Haixun Wang, and Philip S. Yu. δ-cluster: Capturing subspace correlation in a large data set. In *Proc. 2002 Int. Conf. Data Engineering (ICDE'02)*, San Fransisco, CA, April 2002.
24. M. J. Zaki, S. Parthasarathy, M. Ogihara, and W. Li. New algorithms for fast discovery of association rules. In *Proc. 1997 Int. Conf. Knowledge Discovery and Data Mining (KDD'97)*, pages 283–286, Newport Beach, CA, Aug. 1997.

25. L. Zhao and M. Zaki. Tricluster: An effective algorithm for mining coherent clusters in 3d microarray data. In *Proc. 2005 ACM SIGMOD Int. Conf. on Management of Data (SIGMOD'05)*, Baltimore, Maryland, June 2005.

Chapter 4
Role of Human Intelligence in Domain Driven Data Mining

Sumana Sharma and Kweku-Muata Osei-Bryson

Abstract Data Mining is an iterative, multi-step process consisting of different phases such as domain (or business) understanding, data understanding, data preparation, modeling, evaluation and deployment. Various data mining tasks are dependent on the human user for their execution. These tasks and activities that require human intelligence are not amenable to automation like tasks in other phases such as data preparation or modeling are. Nearly all Data Mining methodologies acknowledge the importance of the human user but do not clearly delineate and explain the tasks where human intelligence should be leveraged or in what manner. In this chapter we propose to describe various tasks of the domain understanding phase which require human intelligence for their appropriate execution.

4.1 Introduction

In recent times there has been a call for shift from emphasis on "pattern-centered data mining" to "domain-centered actionable knowledge discovery" [1]. The role of the human user is indispensable in generating actionable knowledge from large data sets. This follows from the fact that the data mining algorithms can only automate the building of models, but there still remain, a multitude of tasks that require human participation and intelligence for their appropriate execution.

Most data mining methodologies describe at least some tasks that are dependent on human actors for their execution [2–6] but do not sufficiently highlight these tasks, or describe the manner in which human intelligence could be leveraged. In Section 4.2 we describe various tasks pertaining to domain-drive data mining (DDDM) that require human input. We specify the role of these tasks in the overall data mining project life cycle with the goal of illuminating the significance of the

---

Sumana Sharma, Kweku-Muata Osei-Bryson
Virginia Commonwealth University, e-mail: sharmas5@vcu.edu, kmuata@isy.vcu.edu

role of the human actor in this process. Section 4.3 presents discussion of directions for future research and Section 4.4 presents the summary of contents of this chapter.

4.2 DDDM Tasks Requiring Human Intelligence

Data Mining is an iterative process comprising of various phases which is turn comprise of various tasks. Examples of these phases include the business (or domain) understanding phase, data understanding and data preparation phases, modeling and evaluation phases and finally the implementation phase whereby discovered knowledge is implemented to lead to actionable results . Tasks pertaining to data preparation and modeling are being increasingly automated, leading to the impression that human interaction and participation is not required for their execution. We believe that this is only partly true. While sophisticated algorithms packaged in data mining software suites can lead to near automatic processing of patterns (or knowledge nuggets) underlying large volumes of data, the search for such patterns needs to be guided using a clearly defined objective. In fact the setting up of a clear objective is only one of the many tasks that are dependent on the knowledge of the human user. We identify at least 11 other tasks that are dependent on human intelligence for their appropriate execution. All these tasks are summarized below. The wording of these statements has been based on the CRISP-DM process model [5], although it must be pointed out that other process models also recommend a similar sequence of tasks for the execution of data mining projects.

1. Formulating Business Objectives
2. Setting up Business Success Criteria
3. Translating Business Objective to Data Mining Objectives
4. Setting up Data Mining Success Criteria
5. Assessing Similarity between Business Objectives of New Projects and Past Projects
6. Formulating Business, Legal and Financial Requirements
7. Narrowing down Data and Creating Derived Attributes
8. Estimating Cost of Data Collection, Implementation and Operating Costs
9. Selection of Modeling Techniques
10. Setting up Model Parameters
11. Assessing Modeling Results
12. Developing a Project Plan

4.2.1 Formulating Business Objectives

The task of formulating business objectives is pervasive to all tasks of the project and provides direction for the entire data mining project. The business objective describes the goals of the project in business terms and it should be completed before

4 Role of Human Intelligence in Domain Driven Data Mining   55

any other task is undertaken or resources committed. Business objectives determination requires discussion among responsible business personnel who interact synchronously or asynchronously to finalize business objective(s). These business personnel may include various types of human actors such as domain experts, project sponsor, key project personnel etc.

The human actors discuss the problem situation at hand, analyze background situation and characterize it by formulating a sound business objective. A good business objective follows the SMART acronym which lays down qualities of a good objective as being Specific, Measurable, Attainable, Relevant and Time-Bound. The human actors must use their judgment to ensure that the business objective set up by them follows these requirements. Not doing to may lead to an inadequate or incorrect objective which in turn may result in failure of the project in the end.

The human actors must also use their intelligence to ensure that the business objective is congruent with the overall objectives of the firm. If this is not the case, then the high level stakeholders are not likely to approve of the project or assign necessary resources for the project's completion. We recommend that the human actors involved in the project formulate the business objective such that the relation between the business objective and the organizational objective becomes clear.

Sometimes, the human actors may have to interact with each other before this relation becomes obvious. For instance, there may be a potential data mining project to find a way to reduce loss rates from a certain segment of a company's credit card customers. However, the objective 'determining a way to reduce loss rates is not a viable one'. Perhaps the biggest drawback is that such as objective does not show how this objective will help achieve the firm's overall objectives. Reformulating the objective such as 'to increase profits from [a particular segment of customers] by lowering charge off rates' may address this issue. Formulating the objective to achieve congruency with organizational objectives as well to satisfy the SMART acronym requires deliberation amongst involved actors. Automated tools cannot help in this case and the execution of the task is dependent on human intelligence.

4.2.2 Setting up Business Success Criteria

Business success criteria are objective and/or subjective criteria that help to establish whether or not the DM project achieved set business objectives and could be regarded as successful. They help to eliminate bias in evaluation of results of DM project. It is important that setting up of business success criteria is preceded by determination of business objectives. The setting up of criteria requires discussion among responsible business personnel who interact synchronously or asynchronously to finalize them.

The success criteria must be congruent with the business objective themselves. For instance, it would be incorrect to set up a threshold for profitability level if the business objective does not aim at increasing profitability. The human actors use their intelligence in setting up criteria that are in sync with the business objectives and which will lead to rigorous evaluation of results at the end of the project.

4.2.3 Translating Business Objective to Data Mining Objectives

The data mining objective(s) is defined as the technical translation of the business objective(s) [5]. However, no automated tools exist that can perform such translation and therefore it is the responsibility of human actors such as business and technical stakeholders to convert the set business objectives into appropriate data mining objective. This is a critical task as one business objective may often be satisfied using various data mining objectives. However the human actors collaborate and decide upon which data mining objective is the most appropriate one. For instance with respect to the example from the credit card firm presented earlier, numerous data mining goals may be relevant. Examples would include increasing profits by increasing approval rates for customers, increasing profits by increasing approval rates but maintaining better or similar loss rates, etc. clearly the selection of one objective over the other cannot be done unless the human actors bring the domain knowledge and background information into play. It is expected that such translation of business objective to data mining objective will involve considerable interaction among actors and utilize their intelligence.

4.2.4 Setting up of Data Mining Success Criteria

The rationale of setting up business success criteria to evaluate the achievement of business objectives also applies to data mining objectives. Data Mining success criteria help to evaluate the technical data mining results yielded by the modeling phase and assess whether or not the data mining objectives will be satisfied by deploying a particular solution. Human actors, often technical stakeholders, may be involved in setting up these criteria. It is important to note that the technical stakeholders may also need to incorporate certain input from the business stakeholders in setting up evaluation criteria. For instance, the business users may be particular about only implementing solutions that have a certain level of simplicity. If the technical stakeholders have decided on a data mining objective that will lead to a classification model, they could incorporate simplicity as a data mining success criteria and assess it using the number of leaves in a classification tree model.

Osei-Bryson [8] provides a method for setting up data mining success criteria. He also recommends setting up threshold values and combination functions. For instance, human actors may agree on accuracy, simplicity and stability as data mining success criteria. Different data mining models built on the same data set may vary with respect to these criteria. Additionally, different criteria may also be weighted differently. In such as case, the human user would need to compare these varying competing models by studying which models satisfy threshold values for all competing criteria and possibly generate a score by summing up the weighted scores for different criteria. Presently no tool exists that can execute this crucial task and it is therefore the responsibility of the human actors to use their judgment to set up the technical evaluation criteria and further details such as their threshold values, weights etc.

4.2.5 Assessing Similarity Between Business Objectives of New and Past Projects

It is important to assess the similarity between business objectives of new projects with past projects so as to avoid duplication of effort and/or to learn from the past project. This task should be performed carefully be the business user as it may require subjective judgment on his part to determine which past project could be regarded as most relevant and if it is possible to use their results in any way in order to increase the efficiency of the new project. A case-based reasoner could be used to help the human user with identifying some of the cases, but still most of the responsibility would still lie with the intelligent human user. The goal of studying past project would be leverage the experience gained in the past and is an example of knowledge re-use the importance of which has been highlighted in the literature [7].

4.2.6 Formulating Business, Legal and Financial Requirements

Requirements formulation is also an important task that must be performed jointly by various human actors. One main type of business requirement that should be assessed by human actors is to determine what kinds of models are needed. Specifically, it must be established whether only an explanatory model is required or if there is no such constraint and non-explanatory models are also applicable. This task requires collaboration and the output is not immediately obvious. In such a case, the human actor, often a business user may discuss with other key stakeholders what kind of model is acceptable. In certain domains such as the credit card industry or insurance industry, a firm may be liable to explaining how and why a certain decision (such as rejecting the loan application of an applicant) was made. If a black box approach such as a standalone neural network model was used to make the decision, the firm is likely to land in trouble. Human actors involved with such domains may use their domain knowledge to ensure that resources are spent in developing only robust explanatory models.

Human actors are also responsibility eliciting and imposing legal requirements in data mining projects. One important legal requirement may be related to the type of data attributes that could be used in building a data mining model. The human users must use their domain knowledge to exclude such variables as sex, race, religion etc from data mining models. Other legal requirements, based on a particular firm's domain should also be clearly established and communicated to relevant stakeholder such as technical personnel involved in setting up the data mining models.

It is also the responsibility of human actors to carefully assess the requirements in form of financial constraints on the data mining project. The human users will need to get an approval of budget and resources from the project sponsor and once more details are established, determine whether the project can be successfully carried out with the granted financial resources. The financial requirements may not always be apparent at the start of the project and it is the duty of the human actors to utilize their domain knowledge to make appropriate recommendations regarding budgetary considerations, if the need arises.

4.2.7 Narrowing down Data and Creating Derived Attributes

Once the business and data mining objectives have been set up, the human actors must narrow down the data attributes that will be used in building the data mining models. This task helps to establish whether or not the required data resources are available and if any additional data needs to be collected or purchased (say from external vendors) by the organization. Key technical stakeholders and technical domain experts participate in this process. Domain experts can be interviewed to determine which data resources are applicable for the project. GSS (group support systems) type tools can be used by technical stakeholders to discuss about and finalize relevant data resources that should be used in the project. Case base of past projects can also be used to data resources used by similar past projects.

The human actors must also participate in creating derived attributes, i.e. attributes created by utilizing two or more data attributes. In some cases, use of a particular data attribute such as debt or income may not yield a good model if used independently. Human actors may agree that a derived attribute such as the debt-to-income ratio makes more business sense and incorporate it in the building of models. The creation of derived attributes requires significant domain knowledge to ensure that variables are only being confined in a meaningful way. This is a critical task as adding derived variables often leads to more accurate data mining models. However as is apparent, this task is dependent on human intelligence for its appropriate execution.

4.2.8 Estimating Cost of Data Collection, Implementation and Operating Costs

Developing estimates of costs of data collection, implementation of solution and operating costs is necessary to ensure that the project meets the budgetary constraints and is feasible to implement. Key technical stakeholders, technical domain experts and technical operational personnel participate in this process. Domain experts can be interviewed to understand costs of data collection and implementation. Project Management cost estimation tools can be used to estimates of each of these costs. Case base of past projects can be used to assess the same costs associated similar past projects.

4.2.9 Selection of Modeling Techniques

Even after the business requirement in form of explanatory or non explanatory model has been identified, there still remains the task of selecting among or using all of the applicable modeling techniques. The human actors are responsible for the task of enumerating applicable techniques and them selecting among these set of techniques. For instance, if the requirement is for an explanatory model, the human actors may agree on classification tree, regression, and k nearest neighbor as being applicable techniques. They may extend this list to also include neural networks, support vector machines etc if non explanatory models were also available. They may also use their knowledge to combine these model to produce ensemble models [2] if applicable.

4.2.10 Setting up Model Parameters

Available data mining software suites such as SAS Enterprise Miner, SPSS Clementine, Angoss Knowledge Seeker etc have simplified the task of searching for patterns using techniques such as decision trees, neural networks, clustering etc. However, in order to efficiently run these data mining algorithms or techniques, the human actor needs to set up the various parameters. He or she will have to use their knowledge to make a choice of parameter values that accurately reflect the objectives and requirements of the project. It should be noted that the multitude of parameter values may be left at their default values. Nearly all data mining software tools update the parameter fields with default values. However this is dangerous and is likely to lead to sub optimal modeling results. Each parameter has some business implication and therefore it is important that the human actor uses his knowledge and judgment in populating the various fields with appropriate values.

4.2.11 Assessing Modeling Results

The assessment of modeling results churned out in form of large number of models by software tools is the responsibility of the intelligent human actors. They must engage in assessing the results generated against technical and business criteria set up earlier and make a decision regarding the appropriate model. For instance, consider the generation of large number of decision tree models generated by varying parameter values. All of the models are built using the same data. In such a case, the human actor must decide on which model is the best one (by studying how each model fares on different evaluation criteria) and select the best among the competing models for actual implementation.

4.2.12 Developing a Project Plan

Developing a project plan is crucial for successful implementation and monitoring of the project. Formal documentation of various tasks associated with the project, actors involved and necessary resources for executing each of the tasks. Both key technical and business stakeholders participate in this process, but a project manager is primarily responsible for formulating the project plan. Work breakdown structuring tools and project management planning and management tools can be used for creation and documentation of project plan. The project plan requires considerable domain expertise to not just correctly estimate the direction for the project but also the resources that will ne necessary to achieve the set objectives. No tool can automate the creation of the project plan for a data mining project and the human user will play an important role in the successful execution of this critical task.

4.3 Directions for Future Research

Future research needs to delve deeper into the role of human intelligence in data mining projects. Data Mining case studies depicting failure of data mining projects are likely to be great lessons and may serve to identify if lack of human involvement was one the main reasons of the failure. Additionally, research also needs to focus on creation of new techniques or utilization of old techniques to assist the human actors in performing the tasks that they are responsible for. With time constraints often being a realistic concern for real world organizations, human actors must be better equipped to execute the tools entitled to them. Some of these tools such as case-based reasoning system, group support system tools, project management tools etc have been highlighted in this chapter. Future research can focus on exploring various other such tools. It is likely to lead to growing recognition of the importance of the human actor and expectedly better results through the increased participation of human actors.

4.4 Summary

The significance of human intelligence in data mining endeavors is being increasingly recognized and accepted. While there has been a strong advancement in area of modeling techniques and algorithms, the role of the human actor and his intelligence have not been sufficiently explored. This chapter aims to bridge this gap by clearly highlighting various data mining tasks that require human intelligence and in what manner. The pitfall of neglecting the role of human intelligence in executing various tasks has also been illuminated. The discussion provided in this chapter will also help renew the focus on the nature of the iterative and interactive Data Mining process which requires the role of the intelligent human in order to lead to valid and meaningful results.

References

1. Cao, L. and C. Zhang (2006). "Domain-Driven Data Mining: A Practical Methodology." International Journal of Data Warehousing and Mining 2(4): 49-65.
2. Berry, M. and G. Linoff (2000). Mastering Data Mining: The Art and Relationship of Customer Relationship Management, John Wiley and Sons
3. Cabena, P., P. Hadjinian, et al. (1998). Discovering Data Mining: From Concepts to Implementation., Prentice Hall.
4. Cios, K. and L. Kurgan (2005). Trends in Data Mining and Knowledge Discovery. Advanced Techniques in Knowledge Discovery and Data Mining. N. Pal and L. Jain, Springer: 1-26.
5. CRISP-DM. (2003). "Cross Industry Standard Process for Data Mining 1.0: Step by Step Data Mining Guide." Retrieved 01/10/07, from http://www.crisp-dm.org/.
6. Fayyad, U., G. Paitetsky-Shapiro, et al. (1996). "The KDD process for extracting useful knowledge from volumes of data." Communications of the ACM 39(11): 27-34.
7. Markus, M. L. (2001). "Toward a Theory of Knowledge Reuse: Types of Knowledge Reuse Situations and Factors in Reuse Success." Journal of Management Information Systems 18(1): 57-94.
8. Osei-Bryson, K.-M. (2004). "Evaluation of Decision Trees." Computers and Operations Research 31: 1933-1945.

Chapter 5
Ontology Mining for Personalized Search

Yuefeng Li and Xiaohui Tao

Abstract Knowledge discovery for user information needs in user local information repositories is a challenging task. Traditional data mining techniques cannot provide a satisfactory solution for this challenge, because there exists a lot of uncertainties in the local information repositories. In this chapter, we introduce ontology mining, a new methodology, for solving this challenging issue, which aims to discover interesting and useful knowledge in databases in order to meet the specified constraints on an ontology. In this way, users can efficiently specify their information needs on the ontology rather than dig useful knowledge from the huge amount of discorded patterns or rules. The proposed ontology mining model is evaluated by applying to an information gathering system, and the results are promising.

5.1 Introduction

In the past decades the information available on the World Wide Web has exploded rapidly. Web information covers a great range of topics and serves a broad spectrum of communities. How to gather needed information from the Web, however, becomes a challenging issue.

Web mining, knowledge discovery in Web data, is a possible direction to answer this challenge. However, the difficulty is that Web information includes a lot of uncertain data. It argues that the key to satisfy an information seeker is to understand the seeker, including her (or his) background and information needs. Usually Web users implicitly use concept models to judge the relevance of a document, although they may not know how to express the models [9]. To obtain such a concept model and rebuild it for a user, most systems use training sets which include both positive and negative samples to obtain useful knowledge for personalized Web search.

Yuefeng Li, Xiaohui Tao
Faculty of Information Technology, Queensland University of Technology, Australia, e-mail: {y2.li,x.tao}@qut.edu.au

The current methods for acquiring training sets can be grouped into three categories: the interviewing (or relevance feedback), non-interviewing and pseudo-relevance feedback strategies. The first category is manual techniques and usually involve great efforts by users, e.g. questionnaire and interview. The downside of such techniques is the cost of time and money. The second category, non-interviewing techniques, attempts to capture a user's interests by observing the user behavior or mining knowledge from the records of the user's browsing history. These techniques are automated, but the generated user profiles lack accuracy, as too many uncertainties exist in the records. The third category techniques perform a search using a search engine and assume the top-K retrieved documents as positive samples. However, not all top-K documents are real positive. In summary, these current techniques need to be improved.

In this paper, we propose an ontology mining model to find perfect training sets in user *local instance (information) repositories* (LIR) using an ontology, a *world knowledge* base. World knowledge is the commonsense knowledge possessed by humans [20], and is also called user background knowledge. An LIR is a personal collection of information items that were frequently visited by a user in a period of time. These information items could be a set of text documents, emails, or Web pages, that implicitly cite the concepts specified in the world knowledge base. The proposed model starts to ask users provide a query to access the ontology in order to capture their information needs at the concept level. Our model aims to better interpret knowledge in LIRs in order to improve the performance of personalized search. It contributes to data mining for the discovery of interesting and useful knowledge to meet what users want. The model is evaluated by applying to a Web information gathering system, against several baseline models. The evaluation results are promising.

The paper is organized as follows. Section 5.2 presents related work. The architecture of our proposed model is presented in Section 5.3. Section 5.4 presents the background information including the world knowledge base and LIRs. In Section 5.5, we describe how to discover knowledge from data and construct an ontology, and in Section 5.6 we present how to mine the topics of user interests from the ontology. The related experiment designs are described in Section 5.7, and the related results are discussed in Section 5.8. Finally, Section 5.9 makes conclusions.

5.2 Related Work

Much effort has been invested in semantic interpretation of user topics and concept models for personalized search. Chirita et al. [1] and Teevan et al. [16] used a collection of a user's desktop text documents, emails, and cached Web pages, to explore user interests. Many other works are focused on using user profiles. A user profile is defined by Li and Zhong [9] as the topics of interests related to a user information need. They also classified Web user profiles into two diagrams: the data diagram and information diagram. A data diagram profile is usually generated by analyzing a database or a set of transactions, e.g. user log data [2, 9, 10, 13]. An information diagram profile is generated by using manual techniques such as questionnaires and interviews or by using the information retrieval techniques and machine-learning methods [10, 17]. These profiles are largely used in personalized search by [2, 3, 9, 17, 21].

Ontologies represent information diagram profiles by using a predefined taxonomy of concepts. Ontologies can provide a basis for the match of initial behavior information and the existing concepts and relations [2, 17]. Li, et al. [7–9, 19] used ontology mining techniques to discover interesting patterns from positive samples and to generate user profiles. Gauch et al. [2] used Web categories to learn personalized ontology for users. Sieg et al. [12] modelled a user's context as an ontological profile with interest scores assigned to the contained concepts. Developed by King et al. [4], *IntelliOnto* is built based on the Dewey Decimal Classification to describe a user's background knowledge. Unfortunately, these aforementioned works cover only a small volume of concepts, and do not specify the semantic relationships of *partOf* and *kindOf* existing in the concepts but only *superClass* and *subClass*.

In summary, there still remains a research gap in semantic study of a user's interests by using ontologies. Filling this gap in order to better capture a user information need motivates our research work presented in this paper.

5.3 Architecture

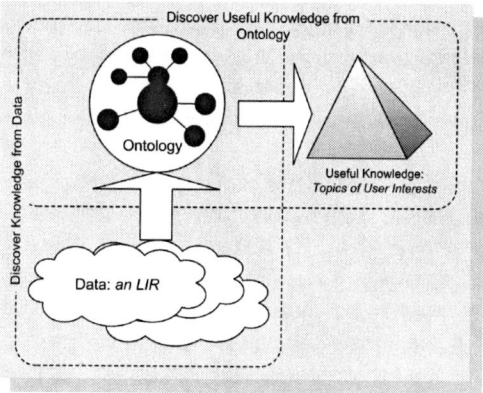

Fig. 5.1 The Architecture of Ontology Mining Model

Our proposed ontology mining model aims to discover the useful knowledge from a set of data by using an ontology. In order to better interpret a user information need, we need to capture the user's interests and preferences. These knowledge underly from a user's LIR and can be interpreted in a high level representation, like ontologies. However, how to explore and discover the knowledge from an LIR remains a challenging issue. Firstly, an LIR is just a collection of unstructured or semi-structured text data. There are many noisy data and uncertainties in the collection. Secondly, not all the knowledge contained in an LIR are useful for user information need interpretation. Only the knowledge relevant to the information need are needed. The ontology mining model is adopted to discover interesting and useful knowledge in LIRs in order to meet the constraints specified for user information needs on the ontology.

The architecture of the model is presented in Fig. 5.1, which shows the process of finding what users want in LIRs. A user first expresses her (his) information need using some concepts in the ontology. We can then label the useful knowledge in the ontology against the queries and generate a personalized ontology for the user. In addition, the relationship between the personalized ontology and LIRs can also be specified to find positive samples in LIRs.

5.4 Background Definitions

5.4.1 World Knowledge Ontology

A world knowledge base is a general ontology that formally describes and specifies world knowledge. In our experiments, we use the Library of Congress Subject Headings[1] (LCSH) for the base. The LCSH ontology is a taxonomic classification developed for organizing the large volumes of library collections and for retrieving information from the library. It aims to facilitate users' perspectives in accessing the information items stored in a library. The system is comprised of a thesaurus containing about 400,000 subject headings that cover an exhaustive range of topics. The LCSH is ideal for a world knowledge base as it has semantic subjects and relations specified.

A subject (or called concept) heading in the LCSH is transformed into a primitive knowledge unit, and the LCSH structure forms the backbone of the world knowledge base. The *BT* and *NT* references defined in the LCSH are to specify two subjects describing the same entity but at different levels of abstraction (or concretion). These references are transformed into the *kindOf* relationships in the world knowledge base. The *UF* references specify the compound subjects and the subjects subdivided by others, and are transformed into the *partOf* relationships. *KindOf* and *partOf* are both transitive and asymmetric. The world knowledge base is formalized as follows:

Definition 5.1. Let $\mathbb{WKB}$ be a taxonomic world knowledge base. It is formally defined as a 2-tuple $\mathbb{WKB} :=< \mathbb{S}, \mathbb{R} >$, where

---

[1] Library of Congress: Classification Web, http://classificationweb.net/.

- $\mathbb{S}$ is a set of subjects $\mathbb{S} := \{s_1, s_2, \cdots, s_m\}$, in which each element is a 2-tuple $s :=\, <label, \sigma>$, where *label* is a label assigned by linguists to subject s and is denoted by $label(s)$, and $\sigma(s)$ is a signature mapping defining a set of relevant subjects to s and $\sigma(s) \subseteq \mathbb{S}$;
- $\mathbb{R}$ is a set of relations $\mathbb{R} := \{r_1, r_2, \cdots, r_n\}$, in which each element is a 2-tuple $r := \,<type, r_v>$, where *type* is a relation type of *kindOf* or *partOf*, and $r_v \subseteq \mathbb{S} \times \mathbb{S}$. For each $(s_x, s_y) \in r_v$, s_y is the subject who holds the *type* of relation to s_x, e.g. s_x is *kindOf* s_y.

5.4.2 Local Instance Repository

A local instance repository (LIR) is a collection of information items (e.g., Web documents) that are frequently visited by a user during a period of time. These items implicitly cite the knowledge specified in the world knowledge base. In this demonstrated model, we use a set of the library catalogue information items that were accessed by a user recently to represent a user's LIR. Each item in the catalogue has a title, a table of contents, a summary, and a list of subjects assigned based on the LCSH. The subjects build the bridge connecting an LIR to the world knowledge base.

We call an element in an LIR as an *instance*, which is a set of terms generated from these information after text pre-processing including stopword removal and word stemming. For a given query q, let $I = \{i_1, i_2, \cdots, i_p\}$ be an LIR where i denotes an instance, and $\mathscr{S} \subseteq \mathbb{S}$ be a set of subjects (denoted by s) corresponding to I. Their relationships can be described as the following mappings:

$$\eta : I \to 2^{\mathscr{S}}, \quad \eta(i) = \{s \in \mathscr{S} | s \text{ is used to describe } i\} \subseteq \mathscr{S}; \tag{5.1}$$

$$\eta^{-1} : \mathscr{S} \to 2^{I}, \quad \eta^{-1}(s) = \{i \in I | s \in \eta(i)\} \subseteq I; \tag{5.2}$$

where $\eta^{-1}(s)$ is a reverse mapping of $\eta(i)$. These mappings aim to explore the semantic matrix existing between the subjects and instances.

Based on these mappings, we can measure the belief of an instance $i \in I$ to a subject $s \in \mathscr{S}$. The listed subjects assigned to an instance are indexed by their importance. Hence, the number of and the indexes of assigned subjects affect the belief of an instance to a subject. Thus, let $\xi(i)$ be the number of subjects assigned to i, $\iota(s)$ be the index of s on the assigned subject list (starting from 1), the belief of i to s can be calculated by:

$$bel(i, s) = \frac{1}{\iota(s) \times \xi(i)} \tag{5.3}$$

Greater $bel(i, s)$ indicates stronger belief of i to s.

5.5 Specifying Knowledge in an Ontology

A personalized ontology is built based on the world knowledge base and focused on a user information need. In Web search, a query is usually a set of terms generated by a user as a brief description of an information need. For an incoming query q, the relevant subjects are extracted from the $\mathbb{S}$ in $\mathbb{WKB}$ using the syntax-matching mechanism. We use $sim(s,q)$ to specify the relevance of a subject $s \in \mathbb{S}$ to q, which is counted by the size of overlapping terms between $label(s)$ and q. If $sim(s,q) > 0$, s is deemed as a positive subject. To construct a personalized ontology, the s's ancestor subjects in $\mathbb{WKB}$, along with their associated semantic relationships $r \in \mathbb{R}$, are extracted. By:

$$\mathscr{S}^+ = \{s | sim(s,q) > 0, s \in \mathscr{S}\}; \tag{5.4}$$

$$\mathscr{S}^- = \{s | sim(s,q) = 0, s \in \mathscr{S}\}; \tag{5.5}$$

$$\mathscr{R} = \{<r,(s_1,s_2)> | <r,(s_1,s_2)> \in \mathbb{R}, (s_1,s_2) \in \mathscr{S} \times \mathscr{S}\}; \tag{5.6}$$

we can construct an ontology $\mathscr{O}(q)$ against the given q. The formalization of a subject ontology $\mathscr{O}(q)$ is as follows:

Definition 5.2. The structure of a personalized ontology that formally describes and specifies query q is a 3-tuple $\mathscr{O}(q) := \{\mathscr{S}, \mathscr{R}, tax^{\mathscr{I}}\}$, where

- $\mathscr{S}$ is a set of subjects ($\mathscr{S} \subseteq \mathbb{S}$) which includes a subset of positive subjects $\mathscr{S}^+ \subseteq \mathscr{S}$ relevant to q, and a subset of negative subjects $\mathscr{S}^- \subseteq \mathscr{S}$ non-relevant to q;
- $\mathscr{R}$ is a set of relations and $\mathscr{R} \subseteq \mathbb{R}$;
- $tax^{\mathscr{I}} : tax^{\mathscr{I}} \subseteq \mathscr{S} \times \mathscr{S}$ is called the backbone of the ontology, which is constructed by two directed relationships $kindOf$ and $partOf$.

A sample ontology is constructed corresponding to a query "Economic espionage"[2]. A part of the ontology is illustrated in Fig 5.2, where the nodes in dark color are the positive subjects in $\mathscr{S}^+$, and the rest (white and grey) are the negatives in $\mathscr{S}^-$.

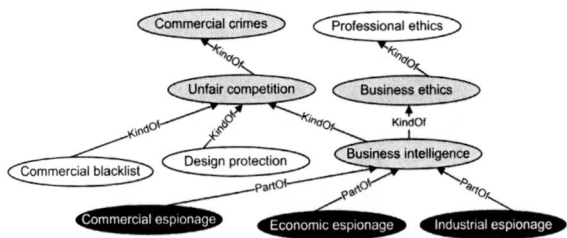

Fig. 5.2 A Constructed Ontology (Partial) for "*Economic Espionage*"

---

[2] A query generated by the linguists in Text REtrieval Conference (TREC), http://trec.nist.gov/.

5 Ontology Mining for Personalized Search

In order to capture a user information need, the constructed ontology needs to be personalized, since a user information need is individual. For this, we use a user's LIR to discover the topics related to the user's interests, and further personalize the user's constructed ontology.

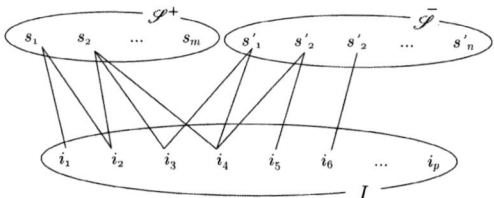

Fig. 5.3 The Relationships Between Subjects and Instances

An assumption exists for personalizing an ontology. Two subjects can be considered specifying the same semantic topic, if they map to the same instances. Similarly, if their mapping instances overlap, the semantic topics specified by the two subjects overlap as well. This assumption can be illustrated using Fig. 5.3. Let $\bar{\mathscr{S}}$ be the compliment set of $\mathscr{S}^+$ and $\bar{\mathscr{S}} = \mathscr{S} - \mathscr{S}^+$. Based on the mapping Eq. (5.1), each i maps to a set of subjects. As a result, a set of instances map to the subjects in different sets of $\mathscr{S}^+$ and $\bar{\mathscr{S}}$. As shown on Fig. 5.3, $s'_1 \in \bar{\mathscr{S}}$ overlaps $s_2 \in \mathscr{S}^+$ by its entire mapping instance set ($\{i_3, i_4\}$), and $s'_2 \in \bar{\mathscr{S}}$ overlaps s_2 by part of the instance set ($\{i_4\}$) only. Based on the aforementioned assumption, we can thus refine the $\mathscr{S}^+$ and $\mathscr{S}^-$, and have $\mathscr{S}^+$ expanded:

$$\mathscr{S}^+ = \mathscr{S}^+ \cup \{s' | s' \in \bar{\mathscr{S}}, \eta^{-1}(s') \cap (\bigcup_{s \in \mathscr{S}^+} \eta^{-1}(s)) \neq \emptyset\};$$

$$\mathscr{S}^- = \bar{\mathscr{S}} - \mathscr{S}^+. \tag{5.7}$$

This expansion is also illustrated in the example displayed in Fig. 5.2, in which the gray subjects are transferred from $\bar{\mathscr{S}}$ to $\mathscr{S}^+$ by having instances referred by some of the dark subjects. This expansion is based on the semantic study of a user's LIR, and thus personalizes the constructed ontology for the user.

As $sim(s,q)$ is used to specify the relevance to q of s, we also need to measure the relevance of the expanded positive subjects to q. The measure starts with calculating the instances' coversets:

$$coverset(i) = \{s | s \in \mathscr{S}, sim(s,q) > 0, s \in \eta(i)\} \tag{5.8}$$

We then have sim_{exp} for the relevance of an expanded positive subject s' by:

$$sim_{exp}(s',q) = \sum_{i \in \eta^{-1}(s')} \sum_{s \in coverset(i)} \frac{sim(s,q)}{|\eta^{-1}(s)|} \tag{5.9}$$

where s is a subject in the initialized but not expended $\mathscr{S}^+$. The value of $sim_{exp}(s',q)$ largely depends on the *sim* values of subjects in $\mathscr{S}^+$ that overlap with s' in their mapping instances.

5.6 Discovery of Useful Knowledge in LIRs

The personalized ontology describes the implicit concept model possessed by a user corresponding to an information need. The topics of user interests can be discovered from the ontology in order to better capture the information need.

We use the ontology mining method of *Specificity* introduced in [15] for the semantic study of a subject in an ontology. Specificity describes a subject's semantic focus on an information need. The specificity value *spe* of a subject s increases if the subject is located toward the leave level of an ontology's taxonomic backbone. In contrast, $spe(s)$ decreases if s is located toward the root level. Algorithm 7 presents a recursive method $spe(s)$ for assigning the specificity value to a subject in an ontology. Specificity aims to assess the strength of a subject in a user's personalized ontology.

---

input : the ontology $\mathscr{O}(q)$; a subject $s \in \mathbb{S}$; a parameter θ between $(0,1)$.
output: the specificity value $spe(s)$ of s.

1 If s is a leaf then let $spe(s) = 1$ and then return;
2 Let S_1 be the set of direct child subjects of s such that $\forall s_1 \in S_1 \Rightarrow type(s_1,s) = kindOf$;
3 Let S_2 be the set of direct child subjects of s such that $\forall s_2 \in S_2 \Rightarrow type(s_2,s) = partOf$;
4 Let $spe_1 = \theta, spe_2 = \theta$;
5 if $S_1 \neq \emptyset$ then calculate $spe_1 = \theta \times min\{spe(s_1) | s_1 \in S_1\}$;
6 if $S_2 \neq \emptyset$ then calculate $spe_2 = \frac{\sum_{s_2 \in S_2} spe(s_2)}{|S_2|}$;
7 $spe(s) = min\{spe_1, spe_2\}$.

Algorithm 1: Assigning Specificity to a Subject

Based on the specificity analysis of a subject, the strength *sup* of an instance i supporting a given query q can be measured by:

$$sup(i,q) = \sum_{s \in \eta(i)} bel(i,s) \times sim(s,q) \times spe(s). \quad (5.10)$$

If s is an expended positive subject, we use sim_{exp} instead of *sim*. If $s \in \mathscr{S}^-$, $sim(s,q) = 0$. The $sup(i,q)$ value increases if i maps to more positive subjects and these positive subjects hold stronger belief to q. The instances with their $sup(i,q)$ greater than a minimum value sup_{min} refer to the topics of the user's interests, whereas the instances with $sup(i,q)$ less than sup_{min} refer to the non-relevant topics. Therefore, we can have two instance sets I^+ and I^-, which satisfy

$$I^+ = \{i|sup(i,q) > sup_{min}, i \in I\}; \tag{5.11}$$

$$I^- = \{i|sup(i,q) < sup_{min}, i \in I\}. \tag{5.12}$$

Let $R = \sum_{i \in I^+} sup(i,q)$, $r(t) = \sum_{i \in I^+, t \in i} sup(i,q)$, $N = |I|$, and $n(t) = |\{i|i \in I, t \in i\}|$. We have the following modified probabilistic formula to choose a set of terms from the set of instances to represent a user's topics of interests:

$$weight(t) = \log \frac{\frac{r(t)+0.5}{R-r(t)+0.5}}{\frac{n(t)-r(t)+0.5}{(N-n(t))-(R-r(t))+0.5}} \tag{5.13}$$

The present ontology mining method discovers the topics of a user's interests from the user's personalized ontology. These topics reflect a user's recent interests and become the key to generate the user's profile.

5.7 Experiments

5.7.1 Experiment Design

A user profile is used in personalized Web search to describe a user's interests and preferences. The techniques that are used to generate a user profile can be categorized into three groups: the interviewing, the non-interviewing, and the pseudo-relevance feedback. A user profile generated by using the interviewing techniques can be called a "perfect" profile, as it is generated manually, and perfectly reflects a user's interests. One representative of such "perfect" profiles is the training sets used in the TREC-11 2002 Filtering Track (see: http://trec.nist.gov/). They are generated by linguists reading each document through and providing a judgement of positive or negative to the document against a topic [11]. The non-interviewing techniques do not involve user efforts directly. Instead, they observe and mine knowledge from a user's activity and behavior in order to generate a training set to describe the user's interests [17]. One representative is the OBIWAN model proposed by Gauch et al [2]. Different from the interviewing and non-interviewing techniques, the pseudo-relevance feedback profiles are generated by semi-manual techniques. These group of techniques perform a search first and assume the top-K returned documents as the positive sample feedback by a user. The Web training set acquisition method introduced by [14] is a typical model of such techniques, which analyzes the retrieved URLs using a belief based method to obtain approximation training sets.

Our proposed model is to compare with the baselines implemented for these typical models in the experiments. The implementation of our proposed model is called "Onto-based", and the three competitors are: (i) the TREC model generating the "perfect" user profiles and representing the manual interviewing techniques. It sets a golden model to mark the achievement of our proposed model; (2) the Web model for the Web training set acquisition method [14] and representing the semi-

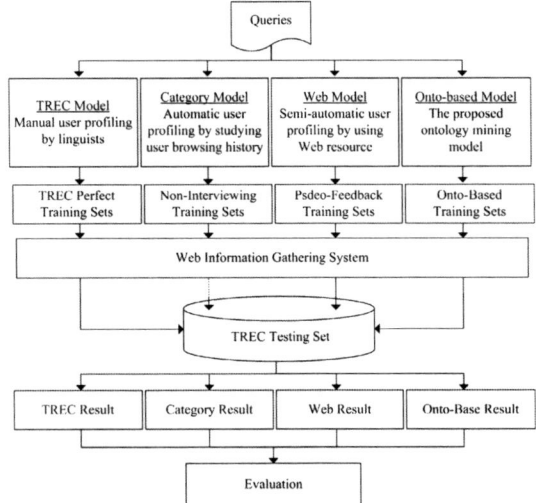

Fig. 5.4 The Dataflow of the Experiments

automated pseudo-relevance feedback mechanism; and (iii) the Category model for the OBIWAN [2] and representing the automated non-interviewing profiling techniques. Fig. 5.4 illustrates the experiment design. The queries go into the four models, and produce different profiles, represented by a training sets. The user profiles are used by the same Web information gathering system to retrieve relevant documents from the testing set. The retrieval results are compared and analyzed for evaluation of the proposed model.

5.7.1.1 Competitor Models

TREC Model: The training sets are manually generated by the TREC linguists. For an incoming query, the TREC linguists read a set of documents and marked each document either positive or negative against the query [11]. Since the queries are also generated by these linguists, the TREC training sets perfectly reflect a user's concept model. The support value of each positive document is assigned with 1, and negative with 0. These training sets are thus deemed as the "perfect" training sets.

The "perfect" model marks the research goal that our proposed model attempts to achieve. A successful retrieval of user interests and preferences can be confirmed if the performance achieved by the proposed model can match or be close to the performance of the "perfect" TREC model.

Category Model: In this model, a user profile is a set of topics related to the user's interests. Each topic is represented by a vector of terms trained from a user's browsing history using the $tf \cdot idf$ method. When searching, the cosine similarity value of an incoming document to a user profile is calculated, and a higher similarity value indicates that the document is more interesting to the user. In order to make the comparison fair, we used the same LIRs in the Onto-based model as the collection of a user's Web browsing history in this model.

Web Model: In this experimental model, the user profiles (training sets) are automatically retrieved from the Web by employing a Web search engine. For each incoming query, a set of positive concepts and a set of negative concepts are identified manually. By using *Google*, we retrieved a set of positive and a set of negative documents (100 documents in each set) using the identified concepts (the same Web URLs are also used by the Onto-based model). The support value of a document in a training set is defined based on (i) the precision of the chosen search engine; (ii) the index of a document on the result list, and (iii) the belief of a subject supporting or against a given query. This model attempts to use Web resources to benefit information retrieval. The technical details can be found in [14].

5.7.1.2 Our Model: Onto-based Model

The taxonomic world knowledge base is constructed based on the LCSH, as described in Section 5.4.1. For each query, we extract an LIR through searching the subject catalogue of Queensland University of Technology Library[3] using the query, as described in Section 5.4.2. These library information are available to the public and can be accessed for free.

We treat each incoming query as an individual user, as a user may come from any domain. For a given query, the model constructs an ontology first. Only the ancestor subjects away from a positive subject within three levels are extracted, as we believe that any subjects in more than that distance are no longer significant and can be ignored. The Onto-based model then uses Eq. (5.7) and (5.9) to personalize the ontology, and mines the ontology using the specificity method. The support values of the corresponding instances are calculated by Eq. (5.10), and the sup_{min} is set as zero and all the positive instances are used in the experiments. The modified probabilistic method (Eq. (5.13)) is then used to choose 150 terms to represent the user's topics of interests. Using the 150 terms, the model generates the training set by filtering the same Web URLs retrieval in the Web model. The positive documents are the top 20 ones that are weighted by the total probability function, and the rest URLs form the negative document set.

5.7.1.3 Information Gathering System

The common information gathering system is implemented, based on a model that tends to effectively gathering information by using user profiles [9]. We choose this model in this paper because it is suitable for both perfect training sets and approximation training sets.

---

[3] http://library.qut.edu.au

Each document in this model is represented by a pattern P which consists of a set of terms (T) and the distribution of term frequencies (w) in the document ($\beta(P)$).

Let PN be the set of discovered patterns. Using these patterns, we can have a probability function:

$$pr_\beta(t) = \sum_{P \in PN, (t,w) \in \beta(P)} support(P) \times w \qquad (5.14)$$

for all $t \in T$, where $support(P)$ is used to describe the percentage of positive documents that can be represented by the pattern for the perfect training sets, or the sum of the supports that are transferred from documents in the approximation training sets, respectively.

In the end, for an incoming document d, its relevance can be evaluated as

$$\sum_{t \in T} pr_\beta(t)\tau(t,d), \text{ where } \tau(t,d) = \begin{cases} 1 & \text{if } t \in d \\ 0 & \text{otherwise.} \end{cases} \qquad (5.15)$$

5.7.2 Other Experiment Settings

The Reuters Corpus Volume 1 (RCV1) [6] is used as the testbed in the experiments. The RCV1 is a large data set of 806,791 documents with great topic coverage. The RCV1 is also used in the TREC-11 2002 Filtering track for experiments. TREC-11 provides a set of topics defined and constructed by linguists. Each topic is associated with some positive and negative documents judged by the same group of linguists [11]. We use the titles of the first 25 topics (R101-125) as the queries in our experiments.

The performance is assessed by two methods: the precision averages at eleven standard recall levels, and F_1 Measure. The former is used in TREC evaluation as the standard for performance comparison of different information filtering models [18]. A recall-precision average is computed by summing the interpolated precisions at the specified recall cutoff and then dividing by the number of queries:

$$\frac{\sum_{i=1}^{N} precision_\lambda}{N}. \qquad (5.16)$$

N denotes the number of experimental queries, and $\lambda = \{0.0, 0.1, 0.2, \ldots, 1.0\}$ indicates the cutoff points where the precisions are interpolated. At each λ point, an average precision value over N queries is calculated. These average precisions then link to a curve describing the precision-recall performance. The other method, F_1 Measure [5], is well accepted by the community of information retrieval and Web information gathering. F_1 Measure is calculated by:

$$F_1 = \frac{2 \times precision \times recall}{precision + recall} \qquad (5.17)$$

Precision and recall are evenly weighted in F_1 Measure. The *macro-F_1* Measure averages each query's precision and recall values and then calculates F_1 Measure, whereas the *micro-F_1* Measure calculates the F_1 Measure for each returned result in a query and then averages the F_1 Measure values. The greater F_1 values indicate the better performance.

5.8 Results and Discussions

Table 5.1 The Average F-1 Measure Results and the Related Comparisons

Model	Macro-F1 Measure			Micro-F1 Measure		
	Average	Improvement	% Change	Average	Improvement	% Change
TREC	0.3944	-0.0061	-1.55%	0.3606	-0.0062	-1.72%
Web	0.382	0.0063	1.65%	0.3493	0.0051	1.46%
Category	0.3715	0.0168	4.52%	0.3418	0.0126	3.69%
Onto-based	0.3883	-	-	0.3544	-	-

The experimental precision and recall results are displayed in Fig. 5.5, where a chart for the precision averages at eleven standard recall levels is displayed. Table 5.1 presents the F_1 Measure results. The figures in "Improvement" column are calculated by using the average Onto-based model F_1 Measure results to minus the competitors' results. The percentages displayed in "% Change" column present the significance level of improvements achieved by the Onto-based model over the competitors, which is calculated by:

$$\% \ Change = \frac{\mathbb{F}_{Onto-based} - \mathbb{F}_{baseline}}{\mathbb{F}_{baseline}} \times 100\%. \tag{5.18}$$

where $\mathbb{F}$ denotes the average F_1 Measure result of an experimental model.

The comparison between the Onto-based and TREC models is to evaluate the user interests discovered by our proposed model to the knowledge specified by linguists completely manually. According to the results illustrated in Fig. 5.5, the Onto-based model has achieved the same performance as the perfect TREC model at most of the cutoff points (0-0.2, 0.4-0.6, 0.9-1.0). Considering that the "perfect" TREC training sets are generated manually, they are more precise than the Onto-based training sets. However, the TREC training sets may not covers the substantial relevant semantic space than the Onto-based training sets. The Onto-based model has about average 1000 documents in an LIR/per query for the discovery of interest topics. In contrast, The number of documents included in each TREC training set is very limited (about 60 documents per query on average). As a result, some semantic meanings referred by a given query are not fully covered by the TREC training sets. In comparison, the Onto-based model training sets cover much broader semantic extent. Consequently, although the expert knowledge contained by TREC sets is more precise, the Onto-based model's precision-recall performance is still close to the TREC model.

The close performance to the perfect TREC model achieved by the Onto-based model is also confirmed by the F_1 Measure results. As shown on Table 5.1, the TREC model outperforms the Onto-based model slightly by only about 1.55% in *Macro-F_1* and 1.72% in *Micro-F_1* Measure. The performance of proposed model is close to the golden model. Considering that the TREC model employs the human power of linguists to read every single document in the training set, which reflects a user's concept model perfectly, the close performance to the TREC model achieved by the Onto-based model is promising.

The comparison of the Onto-based and Category models are to evaluate our proposed model to the automated user profiling techniques using ontology. As shown on Fig. 5.5 and Table 5.1, the Onto-based model outperforms the Category model. On average, the Onto-based model improves performance from the Category model by 4.52% in *Macro-F_1* and 3.69% in *Micro-F_1* Measure. Comparing to the Category model, the Onto-based model specifies the concepts in the personalized ontology using more comprehensive semantic relations of *kindOf* and *partOf*, and analyzes the subjects by using the ontology mining method. The Category model specifies only the simple relations of *superClass* and *subClass*. Furthermore, the specificity ontology mining method appreciates a subject's locality in the ontology backbone, which is closer to the reality. Based on these, it can be concluded that our proposed model describes knowledge better than the Category model.

The comparison of the Onto-based and Web models are to evaluate the world knowledge extracted by the proposed method to the Web model. As shown in Fig. 5.5 and Table 5.1, the Onto-based model outperforms the Web model slightly. On average, the improvement achieved by the Onto-based model over the Web model are 1.65% in *Macro-F_1* and 1.46% in *Micro-F_1* Measure. After investigation, we found that although the same training sets are used, the Web documents, however, are not formally specified by the Web model. In contrast, the Onto-based

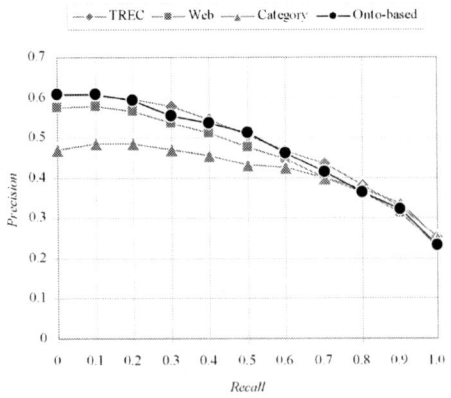

Fig. 5.5 The 11 Standard Recall-Precision Results

training sets integrate the world knowledge and the user interests discovered from the LIRs. Considering both two models using the same Web URLs, the user interests contained in the LIRs leverages the Onto-based model's performance actually. Based on these, we conclude that the proposed model improves the performance of Web information gathering from the Web model.

Based on the experimental results and the related analysis, we can conclude that our proposed ontology mining model is promising.

5.9 Conclusions

Ontology mining is an emerging research field, which aims to discover the interesting and useful knowledge in databases in order to meet some constraints specified on an ontology. In this paper, we have proposed an ontology mining model for personalized search. The model uses a world knowledge ontology, and captures user information needs from a user local information repository. The model has been evaluated using the standard data collection RCV1 with encouraging results.

Compared with several baseline models, the experimental results on RCV1 demonstrate that the performance of personalized search can be significantly improved by ontology mining. The substantial improvement is mainly due to the reducing of uncertainties in information items. The proposed model can reduce the burden of users' evolvement in knowledge discovery. It can also improve the performance of knowledge discovery in databases.

References

1. P. A. Chirita, C. S. Firan, and W. Nejdl. Personalized query expansion for the web. In *Proc. of the 30th intl. ACM SIGIR conf. on Res. and development in inf. retr.*, pages 7–14, 2007.
2. S. Gauch, J. Chaffee, and A. Pretschner. Ontology-based personalized search and browsing. *Web Intelli. and Agent Sys.*, 1(3-4):219–234, 2003.
3. J. Han and K.C.-C. Chang. Data mining for Web intelligence. *Computer*, 35(11):64–70, 2002.
4. J. D. King, Y. Li, X. Tao, and R. Nayak. Mining World Knowledge for Analysis of Search Engine Content. *Web Intelligence and Agent Systems*, 5(3):233–253, 2007.
5. D. D. Lewis. Evaluating and optimizing autonomous text classification systems. In *Proc. of the 18th intl. ACM SIGIR conf. on Res. and development in inf. retr.*, pages 246–254. ACM Press, 1995.
6. D. D. Lewis, Y. Yang, T. G. Rose, and F. Li. RCV1: A New Benchmark Collection for Text Categorization Research. *Journal of Machine Learning Research*, 5:361–397, 2004.
7. Y. Li, W. Yang, and Y. Xu. Multi-tier granule mining for representations of multidimensional association rules. In *Proc. of the intl. conf. on data mining, ICDM06*, pages 953–958, 2006.
8. Y. Li and N. Zhong. Web Mining Model and its Applications for Information Gathering. *Knowledge-Based Systems*, 17:207–217, 2004.
9. Y. Li and N. Zhong. Mining Ontology for Automatically Acquiring Web User Information Needs. *IEEE Transactions on Knowledge and Data Engineering*, 18(4):554–568, 2006.
10. S. E. Middleton, N. R. Shadbolt, and D. C. De Roure. Ontological user profiling in recommender systems. *ACM Trans. Inf. Syst.*, 22(1):54–88, 2004.

11. S. E. Robertson and I. Soboroff. The TREC 2002 filtering track report. In *Text REtrieval Conference*, 2002.
12. A. Sieg, B. Mobasher, and R. Burke. Learning ontology-based user profiles: A semantic approach to personalized web search. *The IEEE Intelligent Informatics Bulletin*, 8(1):7–18, Nov. 2007.
13. K. Sugiyama, K. Hatano, and M. Yoshikawa. Adaptive web search based on user profile constructed without any effort from users. In *Proc. of the 13th intl. conf. on World Wide Web*, pages 675–684, USA, 2004.
14. X. Tao, Y. Li, N. Zhong, and R. Nayak. Automatic Acquiring Training Sets for Web Information Gathering. In *Proc. of the IEEE/WIC/ACM Intl. Conf. on Web Intelligence*, pages 532–535, HK, China, 2006.
15. X. Tao, Y. Li, N. Zhong, and R. Nayak. Ontology mining for personalzied web information gathering. In *Proc. of the IEEE/WIC/ACM intl. conf. on Web Intelligence*, pages 351–358, Silicon Valley, USA, Nov. 2007.
16. J. Teevan, S. T. Dumais, and E. Horvitz. Personalizing search via automated analysis of interests and activities. In *Proc. of the 28th intl. ACM SIGIR conf. on Res. and development in inf. retr.*, pages 449–456, 2005.
17. J. Trajkova and S. Gauch. Improving ontology-based user profiles. In *Proc. of RIAO 2004*, pages 380–389, France, 2004.
18. E.M. Voorhees. Overview of TREC 2002. In *The Text REtrieval Conference (TREC)*, 2002. Retrieved From: http://trec.nist.gov/pubs/trec11/papers/OVERVIEW.11.pdf.
19. S.-T. Wu, Y. Li, and Y. Xu. Deploying approaches for pattern refinement in text mining. In *Proc. of the 6th Intl. Conf. on Data Mining, ICDM'06*, pages 1157–1161, 2006.
20. L.A. Zadeh. Web intelligence and world knowledge - the concept of Web IQ (WIQ). In *Processing of NAFIPS '04.*, volume 1, pages 1–3, 27-30 June 2004.
21. N. Zhong. Toward web intelligence. In *Proc. of 1st Intl. Atlantic Web Intelligence Conf.*, pages 1–14, 2003.

Part II
Novel KDD Domains & Techniques

Chapter 6
Data Mining Applications in Social Security

Yanchang Zhao, Huaifeng Zhang, Longbing Cao, Hans Bohlscheid, Yuming Ou, and Chengqi Zhang

Abstract This chapter presents four applications of data mining in social security. The first is an application of decision tree and association rules to find the demographic patterns of customers. Sequence mining is used in the second application to find activity sequence patterns related to debt occurrence. In the third application, combined association rules are mined from heterogeneous data sources to discover patterns of slow payers and quick payers. In the last application, clustering and analysis of variance are employed to check the effectiveness of a new policy.

Key words: Data mining, decision tree, association rules, sequential patterns, clustering, analysis of variance.

6.1 Introduction and Background

Data mining is becoming an increasingly hot research field, but a large gap remains between the research of data mining and its application in real-world business. In this chapter we present four applications of data mining which we conducted in Centrelink, a Commonwealth government agency delivering a range of welfare services to the Australian community. Data mining in Centrelink involved the application of techniques such as decision trees, association rules, sequential patterns and combined association rules. Statistical methods such as the chi-square test and analysis of variance were also employed. The data used included demographic data, transactional data and time series data and we were confronted with problems

---

Yanchang Zhao, Huaifeng Zhang, Longbing Cao, Yuming Ou, Chengqi Zhang
Faculty of Engineering and Information Technology, University of Technology, Sydney, Australia, e-mail: {yczhao,hfzhang,lbcao,yuming,chengqi}@it.uts.edu.au

Hans Bohlscheid
Data Mining Section, Business Integrity Programs Branch, Centrelink, Australia, e-mail: hans.bohlscheid@centrelink.gov.au

such as imbalanced data, business interestingness, rule pruning and multi-relational data. Some related work include association rule mining [1], sequential pattern mining [13], decision trees [16], clustering [10], interestingness measures [15], redundancy removing [17], mining imbalanced data [11,19], emerging patterns [8], multi-relational data mining [5–7,9] and distributed data mining [4, 12, 14].

Centrelink is one of the largest data users in Australia, distributing approximately $63 billion annually in social security payments to 6.4 million customers. Centrelink administers in excess of 140 different products and services on behalf of 25 Commonwealth government agencies, making 9.98 million individual entitlement payments and recording 5.2 billion electronic customer transactions each year [3]. These statistics reveal not only a very large population, but also a significant volume of customer data. Centrelink's already significant transactional database is further added to by its average yearly mailout of 87.2 million letters and the 32.68 million telephone calls, 39.5 million website hits, 2.77 million new claims, 98,700 field officer reviews and 7.8 million booked office appointments it deals with annually.

Qualification for payment of an entitlement is assessed against a customer's personal circumstances and if all criteria are met, payment will continue until such time as a change of circumstances precludes the customer from obtaining further benefit. However, customer debt may occur when changes of customer circumstances are not properly advised or processed to Centrelink. For example, in a carer/caree relationship, the carer may receive a Carer Allowance from Centrelink. Should the caree pass away and the carer not advise Centrelink of the event, Centrelink may continue to pay the Carer Allowance until such time as the event is notified or discovered through a random review process. Once notified or discovered, a debt is raised for the amount equivalent to the time period for which the customer was not entitled to payment. After the debt is raised, the customer is notified of the debt amount and recovery procedures are initiated. If the customer cannot repay the total amount in full, a repayment arrangement is negotiated between the parties. The above debt prevention and recovery are two of the most important issues in Centrelink and are the target problems in our applications.

In this chapter we present four applications of data mining in the field of social security, with a focus on the debt related issues in Centrelink, an Australia Commonwealth agency. Section 6.2 describes the application of decision tree and association rules to find the demographic patterns of customers. Section 6.3 demonstrates an application of sequence mining techniques to find activity sequences related to debt occurrence. Section 6.4 presents combined association rule mining from heterogeneous data sources to discover patterns of slow payers and quick payers. Section 6.5 uses clustering and analysis of variance to check the effectiveness of a new policy. Conclusions and some discussion will be presented in the last section.

6.2 Case Study I: Discovering Debtor Demographic Patterns with Decision Tree and Association Rules

This section presents an application of decision tree and association rules to discover the demographic patterns of the customers who were in debt to Centrelink [20].

6.2.1 Business Problem and Data

For various reasons, customers on benefit payments or allowances sometimes get overpaid and these overpayments collectively lead to a large amount of debt owed to Centrelink. For example, Centrelink statistical data for the period 1 July 2004 to 30 June 2005 [3] shows that:

- Centrelink conducted 3.8 million entitlement reviews, which resulted in 525,247 payments being cancelled or reduced;
- almost $43.2 million a week was saved and debts totalling $390.6 million were raised as a result of this review activity;
- included in these figures were 55,331 reviews of customers from tip-offs received from the public, resulting in 10,022 payments being cancelled or reduced and debts and savings of $103.1 million; and
- there were 3,446 convictions for welfare fraud involving $41.2 million in debts.

The above figures indicate that debt detection is a very important task for Centrelink staff and we can see from the statistics examined that approximately 14 per cent of all entitlement reviews resulted in a customer debt. However, 86 per cent of reviews resulted in a NIL and therefore it becomes obvious that much effort can be saved by identifying and reviewing only those customers who display a high probability of having or acquiring a debt. Based on the above observation, this application of decision tree and association rules aimed to discover demographic characteristics of debtors; expecting that the results may help to target customer groups associated with a high probability of having a debt. On the basis of the discovered patterns, more data mining work could be done in the near future on developing debt detection and debt prevention systems.

Two kinds of data relate to the above problem: customer demographic data and customer debt data. The data used to tackle this problem have been extracted from Centrelink's database for the period 1/7/2004 to 30/6/2005 (financial year 2004-05).

6.2.2 Discovering Demographic Patterns of Debtors

Customer circumstances data and debt information is organized into one table, based on which the characteristics of debtors and non-debtors are discovered (see

Table 6.1 Demographic data model

Fields	Notes
Customer current circumstances	These fields are from the current customer circumstances in customer data, which are indigenous code, medical condition, sex, age, birth country, migration status, education level, postcode, language, rent type, method of payment, etc.
Aggregation of debts	These fields are derived from debt data by aggregating the data in the past financial year (from 1/7/2004 to 30/06/2005), which are debt indicator, the number of debts, the sum of debt amount, the sum of debt duration, the percentage of a certain kind of debt reason, etc.
Aggregation of history circumstances	These fields are derived from customer data by aggregating the data in the past financial year (from 1/7/2004 to 30/06/2005), which are the number of address changes, the number of marital status changes, the sum of income, etc.

Table 6.2 Confusion matrix of decision tree result

	Actual 0	Actual 1
Predicted 0	280,200 (56.20%)	152,229 (30.53%)
Predicted 1	28,734 (5.76%)	37,434 (7.51%)

Table 6.1). In the data model, each customer has one record, which shows the aggregated information of that customer's circumstances and debt. There are three kinds of attributes in this data model: customer current circumstances, the aggregation of debts, and the aggregation of customer history circumstances, for example, the number of address changes. Debt indicator is defined as a binary attribute which indicates whether a customer had debts in the financial year. In the built data model, there are 498,597 customers, of which 189,663 are debtors.

There are over 80 features in the constructed demographic data model, which proved to be too much for available data mining software to deal with due to the huge search space. The following methods were used to select features: 1) the correlation between variables and debt indicator; 2) the contingency difference of variables to debt indicators with chi-square test ; and 3) data exploration based on the impact difference of a variable on debtors and non-debtors. Based on correlation, chi-square test and data exploration, 15 features, such as ADDRESS CHANGE TIMES, RENT AMOUNT, RENT TYPE, CUSTOMER SERVICE CENTRE CHANGE TIMES and AGE, were selected as input for decision tree and association rule mining.

Decision tree was first used to build a classification model for debtors/non-debtors. It was implemented in Teradata Warehouse Miner (TWM) module of "Decision Tree". In the module, debt indicator was set to dependent column, while customer circumstances variables were set as independent columns. The best result obtained is a tree of 676 nodes, and its accuracy is shown in Table 6.2, where "0" and "1" stand for "no debt" and "debt", respectively. However, the accuracy is poor (63.71%), and the error of false negative is high (30.53%). It is difficult to further improve the accuracy of decision tree on the whole population, however, some leaves of higher accuracy were discovered by focusing on smaller groups.

Association mining [1] was then used to find frequent customer circumstances patterns that were highly associated with debt or non-debt. It was implemented with "Association" module of TWM. In the module, personal ID was set as group column, while item-code was set as item column, where item-code is derived from

Table 6.3 Selected association rules

Association Rule	Support	Confidence	Lift
RA-RATE-EXPLANATION=P and age 21 to 28 ⇒ debt	0.003	0.65	1.69
MARITAL-CHANGE-TIMES =2 and age 21 to 28 ⇒ debt	0.004	0.60	1.57
age 21 to 28 and PARTNER-CASUAL-INCOME-SUM > 0 and rent amount ranging from $200 to $400 ⇒ debt	0.003	0.65	1.70
MARITAL-CHANGE-TIMES =1 and PARTNER-CASUAL-INCOME-SUM > 0 and HOME-OWNERSHIP=NHO ⇒ debt	0.004	0.65	1.69
age 21 to 28 and BAS-RATE-EXPLAN=PO and MARITAL-CHANGE-TIMES=1 and rent amount in $200 to $400 ⇒ debt	0.003	0.65	1.71
CURRENT-OCCUPATION-STATUS=CDP ⇒ no debt	0.017	0.827	1.34
CURRENT-OCCUPATION-STATUS=CDP and SEX=male ⇒ no debt	0.013	0.851	1.38
HOME-OWNERSHIP=HOM and CUSTOMER-SERVICE-CENTRE-CHANGE-TIMES =0 and REGU-PAY-AMOUNT in $400 to $800 ⇒ no debt	0.011	0.810	1.31

customer circumstances and their values. In order to apply association rule analysis to our customer data, we took each pair of feature and value as a single item. Taking feature DEBT-IND as example, it had 2 values, DEBT-IND-0 and DEBT-IND-1. So DEBT-IND-0 was regarded as an item and DEBT-IND-1 was regarded as another. Due to the limitation of spool space, we conducted association rule analysis on a 10 per cent sample of the original data, and the discovered rules were then tested on the whole customer data. We selected the top 15 features to run association rule analysis with minimum support as 0.003, and some selected results are shown in Table 6.3. For example, the first rule shows that 65 per cent of customers with RA-RATE-EXPLANATION as "P" (Partnered) and aged from 21 to 28 had debts in the financial year, and the lift of the rule was 1.69.

6.3 Case Study II: Sequential Pattern Mining to Find Activity Sequences of Debt Occurrence

This section presents an application of impact-targeted sequential pattern mining to find activity sequences of debt occurrence [2]. Impact-targeted activities specifically refer to those activities associated with or leading to specific impact of interest to business. The impact can be an event, a disaster, a government-customer debt, or any other interesting entities. This application was to find out which activities or activity sequences directly triggered or were closely associated with debt occurrence.

6.3.1 Impact-Targeted Activity Sequences

We designed impact-targeted activity patterns in three forms, *impact-oriented activity patterns*, *impact-contrasted activity patterns* and *impact-reversed activity patterns*.

Impact-Oriented Activity Patterns

Mining frequent debt-oriented activity patterns was used to find out which activity sequences were likely to lead to a debt or non-debt. An impact-oriented activity pattern is in the form of $P \rightarrow T$, where the left hand side P is a sequence of activities and the right side is always the target T, which can be a targeted activity, event or other types of business impact. Positive frequent impact-oriented activity patterns ($P \rightarrow T$, or $\bar{P} \rightarrow T$) refer to the patterns likely lead to the occurrence of the targeted impact, say leading to a debt, resulting from either an appeared pattern (P) or a disappeared pattern ($\bar{P}$). On the other hand, negative frequent impact-oriented activity patterns ($P \rightarrow \bar{T}$, or $\bar{P} \rightarrow \bar{T}$) indicate that the target unlikely occurs ($\bar{T}$), say leading to no debt.

Given an activity data set $D = D_T \bigcup D_{\bar{T}}$, where D_T consists of all activity sequences associated with targeted impact and $D_{\bar{T}}$ contains all activity sequences related to non-occurrence of the targeted impact. The count of debts (namely the count of sequences enclosing P) resulting from P in D is $Cnt_D(P)$. The *risk* of pattern $P \rightarrow T$ is defined as $Risk(P \rightarrow T) = \frac{Cost(P \rightarrow T)}{TotalCost(P)}$, where $Cost(P \rightarrow T)$ is the sum of the cost associated with $P \rightarrow T$ and $TotalCost(P)$ is the total cost associated with P. The *average cost* of pattern $P \rightarrow T$ is defined as $AvgCost(P \rightarrow T) = \frac{Cost(P \rightarrow T)}{Cnt(P \rightarrow T)}$.

Impact-Contrasted Activity Patterns

Impact-contrasted activity patterns are sequential patterns having contrasted impacts, and they can be in the following two forms.

- $Supp_{D_T}(P \rightarrow T)$ is high but $Supp_{D_{\bar{T}}}(P \rightarrow \bar{T})$ is low,
- $Supp_{D_T}(P \rightarrow T)$ is low but $Supp_{D_{\bar{T}}}(P \rightarrow \bar{T})$ is high.

We use FP_T to denote those frequent itemsets discovered in those impact-targeted sequences, while $FP_{\bar{T}}$ stands for those frequent itemsets discovered in non-target activity sequences. We define *impact-contrasted patterns* as $ICP_T = FP_T \backslash FP_{\bar{T}}$ and $ICP_{\bar{T}} = FP_{\bar{T}} \backslash FP_T$. The *class difference* of P in two datasets D_T and $D_{\bar{T}}$ is defined as $Cd_{T,\bar{T}}(P) = Supp_{D_T}(P \rightarrow T) - Supp_{D_{\bar{T}}}(P \rightarrow \bar{T})$. The *class difference ratio* of P in D_T and $D_{\bar{T}}$ is defined as $Cdr_{T,\bar{T}}(P) = \frac{Supp_{D_T}(P \rightarrow T)}{Supp_{D_{\bar{T}}}(P \rightarrow \bar{T})}$.

Impact-Reversed Activity Patterns

An impact-reversed activity pattern is composed of a pair of frequent patterns: an underlying frequent impact-targeted pattern 1: $P \rightarrow T$, and a derived activity pattern

2: $PQ \rightarrow \bar{T}$. Patterns 1 and 2 make a contrasted pattern pair, where the occurrence of Q directly results in the reversal of the impact of activity sequences. We call such activity patterns as *impact-reversed activity patterns*. Another scenario of impact-reversed activity pattern mining is the reversal from negative impact-targeted activity pattern $P \rightarrow \bar{T}$ to positive impact $PQ \rightarrow T$ after joining with a trigger activity or activity sequence Q.

To measure the significance of Q leading to impact reversal from positive to negative or vice versa, a metric *conditional impact ratio* (*Cir*) is defined as $Cir(Q\bar{T}|P) = \frac{Prob(Q\bar{T}|P)}{Prob(Q|P) \times Prob(\bar{T}|P)}$. *Cir* measures the statistical probability of activity sequence Q leading to non-debt given pattern P happens in activity set D. Another metric is *conditional Piatetsky-Shapiro's ratio* (*Cps*), which is defined as $Cps(Q\bar{T}|P) = Prob(Q\bar{T}|P) - Prob(Q|P) \times Prob(\bar{T}|P)$.

6.3.2 Experimental Results

The data used in this case study was Centrelink activity data from 1 January 2006 to 31 March 2006. Extracted activity data included 15,932,832 activity records recording government-customer contacts with 495,891 customers, which lead to 30,546 debts in the first three months of 2006. For customers who incurred a debt between 1 February 2006 and 31 March 2006, the activity sequences were built by putting all activities in one month immediately before the debt occurrence. The activities used for building non-debt baskets and sequences were activities from 16 January 2006 to 15 February 2006 for customers having no debts in the first three months of 2006. The date of the virtual non-debt event in a non-debt activity sequence was set to the latest date in the sequence. After the above activity sequence construction, 454,934 sequences were built, out of which 16,540 (3.6 per cent) activity sequences were associated with debts and 438,394 (96.4 per cent) sequences with non-debt. T and $\bar{T}$ denote debt and non-debt respectively, and a_i represents an activity.

Table 6.4 shows some selected impact-oriented activity patterns discovered. The first three rules, $a_1, a_2 \rightarrow T$, $a_3, a_1 \rightarrow T$ and $a_1, a_4 \rightarrow T$ have high *confidences* and *lifts* but low *supports* (caused by class imbalance). They are interesting to business because their *confidences* and *lifts* are high and their *supports* and *AvgAmts* are not too low. The third rule $a_1, a_4 \rightarrow T$ is the most interesting because it has $risk_{amt}$ as high as 0.424, which means that it accounts for 42.4% of total amount of debts.

Table 6.5 presents some examples of impact-contrasted sequential patterns discovered. Pattern "a_{14}, a_{14}, a_4" has $Cdr_{T,\bar{T}}(P)$ as 4.04, which means that it is 3 times more likely to lead to debt than non-debt. Its $risk_{amt}$ shows that it appears before 41.5% of all debts. According to *AvgAmt* and *AvgDur*, the debts related to the second pattern a_8 have both large average amount (26789 cents) and long duration (9.9 days). Its $Cdr_{T,\bar{T}}(P)$ shows that it is triple likely associated with debt than non-debt.

Table 6.6 shows an excerpt of impact-reversed sequential activity patterns. One is *underlying pattern* $P \rightarrow$ *Impact 1*, the other is *derived pattern* $PQ \rightarrow$ *Impact 2*,

Table 6.4 Selected impact-oriented activity patterns

Patterns $P \rightarrow T$	$Supp_D(P)$	$Supp_D(T)$	$Supp_D(P \rightarrow T)$	Confidence	Lift	AvgAmt (cents)	AvgDur (days)	$risk_{amt}$	$risk_{dur}$
$a_1.a_2 \rightarrow T$	0.0015	0.0364	0.0011	0.7040	19.4	22074	1.7	0.034	0.007
$a_3.a_1 \rightarrow T$	0.0018	0.0364	0.0011	0.6222	17.1	22872	1.8	0.037	0.008
$a_1.a_4 \rightarrow T$	0.0200	0.0364	0.0125	0.6229	17.1	23784	1.2	0.424	0.058
$a_1 \rightarrow T$	0.0626	0.0364	0.0147	0.2347	6.5	23281	2.0	0.490	0.111
$a_6 \rightarrow T$	0.2613	0.0364	0.0133	0.0511	1.4	18947	7.2	0.362	0.370

Table 6.5 Selected impact-contrasted activity patterns

Patterns (P)	$Supp_{D_T}(P)$	$Supp_{D_{\bar{T}}}(P)$	$Cd_{T,\bar{T}}(P)$	$Cdr_{T,\bar{T}}(P)$	$Cd_{\bar{T},T}(P)$	$Cdr_{\bar{T},T}(P)$	AvgAmt (cents)	AvgDur (days)	$risk_{amt}$	$risk_{dur}$
a_4	0.446	0.138	0.309	3.24	-0.309	0.31	21749	3.2	0.505	0.203
a_8	0.176	0.060	0.117	2.97	-0.117	0.34	26789	9.9	0.246	0.245
$a_4.a_{15}$	0.255	0.092	0.163	2.78	-0.163	0.36	21127	3.9	0.280	0.141
$a_{14}.a_{14}.a_4$	0.367	0.091	0.276	4.04	-0.276	0.25	21761	2.9	0.415	0.151

Table 6.6 Selected impact-reversed activity patterns

Underlying sequence(P)	Impact 1	Derivative activity Q	Impact 2	Cir	Cps	Local support of $P \rightarrow$ Impact 1	Local support of $PQ \rightarrow$ Impact 2
a_{14}	$\bar{T}$	a_4	T	2.5	0.013	0.684	0.428
a_{16}	$\bar{T}$	a_4	T	2.2	0.005	0.597	0.147
a_{14}	$\bar{T}$	a_5	T	2.0	0.007	0.684	0.292
a_{16}	$\bar{T}$	a_7	T	1.8	0.004	0.597	0.156
$a_{14}.a_{14}$	$\bar{T}$	a_4	T	2.3	0.016	0.474	0.367
$a_{16}.a_{14}$	$\bar{T}$	a_5	T	2.0	0.006	0.402	0.133
$a_{16}.a_{15}$	$\bar{T}$	a_5	T	1.8	0.006	0.339	0.128
$a_{14}.a_{16}.a_{14}$	$\bar{T}$	a_{15}	T	1.2	0.005	0.248	0.188

where *Impact 1* is opposite to *Impact 2*, and Q is a derived activity or sequence. *Cir* stands for *conditional impact ratio*, which shows the impact of the derived activity on *Impact 2* when the underlying pattern happens. *Cps* denotes *conditional P-S ratio*. Both *Cir* and *Cps* show how much the impact is reversed by the derived activity Q. For example, the first row shows that the appearance of a_4 tends to change the impact from $\bar{T}$ to T when a_{14} happens first. It indicates that, when a_{14} occurs first, the appearance of a_4 makes it more likely to become debtable. This pattern pairs indicate what effect an additional activity will have on the impact of the patterns.

6.4 Case Study III: Combining Association Rules from Heterogeneous Data Sources to Discover Repayment Patterns

This section presents an application of combined association rules to discover patterns of quick/slow payers [18, 21]. Heterogeneous data sources, such as demographic and transactional data, are part of everyday business applications and used for data mining research. From a business perspective, patterns extracted from a single normalized table or subject file are less interesting or useful than a full set of multiple patterns extracted from different datasets. A new technique has been designed to discover combined rules on multiple databases and applied to debt recovery in the social security domain. Association rules and sequential patterns from different datasets are combined into new rules, and then organized into groups. The rules produced are useful, understandable and interesting from business perspective.

6.4.1 Business Problem and Data

The purpose of this application is to present management with customers, profiled according to their capacity to pay off their debts in shortened timeframes. This enables management to target those customers with recovery and amount options suitable to their own circumstances and increase the frequency and level of repayment. Whether a customer is a quick or slow payer is believed by domain experts to be related to demographic circumstances, arrangements and repayments.

Three datasets containing customers with debts were used: customer demographic data, debt data and repayment data. The first data contains demographic attributes of customers, such as customer ID, gender, age, marital status, number of children, declared wages, location and benefit. The second dataset contains debt related information, such as the date and time when a debt was raised, debt amount, debt reason, benefit or payment type that the debt amount is related to, and so on. The repayments dataset contains arrangement types, repayment types, date and time of repayment, repayment amount, repayment method (e.g., post office, direct debit, withholding payment), etc. Quick/moderate/slow payers are defined by domain experts based on the time taken to repay the debt, the forecasted time to repay and the frequency/amount of repayment.

6.4.2 Mining Combined Association Rules

The idea was to firstly derive the criterion of quick/slow payers from the data, and then propagate the tags of quick/slow payers to demographic data and to the other data to find frequent patterns and association rules. Since the pay-off timeframe is decided by arrangement and repayment, customers were partitioned into groups according to their arrangement and repayment type. Secondly, pay-off timeframe distribution and statistics for each group were presented to domain knowledge experts, who then decided who were quick/slow payers by group. The criterion was applied to the data to tag every customer as quick/slow payer. Thirdly, association rules were generated for quick/slow payers in each single group. And lastly, the association rules from all groups were organized together to build potentially business-interesting rules.

To address the business problem, there are two types of rules to discover. The first type are rules with the same arrangement and repayment pattern but different demographic patterns leading to different customer classes (see Formula 6.1). The second type are rules with the same demographic pattern but different arrangement and repayment pattern leading to different customer classes (see Formula 6.2).

$$\text{Type A:} \begin{cases} A_1 + D_1 \rightarrow \text{quick payer} \\ A_1 + D_2 \rightarrow \text{moderate payer} \\ A_1 + D_3 \rightarrow \text{slow payer} \end{cases} \quad (6.1)$$

$$\text{Type B:} \begin{cases} A_1 + D_1 \rightarrow \text{quick payer} \\ A_2 + D_1 \rightarrow \text{moderate payer} \\ A_3 + D_1 \rightarrow \text{slow payer} \end{cases} \quad (6.2)$$

where A_i and D_i denotes respectively arrangement patterns and demographic patterns.

6.4.3 Experimental Results

The data used was debts raised in calendar year 2006 and the corresponding customers and repayments in the same year. Debts raised in calendar year 2006 were first selected, and then the customer data and repayment data in the same year related to the above debt data were extracted. The extracted data was then cleaned by removing noise and invalid values. The cleansed data contained 479,288 customers with demographic attributes and 2,627,348 repayments.

Selected combined association rules are given in Tables 6.7 and 6.8. Table 6.7 shows examples of rules with the same demographic characteristics. For those customers, different arrangements lead to different results. It shows that male customers with CCC benefit repay their debts fastest with "*Arrangement=Cash, Repayment=Agent recovery*", while slowest with "*Arrangement=Withholding and Voluntary Deduction, Repayment= Withholding and Direct Debit*" or "*Arrangement=Cash and Irregular, Repayment=Cash or Post Office*". Therefore, for a male customer with a new debt, if his benefit type is CCC, Centrelink may try to encourage him to repay under "*Arrangement=Cash, Repayment=Agent recovery*", and not to pay under "*Arrangement=Withholding and Voluntary Deduction, Repayment=Withholding and Direct Debit*" or "*Arrangement =Cash and Irregular, Repayment=Cash or Post Office*", so that the debt will likely be repaid quickly.

Table 6.7 Selected Results with the Same Demographic Patterns

Arrangement	Repayment	Demographic Pattern	Result	Confidence(%)	Count
Cash	Agent recovery	Gender:M & Benefit:CCC	Quick Payer	37.9	25
Withholding & Irregular	Withholding & Cash or Post Office	Gender:M & Benefit:CCC	Moderate Payer	75.2	100
Withholding & Voluntary Deduction	Withholding & Direct Debit	Gender:M & Benefit:CCC	Slow Payer	36.7	149
Cash & Irregular	Cash or Post Office	Gender:M & Benefit:CCC	Slow Payer	43.9	68
Withholding & Irregular	Cash or Post Office	Age:65y+	Quick Payer	85.7	132
Withholding & Irregular	Withholding & Cash or Post Office	Age:65y+	Moderate Payer	44.1	213
Withholding & Irregular	Withholding	Age:65y+	Slow Payer	63.3	50

Table 6.8 Selected Results with the Same Arrangement-Repayment Patterns

Arrangement	Repayment	Demographic Pattern	Result	Expected Conf(%)	Conf (%)	Support (%)	Lift	Count
Withholding & Irregular	Withholding	Age:17y-21y	Moderate Payer	39.0	48.6	6.7	1.2	52
Withholding & Irregular	Withholding	Age:65y+	Slow Payer	25.6	63.3	6.4	2.5	50
Withholding & Irregular	Withholding	Benefit:BBB	Quick Payer	35.4	64.9	6.4	1.8	50
Withholding & Irregular	Withholding	Benefit:AAA	Moderate Payer	39.0	49.8	16.3	1.3	127
Withholding & Irregular	Withholding	Marital:married & Children:0	Slow Payer	25.6	46.9	7.8	1.8	61
Withholding & Irregular	Withholding	Weekly:0 & Children:0	Slow Payer	25.6	49.7	11.4	1.9	89
Withholding & Irregular	Withholding	Marital:single	Moderate Payer	39.0	45.7	18.8	1.2	147

Table 6.8 shows examples of rules with the same arrangements but different demographic characteristics. The tables indicates that "*Arrangement=Withholding and Irregular, Repayment=Withholding*" arrangement is more appropriate for customers with BBB benefit, while they are not suitable for mature age customers, or those with no income or children. For young customers with a AAA benefit or single, it is not a bad choice suggesting to them, to repay their debts under "*Arrangement=Withholding and Irregular, Repayment=Withholding*".

6.5 Case Study IV: Using Clustering and Analysis of Variance to Verify the Effectiveness of a New Policy

This section presents an application of clustering and analysis of variance to study whether a new policy works or not. The aim of this application was to examine earnings related transactions and earnings declarations in order to ascertain, whether significant changes occurred after the implementation of the "welfare to work" initiative on 1st July 2006. The principal objective was to verify whether customers declare more earned income after the changes, the rules of which allowed them to keep more earned income and still keep part or all of their income benefit.

The population studied in this project were customers who had one or more non-zero declarations and were on the 10 benefit types affected by the "Welfare to Work" initiative across two financial years from 1/7/2005 to 30/6/2007. Three datasets were available and each of them contained 261,473 customer records. Altogether there were 13,596,596 declarations (including "zero declarations"), of which 4,488,751 were non-zero declarations. There are 54 columns in transformed earnings declaration data. Columns 1 and 2 are respectively customer ID and benefit type. The other 52 columns are declaration amounts over 52 fortnights.

6.5.1 Clustering Declarations with Contour and Clustering

At first we employed histograms, candlestick charts and heatmaps to study whether there were any changes between the two years. The result from histograms, candlestick charts and heatmaps all show that there was an increase of the earnings declaration amount for the whole population.

For the whole population, scatter plot indicates no well-separated clusters, while contour shows that some combinations of fortnights and declaration amounts had more customers than others (see Figure 6.1). It's clear that the densely populated areas shifted from low amounts to large amounts from financial year 2005-2006 to financial year 2006-2007. Moreover, the sub-cluster of declarations ranging from $50 to $150 reduced over time, while the sub-cluster ranging from $150 to $250 expanded and shifted towards higher amounts.

The clustering with k-means algorithm did not generate any meaningful clusters. The declarations were divided into clusters by fortnights when the amount is small, while the dominant factor is not time, but amount, when the amount is high. A density-based clustering algorithm, DBSCAN [10], was then used to cluster the declarations below $1000, and due to limited time and space, a random sample of 15,000 non-zero declarations was used as input into the algorithm. The clusters found for all benefit types are shown in Figure 6.2. There are four clusters, separated by the beginning of new year and financial year. From left to right, the four clusters shift towards larger amounts as time goes on, which shows that the earnings declarations increase after the new policy.

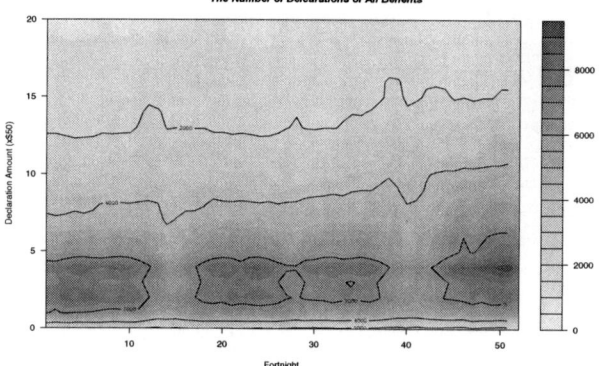

Fig. 6.1 Contour of Earnings Declaration

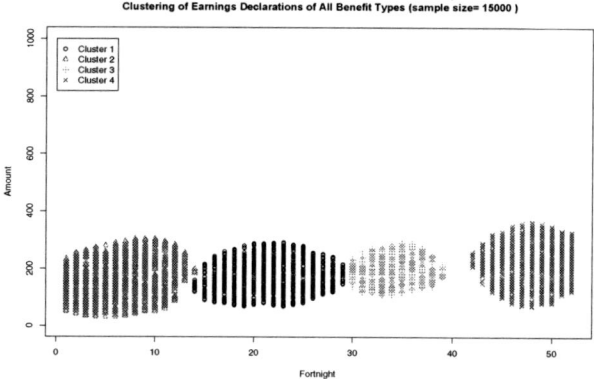

Fig. 6.2 Clustering of Earnings Declaration

The clustering with k-means algorithm does not generate any meaningful clusters. The declarations are divided into clusters by fortnights when the amount is small, while the dominant factor is not time but amount when the amount is high. A density-based clustering algorithm, DBSCAN [10], is then used to cluster the declarations below $1000, and due to limited time and space, a random sample of 15,000 non-zero declarations is used as input into the algorithm. The clusters found for all benefit types are shown in Figure 6.2. There are four clusters, separated by the beginning of new year and financial year. From left to right, the four clusters shift towards larger amounts as time goes on, which shows that the earnings declarations increase after the new policy.

Table 6.9 Hypothesis Test Results Using Mixed Model

Benefit Type	DenDF	FValue	ProbF
APT	1172	4844.50	<.0001
AUS	4872	2.94	0.0863
JSK	9351	2413.06	<.0001
NMA	317	5.67	0.0178
NSA	77801	1448.89	<.0001
PPP	11579	1782.00	<.0001
PTA	3102	421.04	<.0001
SKA	425	2.55	0.1112
STU	41623	16475.2	<.0001
WDA	3398	126.28	<.0001

6.5.2 Analysis of Variance

Hypothesis test was also used to study whether there were significant changes of the earnings declaration before/after the initiative. We employed mixed models to test the changes of earnings declaration. The Mixed Procedure in SAS was used and tests were conducted for every benefit type. The results are shown in Table 6.9, where "DenDF" shows the sample size, "Fvalue" gives an understanding on the difference before/after initiative. The greater this value is, the bigger the difference. "ProbF<0.0001" means there is significant change before/after initiative. So the customers with payment/benefit types APT, JSK, NSA, PPP, PTA, STU and WDA are all with significant changes. "ProbF>0.0001" has two meanings: 1) the difference is not significant, or 2) the sample size is not large enough. Because the sample size of NMA and SKA are very small, the hypothesis test cannot give reliable result on whether there are significant changes or not. The changes on the AUS customers are not significant. The results show that there are significant change for most benefit types, which suggests that the new policy is effective.

6.6 Conclusions and Discussion

Data mining techniques have been used in a social security environment to check the effectiveness of a new policy and discover demographic patterns, activity sequence patterns and debt recovery patterns. The demographic patterns discovered may be used to identify customer groups with a high probability of debt, so that reviews can be conducted to assist in the reduction of debts. Moreover, by discovering activity sequence patterns associated with debt/non-debt, appropriate actions can be suggested to reduce the probability of customers' acquiring a debt. We have also presented effective and practical techniques for discovering rare but significant *impact-targeted activity patterns* in imbalanced activity data and a framework for mining combined association rules from multiple datasets. The above is part of our initial effort to tackle business problems using data mining techniques, and it shows promising applications of data mining to solve real-life problems in near future.

However, there are still many open problems. Firstly, given the likelihood that hundreds or possibly thousands of rules are identified after pruning redundant patterns, how can we efficiently select interesting patterns from them? Secondly, how can domain knowledge be effectively incorporated in data mining procedure to reduce the search space and running time of data mining algorithms? Thirdly, given that the business data is complicated and a single debt activity may be linked to several customers, how can existing approaches for sequence mining be improved to take into consideration the linkage and interaction between activity sequences? And lastly and perhaps most importantly, how can these discovered rules be used to build an efficient debt prevention system to effectively detect debt in advance and give appropriate suggestions to reduce or prevent debt? The above will be part of our future work.

Acknowledgments

We would like to thank Mr Fernando Figueiredo, Mr Peter Newbigin and Mr Brett Clark from Centrelink, Australia, for their support of domain knowledge and helpful suggestions.

This work was supported by the Australian Research Council (ARC) Linkage Project LP0775041 and Discovery Projects DP0667060 & DP0773412, and by the Early Career Researcher Grant from University of Technology, Sydney, Australia.

References

1. R. Agrawal and R. Srikant. Fast algorithms for mining association rules in large databases. In J. B. Bocca, M. Jarke, and C. Zaniolo, editors, *Proceedings of the 20th International Conference on Very Large Data Bases, VLDB*, pages 487–499, Santiago, Chile, September 1994.
2. L. Cao, Y. Zhao, and C. Zhang. Mining impact-targeted activity patterns in imbalanced data. Accepted by *IEEE Transactions on Knowledge and Data Engineering* in July 2007. IEEE Computer Society Digital Library. IEEE Computer Society, http://doi.ieeecomputersociety.org/10.1109/TKDE.2007.190635.
3. Centrelink. Centrelink fraud statistics and centrelink facts and figures, url: http://www.centrelink.gov.au/internet/internet.nsf/about_us/fraud_stats.htm, http://www.centrelink.gov.au/internet/internet.nsf/about_us/facts.htm. Accessed in May 2006.
4. J. Chattratichat, J. Darlington, Y. Guo, S. Hedvall, M. Köler, and J. Syed. An architecture for distributed enterprise data mining. In *HPCN Europe '99: Proceedings of the 7th International Conference on High-Performance Computing and Networking*, pages 573–582, London, UK, 1999. Springer-Verlag.
5. V. Crestana-Jensen and N. Soparkar. Frequent itemset counting across multiple tables. In *PAKDD'00: Proceedings of the 4th Pacific-Asia Conference on Knowledge Discovery and Data Mining, Current Issues and New Applications*, pages 49–61, London, UK, 2000. Springer-Verlag.

6. L. Cristofor and D. Simovici. Mining association rules in entity-relationship modeled databases. Technical report, University of Massachusetts Boston, 2001.
7. P. Domingos. Prospects and challenges for multi-relational data mining. *SIGKDD Explor. Newsl.*, 5(1):80–83, 2003.
8. G. Dong and J. Li. Efficient mining of emerging patterns: discovering trends and differences. In *KDD '99: Proceedings of the fifth ACM SIGKDD international conference on Knowledge discovery and data mining*, pages 43–52, New York, NY, USA, 1999. ACM.
9. S. Dzeroski. Multi-relational data mining: an introduction. *SIGKDD Explor. Newsl.*, 5(1):1–16, 2003.
10. M. Ester, H.-P. Kriegel, J. Sander, and X. Xu. A density-based algorithm for discovering clusters in large spatial databases with noise. In *KDD*, pages 226–231, 1996.
11. H. Guo and H. L. Viktor. Learning from imbalanced data sets with boosting and data generation: the databoost-im approach. *SIGKDD Explor. Newsl.*, 6(1):30–39, 2004.
12. B. Park and H. Kargupta. Distributed data mining: Algorithms, systems, and applications. In N. Ye, editor, *Data Mining Handbook*. 2002.
13. J. Pei, J. Han, B. Mortazavi-Asl, J. Wang, H. Pinto, Q. Chen, U. Dayal, and M.-C. Hsu. Mining sequential patterns by pattern-growth: The prefixspan approach. *IEEE Transactions on Knowledge and Data Engineering*, 16(11):1424–1440, 2004.
14. F. Provost. Distributed data mining: Scaling up and beyond. In *Advances in Distributed and Parallel Knowledge Discovery*. MIT Press, 2000.
15. A. Silberschatz and A. Tuzhilin. What makes patterns interesting in knowledge discovery systems. *IEEE Transactions on Knowledge and Data Engineering*, 8(6):970–974, 1996.
16. Q. Yang, J. Yin, C. X. Ling, and T. Chen. Postprocessing decision trees to extract actionable knowledge. In *ICDM '03: Proceedings of the Third IEEE International Conference on Data Mining*, page 685, Washington, DC, USA, 2003. IEEE Computer Society.
17. M. Zaki. Mining non-redundant association rules. *Data Mining and Knowledge Discovery*, 9:223–248, 2004.
18. H. Zhang, Y. Zhao, L. Cao, and C. Zhang. Combined association rule mining. In T. Washio, E. Suzuki, K. M. Ting, and A. Inokuchi, editors, *PAKDD*, volume 5012 of *Lecture Notes in Computer Science*, pages 1069–1074. Springer, 2008.
19. J. Zhang, E. Bloedorn, L. Rosen, and D. Venese. Learning rules from highly unbalanced data sets. In *ICDM '04: Proceedings of the Fourth IEEE International Conference on Data Mining*, pages 571–574, Washington, DC, USA, 2004. IEEE Computer Society.
20. Y. Zhao, L. Cao, Y. Morrow, Y. Ou, J. Ni, and C. Zhang. Discovering debtor patterns of centrelink customers. In *Proc. of The Australasian Data Mining Conference: AusDM 2006*, Sydney, Australia, November 2006.
21. Y. Zhao, H. Zhang, F. Figueiredo, L. Cao, and C. Zhang. Mining for combined association rules on multiple datasets. In *Proc. of 2007 ACM SIGKDD Workshop on Domain Driven Data Mining (DDDM 07)*, pages 18–23, 2007.

Chapter 7
Security Data Mining: A Survey Introducing Tamper-Resistance

Clifton Phua and Mafruz Ashrafi

Abstract Security data mining, a form of countermeasure, is the use of large-scale data analytics to dynamically detect a small number of adversaries who are constantly changing. It encompasses data- and results-related safeguards; and is relevant across multiple domains such as financial, insurance, and health. With reference to security data mining, there are specific and general problems, but the key solution and contribution of this chapter is still tamper-resistance. Tamper-resistance addresses most kinds of adversaries and makes it more difficult for an adversary to manipulate or circumvent security data mining; and consists of reliable data, anomaly detection algorithms, and privacy and confidentiality preserving results. In this way, organisations applying security data mining can better achieve accuracy for organisations, privacy for individuals in the data, and confidentiality between organisations which share the results.

7.1 Introduction

There is the exceptional progress in networking, storage and processor technology; as well as the increase in data sharing between organisations. As a result, there is the explosive growth in the volume of digital data, a significant portion of which is collected by an organisation for security purposes.

This necessitates the use of security data mining to analyze digital data to discover actionable knowledge. By actionable, we mean that this new knowledge improves the organisation's key performance indicators, enables better decision-making for the organisation's managers, and provides measurable and tangible results. Instead of purely theoretical data-driven data mining, more practical domain-driven data mining is required to discover actionable knowledge.

Clifton Phua, Mafruz Ashrafi
A*STAR, Institute of Infocomm Research, Room 04-21 (+6568748406), 21, Heng Mui Keng Terrace, Singapore 119613, e-mail: {cwphua,mashrafi}@i2r.a-star.edu.sg

This chapter's objective is, as a survey paper, to define the domain of security data mining by organisations using published case studies from various security environments. Although each security environment may have its own unique requirements, this chapter argues that they share similar principles to operate well.

This chapter's main contribution is the focus on ways to engineer tamper-resistance for security data mining applications - mathematical algorithms in computer programs which perform security data mining. With tamper-resistance, organisations applying security data mining can better achieve accuracy for organisations, privacy for individuals in the data, and confidentiality between organisations which share the results.

This chapter is written for the general audience who has little theoretical background in data mining, but interested in practical aspects of security data mining. We assume that the reader knows about or will eventually read up on the data mining process [20] which involves ordered and interdependent steps. These steps consist of data pre-processing, integration, selection, and transformation; use of common data mining algorithms (such as classification, clustering, and association rules); results measurement and interpretation.

The rest of this chapter is organised as follows. We present security data mining's definitions, specific and general issues in Section 7.2. We discuss tamper-resistance in the form of reliable data, anomaly detection algorithms, and privacy and confidentiality preserving results in Section 7.3. We conclude with a summary and future work in Section 7.4.

7.2 Security Data Mining

This section defines terms, presents specfic as well as general problems to security data mining, and offers solutions in the form of successful applications from various security environments.

7.2.1 Definitions

The following definitions (in bold font), which might be highly evident to some readers, are specific to security data mining. An **adversary** is a malicious individual whose aim is to inflict adverse consequences to valuable assets without being discovered. Alternatively, an adversary can be an organisation, and have access to their own data and algorithms. An adversary can create more automated software and/or use more manual means to carry out an attack. Using the relevant, new, and interesting domain of virtual gaming worlds, cheating can be in the form of automated of gold farming. In constrast, cheating can also come in the form of cheap manual labour who game in teams to slaughter high-reward monsters [28].

Internal adversaries work for the organisation, such as employees responsible for data breaches [29, 46]. External adversaries do not have any access rights to the organisation, such as taxpayers who evade tax [13]. Data leak detection uses matching of documents using dictionaries of common terms and keywords, and using fingerprints of sensitive documents, and monitoring locations where sensitive documents are kept. One-class Support Vector Machines (SVM) are trained to rank new taxpayers on known-fraudulent individual and high income taxpayers data. Subsequently, the taxpayers will then be subjected to link analysis using personal data to locate pass-through entities.

Security is the condition of being protected against danger or loss. But a more precise definition of security here is the use of countermeasures to prevent the deliberate and unwarranted behaviour of adversaries [41].

Security data mining, a form of countermeasure, is the use of large-scale data analytics to dynamically detect a small number of adversaries who are constantly changing. It encompasses data- and results-related safeguards. Security data mining is relevant across multiple domains such as financial, insurance, health, taxation, social security, e-commerce, just to name a few. It is a collective term for detection of fraud, crime, terrorism, financial crime, spam, and network intrusion [37]. In addition, there are other forms of adversarial activity such as detection of online gaming [28], data breaches, phishing, and plagarism. The difference between security data mining and fraud data mining is that the former concentrates in the long-term on the adversary, not for short-term profit.

To understand security data mining better, security data mining is compared with database marketing - its opposite domain. A casino can use both domains to increase profit: Non-Obvious Relationship Awareness (NORA) [26] reduces cost, while HARRAH's database marketing [32] increases revenue. In real-time, NORA detects people who are morphing identities. NORA evaluates similarities and differences between people or organisations and shows how current entities are connected to all previous entities. In retrospect, HARRAH cultivates lasting relationships with its core customers. HARRAH discovered that slot players who are retirees are their core customers, and direct resources to develop better customer satisfaction with them.

7.2.2 Specific Issues

The following concepts (in bold font) are specific to security data mining:

- **Resilience**, for security systems, is the ability to degrade gracefully when under most real attacks. The security system needs "defence-in-depth" with multiple, sequential, and independent layers of defence [41] to cover different types of attacks, and to eliminate clearly legitimate examples [24]. In other words, any attack has to pass every layer of defence without being detected.

The security system is a combination of manual approaches; and automated approaches including blacklist matching and security data mining algorithms. The basic automated approaches include hard-coded rules such as matching personal name and address, and setting price and amount limits.

One common automated approach is known fraud matching. Known frauds are usually recorded in a periodically updated blacklist. Subsequently, the current claims/applications/transactions/accounts/sequences are matched against the blacklist. This has the benefit and clarity of hindsight because patterns often repeat themselves. However, there are two main problems in using known frauds. First, they are untimely due to long time delays which provides a window of opportunity for fraudsters. Second, recording of frauds is highly manual.

- **Adaptivity**, for security data mining algorithms, accounts for morphing fraud behaviour, as the attempt to observe fraud changes its behaviour. But what is not obvious, but equally important, is the need to also account for legal (or legitimate) behaviour within a changing environment.

    In practice, for telecommunications superimposed fraud detection [19], there is fraud rule generation from each cloned phone account's labelled data and rule selection to cover most accounts. For anomaly detection, each selected fraud rule is applied in the form of monitors (number and duration of calls) to the daily legitimate usage of each account. StackGuard [8] is a simple compiler which virtually eliminates buffer overflow attacks with only modest speed penalties. To provide an adaptive response to intrusions, StackGuard switches between the more effective MemGuard version and the more efficient Canary version.

    In theory, in spam detection, adversaries learn how to generate more false negatives from prior knowledge, observation, and experimentation [33]. Game theory is adapted to automatically re-learn a cost-sensitive supervised algorithm given the cost-sensitive adversary's optimal strategy [11]. It defines the adversary and classifier optimal strategy by making some valid assumptions.

- **Quality data** is essential for security data mining algorithms through the removal of data errors (or noise). HESPERUS [38] filters duplicates which have been re-entered due to human error or for other reasons. It also removes redundant attributes which have many missing values, and other issues. Data preprocessing for securities fraud detection [18] include known consolidation and link formation techniques to associate people with office locations, infer associations by employment histories, and normalisation techniques by space and time to create a suitable class labels.

7.2.3 General Issues

The following concepts (in bold font) are general to data mining, and are used here to describe security data mining applications.

- **Personal data versus behavioural data**
  Personal data relates to an identified natural people, on the other hand, behavioural data relates to the actions of people under specified circumstances. The data here refers to text form, as image and video data are beyond our scope. Most applications use behavioural data but some, such as HESPERUS [38], use personal data.
  HESPERUS discovers credit card application fraud patterns. It detects sudden and sharp spikes in duplicates within a short time, relative to normal behaviour.

- **Unstructured data versus structured data**
  Unstructured data is not in a tabular or delimited format; while structured data is segmented into attributes where each has an assigned format. In this chapter's subsequent applications, most use structured data but some, such as in software plagarism [40], use unstructured data.
  Unstructured data is transformed into fingerprints - selected and hashed k-grams (using 0 mod p or winnowing) with positional information - to detect software copies. Some issues discussed in the paper include support for a variety of input formats, filter of unnecessary code, and presentation of results.

- **Real-time versus retrospective application**
  A real-time application processes events as they happen, and need to scale up to the arrival and growth of data. In contrast, a retrospective application processes events after they have taken place, and are often used to perform audits and stress tests. A real-time financial crime detection application - Securities Observation, News Analysis, and Regulation (SONAR) [22], and a retrospective managerial fraud detection application - SHERLOCK [5] are described in detail below.
  In real-time, SONAR monitors main stock markets for insider trading by using privileged information of a material nature, and misrepresentation fraud by fabricating news. SONAR mines for explicit and implicit relationships among the entities and events, using text mining, statistical regression, rule-based inference, uncertainty, and fuzzy matching.
  In retrospect, SHERLOCK analyses the general ledger - a formal listing of journal accounts in a business used for financial statement preparation and tax filing - for irregularites which are useful to auditors and investigators. SHERLOCK extracts a few dozen important attributes for outlier detection and classification. Some limitations stated in the paper include data which is hard to pre-process, having a small set of known fraud general ledgers while the major-

ity are unlabelled, and results are hard to interpret.

- **Unsupervised versus supervised application**
  An unsupervised application do not use class labels - usually assignment of records to a particular category - and is more suited for real-time use. A supervised application use class labels and is usually for retrospective use. The following click fraud detection [34] and management fraud detection [47] applications use behavioural, structured data.

  Using user click data on web advertisements, [34] analyses requests using an unsupervised pair-wise analysis with association rules. Using public company account data, use a supervised decision tree to classify time and peer attributes, and apply supervised logistic regression for each leaf time series.

- **Maximum versus no user interaction**
  Maximum user (or domain expert) interaction is required if the consequences for a security breach is severe [25]. User interaction refers to being able to easily annotate, add attributes, or change attribute weights; or to allow better understanding and use of scores (or rules). No user interaction refers to a fully automated application.

  Visual telecommunications fraud detection [9] combines user detection with computer programs. It flexibly encodes data using colour, position, size and other visual characteristics with multiple different views and levels.

7.3 Tamper-Resistance

Figure 7.1 gives a visual overview of tamper-resistance solutions in security data mining. The problems come from data adversaries, internal adversaries, and external adversaries in the form of other organisations sharing the data or results (for example, adversaries always try to look legitimate). The solutions can be summarised as **tamper-resistance**, which addresses most kinds of adversaries and makes it more difficult for an adversary to manipulate or circumvent security data mining. From experience, we recommend reliable data as inputs, anomaly detection algorithms as processes, and privacy and confidentiality preserving results as outputs to enhance tamper-resistance; and we elaborate more on them in the following subsections.

7.3.1 Reliable Data

Reliable data is not just quality data (see previous subsection 7.2.2); but also can be trusted and gives the same results, even with adversary manipulation. By reliable data, we refer to unforgeable, stable, and non-obvious data [43]. To an adversary,

7 Security Data Mining: A Survey Introducing Tamper-Resistance

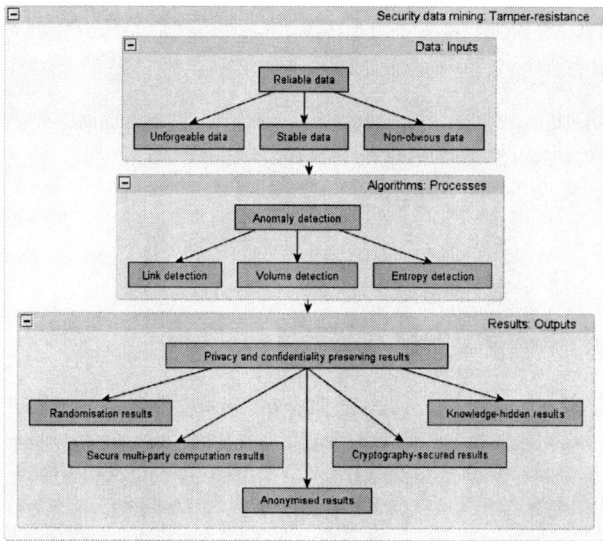

Fig. 7.1 Visual overview

reliable data cannot be replicated with the intent to deceive, has little fluctuation, and is hard to see and understand.

- **Unforgeable data** can be viewed as attributes which are generated subconsciously, such as rhythm-based typing patterns [36] which is based on timing information of username and password. As an authentication factor, rhythm-based typing patterns is cheap, easily accepted by most users, and can be used beyond keyboards. However, there exists policy and privacy issues.

- **Stable data** include communication links between adversaries where links are already available. By linking mobile phone accounts using call quantity and durations to form Communities Of Interest (COI), two distinctive characteristics of fraudsters can be determined. Fraudulent phone accounts are linked as fraudsters call each other or the same phone numbers, and fraudulent call behaviour from known frauds are reflected in some new phone accounts [7].

    Also, stable data can come from attribute extraction where attributes are not directly available or when there are too many attributes. To find new, previously unseen, malicious executables and differentiate them from benign programs, there is attribute extraction of various information such as Dynamically Linked Library (DLL) calls, consecutive printable strings, and byte sequences [42].

- **Non-obvious data** refer to attributes with characteristic distributions. For intrusion detection, these attributes describe the network traffic, such as historical averages of source and destination Internet Protocol (IP) addresses of packets, source and destination port numbers, type of protocol, number of bytes per pacet, and time elapsed between packets. In addition, more of such attributes are from router data, such as Central Processing Unit (CPU), memory usage, and traffic volume [35].

In online identity theft, each phishing attack has several stages starting from delivery of attack to ending of receiving of money [15]. The point here is to collect and mine non-obvious data from stages where adversaries least expect.

7.3.2 Anomaly Detection Algorithms

To detect anomalies early (also known as abnormalities, deviations, or outliers), anomaly detection algorithms are a type of security data mining algorithm which originate from network intrusion detection research [14]. They profile normal behaviour (also known as norm or baseline) by outputting suspicion scores (or rules). Anomaly detection algorithms can be used on data for various security environments, in different stages, at different levels of granularity (such as at the global, account, or individual levels), or for groups of similar things (such as dates, geography, or social groups).

Anomaly detection algorithms require class imbalanced data - plenty of normal compared to anomalous behaviour which is common in security data - to be useful [16]. They are only effective when normal behaviour has high regularity [30]. The common anomaly detection algorithms monitor for changes in links, volume, and entropy:

- **Link detection** is to find good or bad connections between things. For rhythm-based typing where links have to be discovered, [36] uses classifiers to measure the link (or similarity) between an input keystroke timing and a model of normal behaviour of the keystroke timing. Each attribute for a model of normal behaviour is an updated average from a predefined number of keystrokes. For telecommunications fraud detection where links are available, [7] examines temporal evolution of each large dynamic graph for subgraphs called COIs. For professional software plagarism detection, [31] mines Program Dependence Graphs (PDG) which links code statements based on data and control dependencies which reflects developers' thinking when code is written.

    However, for securities fraud detection, although fraud is present when there are links between groups of representatives that pass together through multiple places of employment, these links will also find harmless sets of friends that worked together in the same industry and a multitude of non-fraud links [21].

- **Volume detection** is to monitor the significant increase or decrease in amount of something. For credit card transactional fraud detection, Peer Group Analysis [6] monitors inter-account behaviour by comparison of the cumulative mean weekly amount between a target account and other similar accounts (peer group). Break Point Analysis [6] monitors intra-account behaviour by detecting rapid weekly spending. A neural network trained on a seven day moving window of Automated Teller Machines' (ATM) daily cash output to detect anomalies. For bio-terrorism detection, Bayesian networks [48] observe sudden increases of certain illness symptoms from real emergency department data. Time series analysis [23] tracks daily sales of throat, cough, and nasal medication; and some grocery items such as facial tissues, orange juice, and soup.

However, sudden increases in volume can come from common non-malicious fluctuations.

- **Entropy detection** is to measure the sudden increase of decrease of disorder (or randomness) in a system. Entropy is a function of $k \log p$, where k is a constant and p is the probability of a given configuration. For network intrusion detection, the traffic and router attributes are characterised by distinct curves which uniquely profile the traffic: high entropy is represented by a uniform curve, while low entropy is shown as a spike.

Even if adversaries try to look legitimate and keep usage volume low, entropy detection can still find network intrusions which differ in some way from the network's established usage patterns [30, 35].

7.3.3 Privacy and Confidentiality Preserving Results

Data sharing and data mining can be good (increases accuracy), but data mining can be bad (decreases privacy and confidentiality) [10]. For example, suppose a drug manufacturing company wishes to collect responses (i.e. record) from each of the clients containing their dining habits and adverse effects of a drug. The relationship between dining habits and the drug could give the drug manufacturing company some insight knowledge about its side effects. The clients may not be interested to provide information because of their privacy.

Another example [45] is a car manufacturing company who incorporates several components such as tires, electrical equipments, etc. made by independent producers. Each of these producers has their proprietary databases which it may not be interested to share. However, in practical scenarios sharing those databases is important and we could take the Ford Motors and Firestone Tires provide a real example of this type. Ford Explorers with Firestone tires from a specific factory had tire-thread separation problem which resulted in 800 injuries. As those tires did not cause problems to other vehicles or the other tires in Ford Explorer did not pose such problems, thus neither Ford nor Firestone wants to take responsibility. The delays in identifying the real problem resulted in public concern and eventually led to replacement of 14.1 million tires. In reality, many of those tires were probably fine as Ford Explorer accounted for only 6.5 million of the replacement tires. If both companies had discovered the association between the different attributes of their proprietary databases, then this safety concern can be avoided before it becomes public.

Privacy and confidentiality are important issues in security data mining because organisations use personal, behavioural, and sensitive data. Explicit consent has been given by the people to use their personal data and behavioural data for a specific purpose, and all personal data is protected from unauthorised disclosure or intelligible interception. Privacy laws, non-disclosure agreements, and ethical codes of conduct have to be adhered to. Sometimes, the exchange of raw or summarised results with other organisations may expose personal or sensitive data. Therefore, the following are ways to increase privacy and confidentiality, mainly from association rules literature:

- **Randomisation** is simple probabilistic distortion of user data, employing random numbers generated from a pre-defined distributed function. A centralised environment to maintain privacy and accuracy of resultant rules has been proposed [39]. However, the distortion process employs system resources for a long period when the dataset has large number of transactions. Furthermore, if this algorithm is used in the context of a distributed environment, this needs uniform distortion among various sites in order to generate unambiguous rules. This uniform distortion may disclose confidential inputs of individual site and may also breach the privacy of data (such as exact support of itemsets), and hence it is not suitable for large distributed data mining.

  To discover patterns from distributed datasets, a randomisation technique [17] could be deployed in an environment where a number of clients are connected to a server. Each client sends a set of items to the server where association rules are generated. During the sending process, the client modifies the itemsets according to its own randomisation policies. As a result, the server is unable to find the exact information about the client.

  However, this assumption is not suitable for distributed association rule mining because it generates global frequent itemsets by aggregating support counts of all clients.

- **Secure Multi-party Computation (SMC)**-based [3,27] perform a secure computation at individual site. To discover a global model, those algorithms secretly exchange the statistical measures among themselves. This is more suitable for few external parties.

  A privacy preserving association rule mining is defined for horizontally partitioned data (each site shares a common schema but has different records) [27]. Two different protocols were proposed: secure union of locally large itemsets, and a testing support threshold without revealing support counts. The former protocol uses cryptography to encrypt local support count, and therefore, it is not possible to find which itemset belongs to which site. However, it reveals the local itemsets to all participating sites in where these itemset are also locally frequent. Since the first protocol gives the full set of locally frequent itemsets, then in order to find which of these itemsets are globally frequent, the latter protocol is used. It adds a random number to each support count and finds the excess supports. Finally, these excess supports are sent to the second site where it learns nothing about the first site's actual dataset size or support. The second site adds its excess support and sends the value until it reaches the last site.

This protocol can raise a collusion problem. For example, site i and $i+2$ in the chain can collude to find the exact excess support of site $i+1$. To generate patterns from vertically partitioned distributed dataset, a technique is used to maintain the privacy of resultant patterns in vertically partitioned distributed data sources (across two data sources only) [45]. Each of the parties holds some attributes of each transaction. However, if the number of disjoint attributes among the site is high, this technique incurs huge communication costs. Furthermore, this technique is designed for an environment where there are two collaborating sites, each of them holding some attributes of each transaction. Hence, it may not be applied in an environment where collaborating sites do not possess such characteristics.

- **Anonymity** minimises potential privacy breaches. The above two techniques - randomisation and SMC - focused on how to find frequency of itemsets from large dataset such a way that none of the participants is able to see the exact local frequency each of the individual itemset. Though the patterns discovered using these methods does not reveal exact frequency of an itemset however, the resultant patterns may reveal some information about the original dataset which are not intentionally released. In fact, such inferences represent *per se* a threat to privacy. To overcome such potential threat, k-anonymous patter discovery method is proposed [4]. Unlike the randomisation, the proposed method generates patterns using data mining algorithm from the real dataset. Then these patterns are analysed against several anonymity properties.

  The anonymity properties check whether collection of patterns guarantee the anonymity or not. Based on the outcome of the anonymity, the patterns collection is sanitised in such a way that the anonymity of a given pattern collection is preserved. As the patterns are generated using the real dataset, the main problem of this approach is how to discover patterns from distributed datasets. In fact, if each of the participating sites applies this method at local sites, then the resultant global patterns will have discrepancies which could diminish the goal of distributed pattern mining.

- **Cryptography**-based techniques [27, 49] use the public key cryptography system to generate a global model. This is more suitable for many external parties. Despite cryptography system has computational and communication overhead, recent research argues it is possible to generate privacy preserving patterns and with achieve good performance. For example, a cryptography-based system that performs sufficiently efficient to be useful in the practical data mining scenarios [49]. Their proposed method discovers patterns in a setting where number of participant is large. Each of the participants sends their own private input to a data miner who will generate patterns from these inputs using the homomorphic property of ElGamal cryptography system.

The main problem of cryptography-based approach is the underlying assumptions. For example, all of the cryptography-based methods assume participating parties are semi-honest, that is, each of them executes the protocol exactly the same manner as described in the protocol specification. Unless each of the participants is semi-honest, those methods may not able to preserve the privacy of the each of the participant's private input.

- **Knowledge hiding** is a way to preserve privacy of sensitive knowledge by hiding frequent itemsets from large datasets. Heuristics were applied to reduce the number of occurrences to such a degree that its support is below the user-specified support threshold [2]. This work was extended to confidentiality issues of association rule mining [12]. Both works assume datasets are local and that hiding some itemsets will not affect the overall performance or mining accuracy.

  However, in distributed association rule mining, each site has its own dataset and a similar kind of assumption may cause ambiguities in the resultant global rule model.

7.4 Conclusion

This chapter is titled *Security Data Mining: A Survey Introducing Tamper-Resistance*, that is, motivations, definitions, and problems are discussed and tamper-resistance as an important solution is recommended. The growth of security data with adversaries has to be accompanied by both theory-driven and domain-driven data mining. Inevitably, security data mining with tamper-resistance has to incorporate domain-driven enhancements in the form of reliable data, anomaly detection algorithms, and privacy and confidentiality preserving results. Future work will be to apply tamper-resistance solutions to the detection of data breaches, phishing, and plagarism; for specific results to support the conclusion of this chapter.

References

1. Adams, N.: 'Fraud Detection in Consumer Credit'. Proc. of UK KDD Workshop (2006)
2. Atallah, M., Bertino, E., Elmagarmid, A., Ibrahim, M., Verykios, V.,: 'Disclosure Limitation of Sensitive Rules'. Proc. of KDEX99, pp. 45–52 (1999)
3. Ashrafi, M., Taniar, D., Smith, K.: 'Reducing Communication Cost in a Privacy Preserving Distributed Association Rule Mining'. Proc. of DASFAA04, LNCS 2973, pp. 381–392 (2004)
4. Atzori, M., Bonchi, F., Giannotti, F., Pedreschi, D.: 'k-Anonymous Patterns'. Proc. of PKDD05, pp. 10–21 (2005)
5. Bay, S., Kumaraswamy, K., Anderle, M., Kumar, R., Steier, D: 'Large Scale Detection of Irregularities in Accounting Data'. Proc. of ICDM06, pp. 75–86 (2006)

6. Bolton, R., Hand, D.: 'Unsupervised Profiling Methods for Fraud Detection'. Proc. of CSCC01 (2001)
7. Cortes, C., Pregibon, D., Volinsky, C.: 'Communities of Interest'. Proc. of IDA01. pp. 105–114 (2001)
8. Cowan, C., Pu, C., Maier, D., Walpole, J., Bakke, P., Beattie, S., Grier, A., Wagle, P., Zhang, Q., Hilton, H: 'StackGuard: Automatic Adaptive Detection and Prevention of Buffer-Overflow Attacks'. Proc. of 7th USENIX Security Symposium (1998)
9. Cox, K., Eick, S., Wills, G.: 'Visual Data Mining: Recognising Telephone Calling Fraud'. *Data Mining and Knowledge Discovery* **1**. pp. 225–231 (1997)
10. Clifton, C., Marks, D.: 'Security and Privacy Implications of Data Mining'. Proc. of SIGMOD Workshop on Data Mining and Knowledge Discovery. pp. 15–19 (1996)
11. Dalvi, N., Domingos, P., Mausam, Sanghai, S., Verma, D.: 'Adversarial Classification'. Proc. of SIGKDD04 (2004)
12. Dasseni, E., Verykios, V., Elmagarmid, A., Bertino, E.: 'Hiding Association Rules by Using Confidence and Support'. LNCS 2137, pp. 369-379 (2001)
13. DeBarr, D., Eyler-Walker, Z.: 'Closing the Gap: Automated Screening of Tax Returns to Identify Egregious Tax Shelters'. *SIGKDD Explorations.* **8**(1), pp. 11–16 (2006)
14. Denning, D.: 'An Intrusion-Detection Model'. *IEEE Transactions on Software Engineering.* **13**(2), pp. 222–232 (1987)
15. Emigh, A., 'Online Identity Theft: Phishing Technology, Chokepoints and Countermeasures'. ITTC Report on Online Identity Theft Technology and Countermeasures (2005)
16. Eskin, E., Arnold, A., Prerau, M., Portnoy, L., Stolfo, S.: 'A Geometric Framework for Unsupervised Anomaly Detection: Detecting Intrusions in Unlabeled Data'. Applications of Data Mining in Computer Security, Kluwer (2002)
17. Evfimievski, A., Srikant, R., Agrawal, R., Gehrke, J.: 'Privacy Preserving Mining of Association Rules', *Information Systems*, **29**(4): pp. 343–364 (2004)
18. Fast, A., Friedland, L., Maier, M., Taylor, B., Jensen, D., Goldberg, H., Komoroske, J.: 'Relational Data Pre-Processing Techniques for Improved Securities Fraud Detection'. Proc. of SIGKDD07 (2007)
19. Fawcett, T., Provost, F.: 'Adaptive Fraud Detection'. *Data Mining and Knowledge Discovery.* **1**(3), pp. 291–316 (1997)
20. Fayyad, U., Piatetsky-Shapiro, G., Smyth, P., Uthurusamy, R.: Advances in Knowledge Discovery and Data Mining. AAAI (1996)
21. Friedland, L., Jensen, D.: 'Finding Tribes: Identifying Close-Knit Individuals from Employment Patterns'. Proc. of SIGKDD07 (2007)
22. Goldberg, H., Kirkland, J., Lee, D., Shyr, P., Thakker, D: 'The NASD Securities Observation, News Analysis and Regulation System (SONAR)'. Proc. of IAAI03 (2007)
23. Goldenberg, A., Shmueli, G., Caruana, R.: 'Using Grocery Sales Data for the Detection of Bio-Terrorist Attacks'. *Statistical Medicine* (2002)
24. Hand, D.: 'Protection or Privacy? Data Mining and Personal Data'. Proc. of PAKDD06, LNAI 3918. pp. 1–10 (2006)
25. Jensen, D.: 'Prospective Assessment of AI Technologies for Fraud Detection: A Case Study'. AI Approaches to Fraud Detection and Risk Management. AAAI Press, pp. 34–38 (1997)
26. Jonas, J.: 'Non-Obvious Relationship Awareness (NORA)'. Proc. of Identity Mashup (2006)
27. Kantarcioglu, M., Clifton, C.: 'Privacy-Preserving Distributed Mining of Association Rules on Horizontally Partitioned Data'. *IEEE Transactions on Knowledge and Data Engineering.* **16**(9), pp. 1026–1037 (2004)
28. Kushner, D.: 'Playing Dirty: Automating Computer Game Play Takes Cheating to a New and Profitable Level'. *IEEE Spectrum.* **44**(12) (INT), December 2007, pp. 31–35 (2007)
29. Layland, R.: 'Data Leak Prevention: Coming Soon To A Business Near You'. *Business Communications Review.* pp. 44–49, May (2007)
30. Lee, W., Xiang, D.: 'Information-theoretic Measures for Anomaly Detection'. Proc. of 2001 IEEE Symposium on Security and Privacy (2001)
31. Liu, C., Chen, C., Han, J., Yu, P.: 'GPLAG: Detection of Software Plagiarism by Program Dependence Graph Analysis'. Proc. of SIGKDD06 (2006)

32. Loveman, G.: 'Diamonds in the Data Mine'. *Harvard Business Review*. pp. 109–113, May (2003)
33. Lowd, D., Meek, C.: 'Adversarial Learning'. Proc. of SIGKDD05 (2005)
34. Metwally, A., Agrawal, D., Abbadi, A.: 'Using Association Rules for Fraud Detection in Web Advertising Networks'. Proc. of VLDB05 (2005)
35. Nucci, A., Bannerman, S.: 'Controlled Chaos'. *IEEE Spectrum*. **44**(12) (INT), December 2007, pp. 37–42 (2007)
36. Peacock, A., Ke X., Wilkerson, M.: 'Typing Patterns: A Key to User Identification'. *IEEE Security and Privacy* **2**(5), pp. 40–47 (2004)
37. Phua, C., Lee, V., Smith-Miles, K., Gayler, R.: 'A Comprehensive Survey of Data Mining-based Fraud Detection Research'. Clayton School of Information Technology, Monash University (2005)
38. Phua, C.: 'Data Mining in Resilient Identity Crime Detection'. PhD Dissertation, Monash University (2007)
39. Rizvi, S., Haritsa, J.: 'Maintaining Data Privacy in Association Rule Mining'. Proc. of VLDB02 (2002)
40. Schleimer, S., Wilkerson, D., Aiken, A.: 'Winnowing: Local Algorithms for Document Fingerprinting'. Proc. of SIGMOD03. pp. 76–85 (2003)
41. Schneier, B.: Beyond Fear: Thinking Sensibly about Security in an Uncertain World. Copernicus (2003)
42. Schultz, M., Eskin, E., Zadok, E., Stolfo, S.: 'Data Mining Methods for Detection of New Malicious Executables'. Proc. of IEEE Symposium on Security and Privacy. pp. 178–184 (2001)
43. Skillicorn, D.: Knowledge Discovery for Counterterrorism and Law Enforcement. CRC Press, in press (2008)
44. Sweeney, L.: 'Privacy-Preserving Surveillance using Databases from Daily Life'. *IEEE Intelligent Systems*. **20**(5): pp. 83—84 (2005)
45. Vaidya, J., Clifton C.: 'Privacy Preserving Association Rule Mining in Vertically Partitioned Data'. Proc. of SIGKDD02.
46. Viega, J.: 'Closing the Data Leakage Tap'. *Sage*. **1**(2): Article 7, April (2007)
47. Virdhagriswaran, S., Dakin, G.: 'Camouflaged Fraud Detection in Domains with Complex Relationships'. Proc. of SIGKDD06 (2006)
48. Wong, W., Moore, A., Cooper, G., Wagner, M.: 'Bayesian Network Anomaly Pattern Detection for Detecting Disease Outbreaks'. Proc. of ICML03 (2003)
49. Yang, Z., Zhong, S., Wright, R.: 'Privacy-Preserving Classification of Customer Data without Loss of Accuracy'. Proc. of SDM05 (2005)

Chapter 8
A Domain Driven Mining Algorithm on Gene Sequence Clustering

Yun Xiong, Ming Chen, and Yangyong Zhu

Abstract Recent biological experiments argue that similar gene sequences measured by permutation of the nucleotides do not necessarily share functional similarity. As a result, the state-of-the-art clustering algorithms by which to annotate genes with similar function solely based on sequence composition may cause failure. The recent study of gene clustering techniques that incorporate prior knowledge of the biological domain is deemed to be an essential research subject of data mining, specifically aiming at one for biological sequences. It is now commonly accepted that co-expressed genes generally belong to the same functional category. In this paper, a new similarity metric for gene sequence clustering based on features of such co-expressed genes is proposed, namely 'Tendency Similarity on N-Same-Dimensions', in terms of which a domain driven algorithm 'DD-Cluster' is designed to group together gene sequences into 'Similar Tendency Clusters on N-Same-Dimensions', i.e., co-expressed gene clusters. Compared with earlier clustering methods considering composition of gene sequences alone, the resulting 'Similar Tendency Clusters on N-Same-Dimensions' proved more reliable for assisting biologists in gene function annotation. The algorithm has been tested on real data sets and has shown high performance, the clustering results having demonstrated effectiveness.

8.1 Introduction

The study of biological functions of gene coded product and their effect on life activity presents one of the biggest challenges in the post genomic era. Sequence alignment is a well-known method widely used for searching similar gene sequences, whose function information is hopefully considered to be useful for re-

---

Yun Xiong, Ming Chen, Yangyong Zhu
Department of Computing and Information Technology, Fudan University, Shanghai 200433, China, e-mail: {yunx,chenming,yyzhu}@fudan.edu.cn

search on gene functions. However, biological experiments in recent years argue that similar gene sequences measured by permutation of the nucleotides do not necessarily exhibit functional similarity. As a result, the state-of-the-art clustering algorithms based on sequence information alone may not perform well. It is now possible for researchers to incorporate biological domain knowledge for gene clustering, such as gene expression data, co-expression profiles, or additional helpful information. Because genes from the same co-expressed gene cluster are more likely to co-express during the same cell cycle, particular attention paid to the conserved sequence pattern from the upstream of co-expressed gene clusters is expected to improve the accuracy of transcriptional regulatory elements prediction [1], which would in turn assist future study of functional elements and association rules, if any, between regulatory elements and corresponding physiological conditions or environmental information. This strategy will perhaps eventually become an essential research subject of transcriptional regulatory networks in the future.

The gene expression data are usually measured under several physiological conditions, involving organisms, distinct phases of the cell cycle, drug effect-time, etc., often resulting in high dimensional features. As a result, the distance metrics defined in total space can not be directly applied for clustering such high dimensional data. However, a biological process of interest usually involves only a small portion of genes, and happens under a minority of conditions within a certain time interval. It is meaningful to reveal gene transcriptional regulatory mechanisms by some genes under certain conditions with a preserving tendency but different expression levels. Some biclustering algorithms for gene expression data in subspace proposed by some earlier researchers, such as BiClustering [2], pCluster [3], MaPle(Maximal Pattern-based Clustering) [4], OPSM (Order-Preserving Submatrix) [5], are subject to strong constraints on both expression level rate and the order of conditions. Due to this property, they can hardly be applied for genes with similar expression patterns but different expression levels. Liu et al. proposed the algorithm OP-Cluster (Order Preserving Cluster) [6] to solve this problem. As the pruning condition of the OPC tree depends on the user specified threshold of dimensionality, it cannot be satisfied when the dimensionality is relatively low compared to that of the total space. In that case, the cost of constructing such a tree will increase dramatically, thereby resulting in lower efficiency. In addition, OP-Cluster cannot properly handle patterns containing "order equivalent class" (i.e., a sequential pattern that contains itemset). Apart from previous work, we propose a new similarity metric for gene clustering and design a biological domain-driven algorithm.

8.2 Related Work

Previous clustering algorithms, such as hierarchy based complete-link [7], partitioning around K-medoid [8], etc., can hardly be applied to clustering of large scale data sets of biological sequences. This is because they require global pairwise comparison to compute the sequence similarity, which often leads to prohibitive time cost. It has been realized that biological sequences can exhibit their similarity by means of common patterns they share. Correspondingly, researchers have proposed several clustering algorithms based upon feature extraction of sequential patterns. This kind of algorithms performs very well on protein family clustering, since they share like function domain. But for gene sequences of remote homologies, sequence composition of linear permutation may not be sufficient for extracting their common patterns, and so any clustering algorithms of this kind cannot be applied to clustering of gene sequences with similar functions. Two important features for gene sequence clustering are known to be gene expression data and co-expression profiles.

The gene expression data is of high dimensionality for which clustering is often performed in high dimensional space. Thorough research on L_k-norm by Aggarwal et al. [9] suggested that in most cases it cannot be used as distance metric in such space because the distance from one point to its nearest neighbor is almost equivalent to that from the same point to its farthest neighbor in high dimensional space. The fact that each cluster in a data set often refers to certain subspace makes it reasonable to cluster together such data in a subspace rather than in the original space [10]. Aiming at properties of gene expression data, the clustering method should include clustering according to expression levels under certain part of conditions, or clustering of conditions in terms of expression values of several genes. Obviously, one needs to consider gene expression data in subspaces, one typical algorithm being BiCluster [2] introduced by Cheng et al..

Yang et al. proposed a pattern-based subspace clustering algorithm pCluster, on the basis of which Pei et al. introduced an improved version—MaPle [4]. However, good results from these algorithms require not only coherent tendency but also their expression values being strictly proportional. Ben-Dor et al. proposed the OPSM model [5] to cluster together genes with coherent tendency free from considering the expression values of related columns. The major limitation of the model is that conditions must be placed in a strict order. In order to search gene sequences with a similar expression profile but distinct expression level, Liu et al. proposed a more general model, called the OPC model, to cluster together gene sequences with coherent tendency [6]. The model trickily transforms the problem of sequential pattern mining into clustering in such a way that gene expression data are represented by sequential patterns. Although OPC treats OPSM and pCluster as its two special cases, and allows genes with similar expression level to constitute one "order equivalence class", it may not work well when itemsets are involved. Different from [6], this paper defines a new similarity metric, the 'Tendency Similarity on N-Same-Dimensions', to measure the coherent tendency of gene expression profiles. It follows that the similarity of gene sequences can be effectively measured in an alternative way. Accordingly, we propose a new gene clustering algorithm 'DD-Cluster', which can produce gene clusters with different expression values but coherent tendency. It also takes into account the sequential pattern mining problem with itemset brought by 'order equivalent class'. The clustering results of our algorithm can cover that of 'OP-Cluster'.

8.3 The Similarity Based on Biological Domain Knowledge

One main purpose of data mining on biological sequences is to identify putative functional elements and reveal their interrelationships. The availability of the mining results plays a key role in biological sequence analysis and has been the subject of much concern. For biological sequence mining in a specific field, a well-defined similarity metric is essential to guarantee that the mining results meet practical needs. Due to the specificity of biological data and the diversity of biological demands, the incorporation of biological domain knowledge is often required to help define such metrics. In some present applications of biological sequence analysis, it is the improper similarity metrics that cause the state-of-the-art clustering algorithm to result in inconsistency with real biological observations.

The modern technology of gene chips has been accumulating considerable gene expression data containing rich biological knowledge for the study of bioinformatics. Gene expression is the result of the cell, organisms and organs being affected by inheritance and environment. It can reflect expression levels of specific organisms or genes in cells under different physiological conditions, or on different time points in certain continuous time intervals [11]. Biological experiments reveal that gene sequences exhibiting similar expression levels generally belong to the same functional category [12]. Cases in point are signal transmission and cut injury recovery. Therefore, gene co-expression profiles can reflect the similarity between gene sequences, providing a basis on which researchers can define new similarity metrics for gene sequences. Gene clustering algorithms based on such similarity metrics can produce gene clusters with co-expressed functions in certain conditions or environments which will hopefully reveal gene regulatory mechanisms.

8.4 Problem Statement

Table 8.1 Gene expression data set

Gene	Condition			
	D_1	D_2	D_3	D_4
G_1	942	1000	845	670
G_2	992	340	865	620
G_3	962	820	855	640
G_4	982	78	825	660
G_5	972	130	875	650

Table 8.2 Sample data set

Gene	Condition			
	D_1	D_2	D_3	D_4
G_1	3942	845	3770	754
G_2	1392	865	620	814
G_3	6392	3765	1513	4887
G_4	3918	2532	1156	3180
G_5	3679	1421	1368	220

Example 1. Table 8.1 shows a gene expression data set of five genes under four conditions. Generally, let m be the numbers of genes, n be the numbers of condi-

8 A Domain Driven Mining Algorithm on Gene Sequence Clustering

tions. In Table 8.1, $m=5$, $n=4$; each row corresponds to a gene; the values in each row represent expression situations of a gene under different conditions; the values in each column represent the expression values of different genes under one condition; and each element represents the expression value of the ith gene G_i on the jth condition D_j. We call the collection of all m genes $G_i(1 \leq i < m)$ the set of gene sequences, denoted as $\hat{G}$; and the set of all n conditions D_j the set of conditions, denoted as $\hat{D}$. Let O be an object composed of gene and corresponding conditions, denoted as $O(D_1, D_2, \ldots, D_j, \ldots, D_n)$.

It is difficult to measure the similarity between genes using L_k distance metric in a space full of all conditions, because their expression values on condition D_2 are far apart from each other in distance (shown in figure8.1a). However, if we only consider three conditions (D_1, D_3, D_4), then similar relationship among them can be found (shown in figure 8.1b).

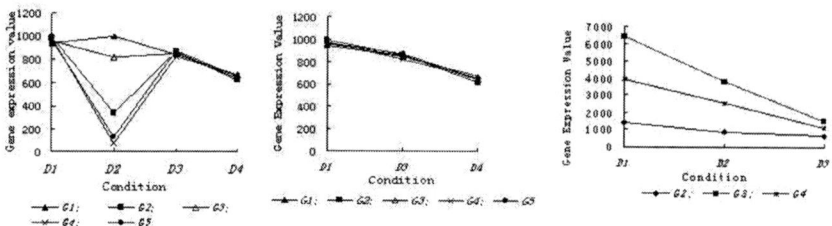

Fig. 8.1 Similarity among genes in (a)total space (b)subspace (c)subspace

Figure8.1b illustrates that if expression values of one gene are similar to others on $N(N \leq n)$ same conditions, then these genes are supposed to be similar under such N conditions. If N is not less than a certain threshold (specified by experts), then these genes can be regarded to be similar. Furthermore, gene expression data naturally exhibit fluctuate property, therefore, though the expression values of different gene sequences are far apart, they may share coherent tendency (see table 8.2 and figure 8.1c).

Definition 8.1. (n-Dimensional Object Space) An object with n attribution values $O(x_1, x_2, \ldots, x_n)$, we define O as a n- dimensional object. All such objects $O(x_1, x_2, \ldots, x_n)$ form a n-dimensional object space Ω, $\Omega = D_1 \times D_2 \times \cdots \times D_n$, where each $D_i(i = 1, 2, \ldots, n)$ represents domain of values on one dimension.

Let a gene sequence be an object in Ω, then gene expression data set $\hat{A}$ can be regarded as one instance of Ω, called n dimension gene object set, denoted as Σ, where each object O corresponds to one object O in set $\hat{A}$, and each dimension D represents one condition $D(D \in \hat{D})$.

Definition 8.2. (Similar Dimension Group ItemSet) Let O be an object in n-dimensional object space $\Omega, x_1, x_2, \ldots, x_n$ be a series of attribute values in a non-decreasing order, δ be the user specified threshold. If $(x_{i+j} - x_i) < \delta$, we say that attributes

$D_i, D_{i+1}, \ldots, D_{i+j} (n \geq i > 0, n \geq j > 0)$ are close to each other and call the set of $\langle D_i, D_{i+1}, \ldots, D_{i+j} \rangle$ similar dimension group Itemset.

The purpose of giving the definition is to represent the insignificant difference between the values of two or more attributes [6].

Definition 8.3. (Object Dimensional Order) Given an object $O(x_1, x_2, \ldots, x_n)$, if $x_l \leq x_k$, we say that O possesses dimensional order $D_l \prec D_k$, i.e., D_l is said to be ahead of D_k. If the value of x_l is similar to that of x_k, we say D_l is order irrelevant to D_k.

Definition 8.4. (Object Dimensional Sequence) The attribute values of $O(x_1, x_2, \ldots, x_n)$ as $x_{k1} \leq x_{k2} \leq \ldots \leq x_{kn}$ sorted in ascending order, whose corresponding dimension series $D_{k1}, D_{k2}, \ldots, D_{ki}, \ldots, D_{kn}$ are defined as object dimensional sequence, denoted as $O(D_{k1}, D_{k2}, \ldots, D_{ki}, \ldots, D_{kn})$. if $D_{ki1}, D_{ki2}, \ldots, D_{kij}$ are irrelevant to each other, we denoted them as $(D_{ki1}, D_{ki2}, \ldots, D_{kij})$, thereby object dimensional sequence is denoted as $O\langle D_{k1}, D_{k2}, \ldots, (D_{ki1}, D_{ki2}, \ldots, D_{kij}), \ldots, D_{kn} \rangle$.

An object dimensional sequence corresponds to one object in Σ. In this way, set Σ can be transformed into a n-dimensional sequence database($DSDB$) storing m object dimensional sequences, denoted as $\Delta \{O_1, O_2, \ldots, O_i, \ldots, O_m\}$.

Table 8.3 $DSDB\Delta$ from Tab.8.2

SeqId	Sequence
O_1	$\langle D_4 D_2 (D_1 D_3) \rangle$
O_2	$\langle D_3 (D_2 D_4) D_1 \rangle$
O_3	$\langle D_3 D_2 D_4 D_1 \rangle$
O_4	$\langle D_3 D_4 D_2 D_1 \rangle$
O_5	$\langle D_4 D_3 D_2 D_1 \rangle$

Example 2 Transforming a gene expression data set (shown in table 8.2) into a dimensional sequence database δ (shown in table 8.3)

Definition 8.5. (Subsequence) Given an object dimensional sequence $O_1 = D_1, D_2, \ldots, D_n$ and $O_2 = D'_1, D'_2, \ldots, D'_m (m \leq n)$. If there exsit $1 \leq i_1 < i_2 < \ldots < i_m \leq n$, and $D'_1 \sqsubseteq D_{i2}, D'_2 \sqsubseteq D_{i2}, \ldots, D'_m \sqsubseteq D_{im}$, then we say that O_2 is a subsequence of the object dimensional sequence O_1, or O_1 contains O_2.

Definition 8.6. (Similar Tendency) Given two object dimensional sequences $O_i \langle D_{i1}, D_{i2}, \ldots, D_{in} \rangle$, $O_j \langle D_{j1}, D_{j2}, \ldots, D_{ji} \rangle$, in database Δ, l dimensions $D_{k1}, D_{k2}, \ldots, D_{kl}$. For O_i, O_j, if there exsits $D_{k1} \prec D_{k2} \prec \ldots \prec D_{ki}$, we say that O_i and O_j exhibit similar tendency on l dimensions $D_{k1}, D_{k2}, \ldots, D_{kl}$. Where, $\{D_1, D_2, \ldots, D_n\} = \{D_{i1}, D_{i2}, \ldots, D_{in}\} = \{D_{j1}, D_{j2}, \ldots, D_{ji}\}, \{D_1, D_2, \ldots, D_n\} \supset \{D_{k1}, D_{k2}, \ldots, D_{ki}\}$.

Definition 8.7. (Tendency Similarity on N-Same-Dimensions and Similar Tendency Pattern on N-Same-Dimensions) Given a n-dimensional sequence database, if there exist k object dimensional sequences $O_1, O_2, \ldots, O_i, \ldots, O_k$ in Δ such that they exhibit similar tendency $N (N \leq n)$ same dimensions $(D_{k1}, D_{k2}, \ldots, D_{kN})$, then these k object dimensional sequences are similar tendency on N-same-dimensions, such a similarity measure is defined as tendency similarity on N-same-dimensions. Accordingly the subsequence $\langle D_{k1}, D_{k2}, \ldots, D_{kN} \rangle$ composed of such same N dimensions is regarded as similar tendency pattern on N-same-dimensions, denoted as N-*pattern*.

The length of similar tendency pattern on N-Same-Dimensions is defined as dimension support of the pattern, denoted as *dim_sup*. The number of object dimensional sequences which contain the pattern in a dimensional sequence database is defined as cluster support of similar tendency pattern on N-same-dimensions, denoted as *clu_sup*.

Definition 8.8. (Similar Tendency Cluster on N-Same-Dimensions) Given k object dimensional sequences such that they are similar tendency on N-same-dimensions, let the set C_o of k object dimensional sequences be $(O_1, O_2, \ldots, O_i, \ldots, O_k)$, if k is no less than cluster support threshold, then we say that the set C_o is a similar tendency cluster on N-Same-Dimensions.

Given a dimensional sequence database Δ and a cluster support threshold *clu_sup*. Similar tendency clustering on N-same-dimension is the process, which clustering object dimensional sequences $O_1, O_2, \ldots, O_m$ in Δ into a similar tendency cluster on N-same-dimensions, such that the number of object dimensional sequences in each cluster is no less than *clu_sup*.

8.5 A Domain-Driven Gene Sequence Clustering Algorithm

Domain-driven gene sequence clustering algorithm DD-Cluster is to do similar tendency clustering on N-same-dimensions for m genes in one n dimensional object space Ω according to the expression values on n conditions in order to get all similar tendency clusters on N-same-dimensions, i.e., co-expressed gene sequences clusters. DD-Cluster takes two steps: Step 1(Preprocessing): To transform gene expression data set into one object dimensional sequence database Δ in the way mentioned in example 2. Step 2(Sequential Pattern Mining Phase): To find similar tendency patterns on N-same-dimensions in a dimensional sequence database Δ. Each of the result patterns represents a subspace composed of N dimensions. The gene objects containing such patterns share coherent tendency on the N dimensions, and are grouped into a similar tendency cluster on N-same-dimensions.

Definition 8.9. (Prefix with ItemSet) Given a subsequence of an object dimensional sequence $O_1 = \langle D_1, D_2, \ldots, D_n \rangle, O_2 = \langle D'_1, D'_2, \ldots, D'_n \rangle, (m \leq n).O_2$ is called Prefix with ItemSet of O_1 if and only if 1) $D_i = D'_i$ for $(i \leq m-1)$; 2) $D'_m \sqsubseteq D_m$; and 3) all the items in $(D_m\text{-}D'_m)$ are alphabetically after those in D'_m.

Given object dimensional sequences O_1 and O_2 such that O_2 is a subsequence of O_1, i.e., $O_2 \sqsubseteq O_1$. A subsequence O_3 of sequence O_1 (i.e., $O_3 \sqsubseteq O_1$) is called a projection with itemset of O_1 w.r.t. prefix O_2 if and only if 1) O_3 has prefix O_2; 2) there exists no proper super-sequence O_4 is a subsequence of O_3, (i.e., $O_3 \sqsubseteq O_4$, but $O_3 \neq O_4$) such that O_4 is a subsequence of O_1 and also has prefix O_2.

The collection of projections of sequences in a dimensional sequence database w.r.t. subsequence O is called projected database with itemset, denoted as $DSSD|_O$.

The prefix tree of a dimensional sequence database is a tree such that: i)Each path corresponds to a sequence except for that from root node; ii)Left nodes are alphabetically before right nodes; iii)Child nodes are composed of items, the support of which is no less than the threshold in the projection database of parent nodes.

Many duplicate projection databases will be generated in the sequential pattern mining process using the Prefix-Projection method, thereby resulting in many identical databases to be scanned. To overcome this drawback, [13] presents the SPMDS algorithm without scanning duplicate projection databases. SPMDS computes MD5 for pseudo projection of projection databases, and decides whether the projection databases are equal or not by comparing their MD5 results. SPMDS can decrease the generation and scanning of duplicate projection databases. However, this checking evidence cannot directly be applied to find sequential patterns with itemset.

Table 8.4 Dimensional sequence database with itemset

SeqId	Sequence
O_1	$\langle(D_1D_2)(D_2D_3D_4)\rangle$
O_2	$\langle(D_1D_2)(D_2D_3D_4)\rangle$
O_3	$\langle(D_1D_1)(D_2D_4)\rangle$

Example 3 A dimensional sequence database is shown in table 8.4. Assume that the dimension support is 2.

Suppose a prefix sequence $\langle(D_1D_2)\rangle$ has been mined. When computing the projection database of $\langle D_2 \rangle$, we observe that the projection database of $\langle D_2 \rangle$ and $(\langle D_1D_2 \rangle)$ are the same. In such a case, SPMDS takes the child nodes of $\langle(D_1D_2)\rangle$ as the child nodes of $\langle D_2 \rangle$. But this method is not well adapted to find the sequential patterns with itemset, because some valid patterns such as $\langle(D_2D_3)\rangle$, $\langle(D_2D_3D_4)\rangle$ could never be identified. Therefore, SPMDS cannot be well adapted to mine similar tendency pattern on N-same-dimensions with similar dimension group itemset.

I-EXTEND means to do expansions in a similar dimensional group itemset. For example, the node $\langle(D_1)\rangle$ in a prefix tree can be extended to $\langle(D_1D_2)\rangle$. S-EXTEND means to do expansions among the sets of dimensions. For example, the node $\langle(D_1)\rangle$ in prefix tree can be extended to $\langle(D_1D_2)\rangle$.

Definition 8.10. (Local Pseudo Projection) The projection database of each sequence can be represented as pointers which point to the original database. We define the array of these pointers as local pseudo projection.

8 A Domain Driven Mining Algorithm on Gene Sequence Clustering

Definition 8.11. (Pseudo Projection) As for the sequence with itemset, the projection database should deal with itemset, accordingly we regard the pointers as a set. We define the array of the sets of pointers as Pseudo Projection.

Example 4, The pseudo projection and local pseudo projection of sequence $\langle (D_2) \rangle$ and $\langle (D_1 D_2) \rangle$ are shown in Figure 8.2. Even if their local pseudo projections are identical, their pseudo projections are different.

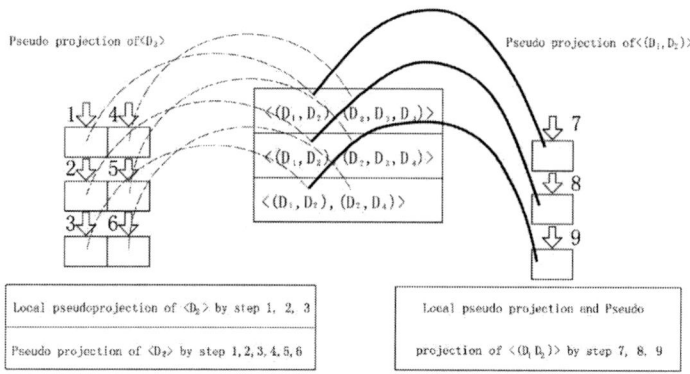

Fig. 8.2 Psedo-projection of sequence with itemset

Theorem 8.1. *Two projection databases are equivalent if and only if their pseudo projections are equal [13].*

Corollary 1 If without regard to confliction from hash function, two projection databases are equivalent if and only if the hash values corresponding to the pseudo projection are equal.

Based on Corollary 1, its hash values are stored by a prefix tree when each projection database is created. In the process of the tree being generated, if hash value of node p is equal to that of current node w, then these projection databases are said to be equivalent. Therefore, we only take child nodes of p in the projection database as that of w instead of scanning the duplicate projection database [13].

Theorem 8.2. *If the pseudo projections of two nodes are equal to each other, then they can be merged. And if local pseudo projections corresponding to two sequences are equal, then the nodes created by S-EXTEND must be identical.*

Proof. Given two sequences O_1, O_2. For each frequent item i to do S-EXTEND, if local pseudo projection of O_1 is equal to that of O_2, then i and the last item in O_1 or O_2 must not be in the same element. Therefore, an item is frequent if and only if it is frequent in one projection database. If pseudo projections of two sequences are equal, then their local pseudo projections must be also identical, so does by S-EXTEND. Since pseudo projection is defined as a set of pointers when each item is extended, from which nodes generated by I-EXTEND and S-EXTEND will be equal, which in turn could be merged. □

According to **theorem 2**, even if the hash value of $\langle D_2 \rangle$'s pseudo projection is not equal to that of $\langle (D_1 D_2) \rangle$, both of them can still share S-EXTEND node. Furthermore, when operating I-EXTEND on D_2, we can find the pseudo projections of $\langle (D_2 D_3) \rangle$, $\langle (D_2 D_4) \rangle$ and that of $\langle (D_1 D_2) D_3 \rangle$, $\langle (D_1 D_2) D_4 \rangle$ to be equal. In this way node D_3 and D_4 can be merged. The compressed prefix-tree for database in table 8.4 is shown in figure 8.3.

If two prefix sequences correspond to the equivalent projection databases, then their sizes will be equal, and the last item in the two prefix sequences are the same [13]. According to the judgment, the efficiency can be further improved.

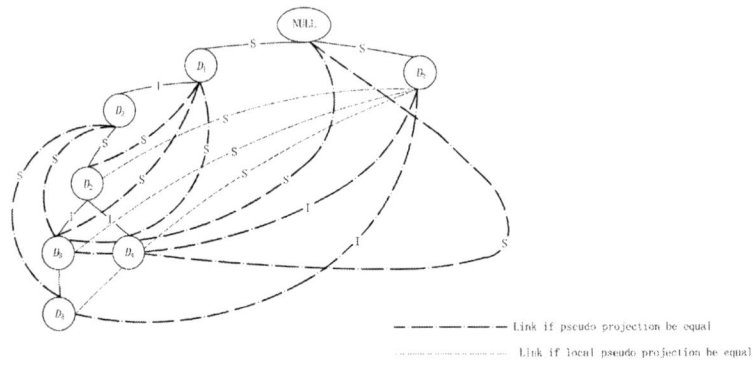

Fig. 8.3 Compressed prefix-tree for table8.4

In summary, our algorithm is illustrated as follows.
Algorithm 1 DD-Cluster(s, D, clu_sup, F)
Input: Dimensional sequence database D, the cluster support threshold clu_sup.
Output: the set of tendency similar patterns on N-same-dimension F.
1) s=NULL;
2) Scanning database D to find the frequent itemset w with length 1.
3) for each w do
4)   Define a pointer from s to w, and let $s.attr$ = S-EXTEND;
5)   Create the pseudo projection P_w of w;
6)   call iSet-GePM(w, D, clu_sup, P_w);
7)   Transfer prefix tree s, and put $s.attr$ into the set of F.

Algorithm 2 iSet-GePM(s, D, clu_sup, P_s) // similar tendency patterns on N-same-dimension with itemset mining algorithm
Input: A tree node s, dimensional sequence database D, cluster support threshold clu_sup and the pseudo projection of s as P_s
Output: Prefix Tree.

1) Scan P_s, Compute local pseudo projection of s using hash function d', and then evaluate the pseudo projection using hash function d;
2) Search the prefix tree;
3) if exist $w.d = s.d$ do
4)   Find the parent node of s p;
5)   Take w as the child node of p, $p.attr$ remains unchanged;
6)   delete s;
7)   return
8) if exist $w.d' = s.d'$ do
9)   for each node such that $w.attr$ = S-EXTEND do
10)   Take r as the child node of s, and let $s.attr$ = S-EXTEND;
11)   Find node r marked I-EXTEND by scanning D according to P_s;
12)   for each r do
13)   Take r as the child node of s, and let $s.attr$ = I-EXTEND;
14)   Create the projection database of r P_r;
15)   call iSet-GePM(r, D, clu_sup, Pr);
16)   return
17) Find all of nodes r marked I-EXTEND and nodes t marked S-EXTEND by scanning D according to P_s;
18) if the length of P_s is less than clu_sup or r, t not exist.
19)   return
20) for each r do
21)   Take r as the child node of s, and let $s.attr$ = I-EXTEND;
22)   Create the projection database of r P_r;
23)   call iSet-GePM(r, D, clu_sup, Pr);
24) for each t do
25)   Take t as the child node of s, and let $s.attr$ = S-EXTEND;
26)   Create the projection of t as P_t;
27)   call iSet-GePM(t, D, clu_sup, Pt);
28) return

In order to find clusters on N-same-dimensions we prune result set F according to user-specified dimension support threshold N. When counting cluster support to decide whether a pattern is frequent, we preserve the sequence id of this pattern.

8.6 Experiments and Performance Study

We tested DD-Cluster algorithm with both real and synthetically generated data sets to evaluate the performance. One real data set is the Breast Tumor microarray data containing expression levels of 3226 genes under 22 tissues [14], the other is the Yeast microarray data containing expression levels of 2884 genes under 17 conditions [15]. Synthetically generated data is a matrix with 5000 rows by 50 columns. The algorithm was implemented with C programming language and executed on a PC with 900Hz CPU and 512M main memory running Linux. We choose OP-Cluster algorithm in order to make comparison. The executed program of OP-Cluster algorithm was provided by [6] [16], which was downloaded from http://www.cs.unc.edu/ liuj/software/.

Experiment 1. Biological Significance Verification

1)Using criteria (Gene Ontology [17] and p-value) to evaluate mining results

We tested the DD-Cluster algorithm on Yeast dataset with $dim\_sup = 40\%, clu\_sup = 20\%$. We use GO (Gene Ontology) and p-value as criteria to verify whether the gene sequences in similar tendency cluster on N-same-dimension exhibit similar functions or share identical biological process, and so on [18] [19]. Gene Ontology Term Finder tool for yeast genome (http://db.yeast genome.org/cgi-bin/GO/goTerm Finder) which provided by the gene ontology (www.geneontology.org) was applied to evaluate the biological significance of the result clusters in terms of three gene function categories: biological processes, cellular components, and gene functions.

Figure 8.4 illustrates the expression profiles for three clusters found by DD-Cluster from the Yeast dataset. The algorithm can successfully find co-expressed genes whose expression level is distinct from each other but which share coherent tendency.

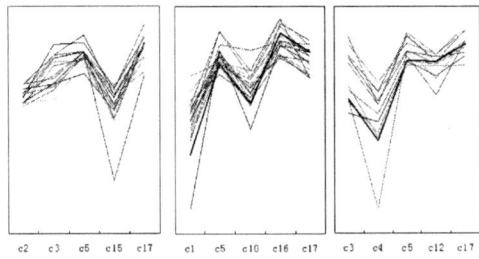

Fig. 8.4 Three similar tendency clusters on 5-same-dimensions found by DD-Cluster

In terms of biological significance, if the majority of genes in one cluster appears in one category, the P-value will be small, that is, the closer the p-value approaches zero, the more significant the correlation between particular GO term and the gene cluster becomes [6] [17] [19]. We submitted clustering results to online tool-Gene Ontology Term Finder. The evaluation results are shown in table 8.5.

Table 8.5 shows the hierarchy of functional annotations in terms of Gene Ontology for each cluster, whose statistical significance is given by p-values. For example, gene sequences in C_9 are involved in a translation process whose function is associated with Structural molecule activity. The experimental results show that gene sequences in similar tendency cluster on N-same-dimension exhibit similar biological function, component and process, and with low P-value.

2) Results comparison

We tested both DD-Cluster and OP-Cluster algorithm on the Breast Tumor datasets by choosing $dim\_sup = 20\%, clu\_sup = 20\%$. Without regard to similar

8 A Domain Driven Mining Algorithm on Gene Sequence Clustering

Table 8.5 GO terms of similar tendency cluster on N-same-dimensions

Cluster	Function	Components	Process
C_1	Structural constituent of ribosome (p-value=0.00250)	Cytosolic small ribosomal subunit(sensu Ekaryota) (p-value=7.60e-06)	Biosynthetic process (p-value=0.00555)
C_5	Phenylalanine-tRNA ligase activity (p-value=0.72e-04)	Phenylalanine-tRNA ligase complex (p-value=3.9e-04)	Phenylalanyl-tRNA aminoacylation (p-value=0.00228)
C_9	Structural molecule activity (p-value=2.4e-04)	Cytosolic ribosome (sensu Ekaryota) (p-value=6.18-06)	Translation (p-value=0.00296)

dimension group itemset, they can mine 28 clusters in which genes possess coherent tendency, i.e., both of their clustering results are the same. Therefore, DD-Cluster can mine all clusters found by OP-Cluster.

Experiment 2. Scalability

1) Response time with respect to cluster support

We tested both DD-Cluster and OP-Cluster on the Breast Tumor datasets. Let $dim\_sup\%$ be 20. Figure 8.5 shows that the response time decreases when the cluster support threshold is increased on both DD-Cluster and OP-Cluster. But with the same cluster support threshold, DD-Cluster shows better performance.

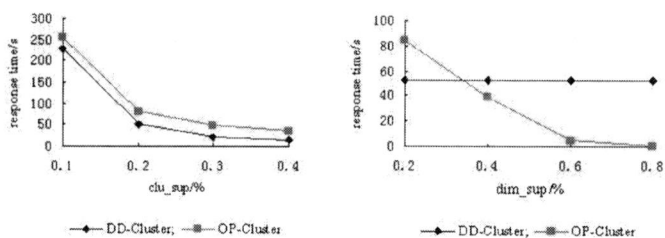

Fig. 8.5 Response time vs. cluster support on breast tumor data

Fig. 8.6 Response time vs. dimension support on breast tumor data

2) Response time with respect to dimension support

Let clu_sup be 20%. Figure 8.6 shows that the response time is stable on DD-Cluster, i.e., DD-Cluster is independent on dimension support, because the pruning step is performed after mining all patterns with various lengths. Therefore, DD-Cluster is more efficient when dimension support is relatively small with respect to total space, which conforms to domain knowledge that biological process of interest usually happens under a minority of conditions. However, the efficiency of OP-Cluster is subject to dimension support. Because OP-Cluster adopts top-down search method, the prune strategy on OPC tree depends on user-specified minimum number of dimensions N. If the threshold is relatively small, pruning demand can hardly be satisfied. Thereby, OPC tree will grow fast and become very large.

3) Response time with respect to the number of sequences

We fixed the dimensionality to 22, and let the number of sequences be 1000, 2000, 3000 respectively. We selected four combinations of the cluster support and the dimension support: a) 20%, 20%; b)80%, 20%; c)20%, 80%; d)80%, 80%.

Figure 8.7 shows that if the number of sequences is increasing and meanwhile dimension support and cluster support are relatively small, DD-Cluster performs better than OP-Cluster.

4) Response time with respect to different dimensionality

We tested algorithms on the data set with 1000 rows and 3000 rows respectively from synthetically generated data sets. We fixed cluster support to 20% and dimension support 20%. The result is shown in figure 8.8.

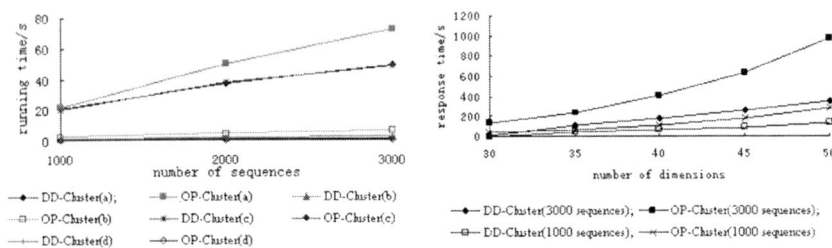

Fig. 8.7 Response time v.s. number of sequences in breast tumor data

Fig. 8.8 Response time v.s. number of dimensions in synthetically generated data

For larger sequences data sets, when the dimensionality increases, the response time of OP-Cluster algorithm increases much faster than DD-Cluster, i.e., DD-cluster is more efficient.

8.7 Conclusion and Future Work

Gene sequence clustering technique plays an important role in Bioinformatics. Based on domain knowledge in biology, a novel similarity measure "Tendency Similarity on N-Same-Dimensions" is proposed in order to clustering gene sequences. The similarity can capture co-expression characteristics among gene sequences. Furthermore, a clustering algorithm DD-Cluster was developed to search similar tendency clusters on N-Same-Dimensions, i.e., co-expressed gene sequences clusters, in which gene sequences exhibit similar function. Compared with other gene sequence clustering methods that merely make use of sequences information, similar tendency clusters on N-Same-Dimensions can give better explanation on gene sequences functions. Experimental results show that the gene sequence with similar function clustering was implemented much more effectively by DD-Cluster compared with that by previous algorithms, and has shown great performance. The accuracy of gene sequence pattern mining results can be further improved by identifying such patterns from co-expressed gene clusters, in the hope of directing the identification of functional elements. In future work, we will develop a mining algorithm to recognize transcriptional factor binding sites by virtue of results from DD-cluster in order to improve its accuracy, and to study corresponding physiological conditions which affect transcriptional regulatory mechanism.

Acknowledgements The research was supported in part by the National Natural Science Foundation of China under Grant No.60573093, the National High-Tech Research and Development Plan of China under Grant No.2006AA02Z329.

References

1. Mao, L. Y., Mackenzie, C., Roh, J. H., Eraso, J. M., Kaplan, S., Resat, H.. Combining microarray and genomic data to predict DNA binding motifs. *Microbiology*, 2005, 151(10): 3197-3213.
2. Cheng, Y., Church, G.. Biclustering of expression data. Bourne, P., Gribskov, M., Altman, R.(Eds.). *Proceedings of the 8th International Conference on Intelligent Systems for Molecular Biology*. San Diego: AAAI Press, 2000: 93-103.
3. Wang, H. X., Wang, W., Yang, J., Yu, P. S.. Clustering by pattern similarity in large data sets. Franklin, M. J., Moon, B., Ailamaki, A.. *Proceedings of the 2002 ACM SIGMOD International Conference on Management of Data*. Madison, Wisconsin: ACM, 2002:394-405.
4. Pei, J., Zhang, X. L., Cho M. J., Wang, H. X., Yu, P. S.. MaPel: A fast algorithm for maximal pattern-based clustering. *Proceedings of the 3rd IEEE International Conference on Data Mining (ICDM)*. Melbourne, Florida, USA: IEEE Computer Society, 2003: 259-266.
5. Ben-Dor, A., Chor, B., Karp, R., Yakhini, Z.. Discovering local structure in gene expression data: The order-preserving submatrix problem. *Proceedings of the 6th Annual International Conference on Computational Biology*. Washington, DC, USA: ACM, 2002: 49-57.
6. Liu, J. Z., Wang, W.. OP-Cluster: Clustering by tendency in high dimensional space. *Proceedings of the 3rd IEEE International Conference on Data Mining (ICDM)*. Melbourne, Florida, USA: IEEE Computer Society, 2003:187-194.
7. Day, W. H. E., Edelsbrunner, H.. Efficient algorithms for agglomerative hierarchical clustering methods. *Journal of Classification*, 1984, 1(1): 7-24.
8. Kaufman, L., Rousseeuw, P. J.. Finding groups in data: An introduction to cluster analysis. New York: Johh Wiley and Sons, 1990.
9. Aggarwal, C. C., Hinneburg, A., Keim1, D.. On the surprising behavior of distance metrics in high dimensional space. Bussche, J. V., Vianu, V.(Eds.). *The 8th International Conference on Database Theory*. London, UK: Lecture Notes in Computer Science, 2001: 420-434.
10. Agrawal, R., Gehrke, J., Gunopulos, D., Raghavan, P.. Automatic subspace clustering of high dimensional data for data mining applications. Haas, L. M., Tiwary, A.(Eds.). *Proceedings of the ACM SIGMOD International Conference on Management of Data*, Seattle, WA, USA: ACM Press, 1998: 94-105.
11. Moreau, Y., Smet, F. D., Thus, G., Marchal, K., Moor, B. D.. Functional bioinformatics of microarray data: From expression to regulation. *Proceedings of the IEEE*, 2002, 90(11): 1722-1743.
12. Eisen, M. B., Spellman, P. T., Brown, P. O., Botstein, D.. Cluster analysis and display of genome-wide expression patterns. *Proceedings of the National Academy of Sciences of the United States of America (PNAS)*, 1998, 95(25): 14863-8.

13. Zhang, K., Zhu, Y. Y.. Sequence pattern mining without duplicate project database scan. *Journal of Computer Research and Development*, 2007, 44(1): 126-132.
14. Hedenfalk, I., Duggan, D., Chen, Y. D.. Gene-expression profiles in hereditary breast cancer. *The New England Journal of Medicine*, 2001, 344(8): 539-548.
15. Tavazoie, S., Hughes, J. D., Campbell, M. J., Cho, R. J., Church, G. M.. Systematic determination of genetic network architecture. *Nature Genetics*, 1999, 281-285.
16. Liu, J. Z., Yang, J., Wang, W.. Biclustering in gene expression data by tendency. *Proceedings of the 2004 IEEE Computational Systems Bioinformatics Conference*. United States: IEEE Computer Society, 2004: 182-193.
17. Ashburner, M., Ball, C. A., Blake, J. A.. Gene ontology: Tool for the unification of biology. *Nature Genetics*, 2000:25(1), 25-29.
18. Xu, X., Lu, Y., Tung, A. K. H.. Mining shifting-and-scaling co-regulation patterns on gene expression profiles. In: Liu, L., Reuter, A., Whang, K. Y. (Eds.). *Proceedings of the 22nd International Conference on Data Engineering(ICDE 2006)*, Atlanta, GA, USA. IEEE Computer Society, 2006: 89-100.
19. Zhao, Y. H., Yu, J. X., Wang, G. R., Chen, L. Wang, B., Yu, G.. Maximal subspace co-regulated gene clustering. *IEEE Transactions on Knowledge and Data Engineering*. 2008: 83-98.

Chapter 9
Domain Driven Tree Mining of Semi-structured Mental Health Information

Maja Hadzic, Fedja Hadzic, and Tharam S. Dillon

Abstract The World Health Organization predicted that depression would be the world's leading cause of disability by 2020. This is calling for urgent interventions. As most mental illnesses are caused by a number of genetic and environmental factors and many different types of mental illness exist, the identification of a precise combination of genetic and environmental causes for each mental illness type is crucial in the prevention and effective treatment of mental illness. Sophisticated data analysis tools, such as data mining, can greatly contribute in the identification of precise patterns of genetic and environmental factors and greatly help the prevention and intervention strategies. One of the factors that complicates data mining in this area is that much of the information is not in strictly structured form. In this paper, we demonstrate the application of tree mining algorithms on semi-structured mental health information. The extracted data patterns can provide useful information to help in the prevention of mental illness, and assist in the delivery of effective and efficient mental health services.

9.1 Introduction

The World Health Organization predicted that depression would be the worlds leading cause of disability by 2020 [16]. This is calling for urgent interventions. Research into mental health has increased and has resulted in a wide range of results and publications covering different aspects of mental health and addressing a variety of problems. The most frequently covered topics include the variations between different illness types, descriptions of illness symptoms, discussing the different kind of treatments and their effectiveness, explaining the disease causing fac-

---

Maja Hadzic, Fedja Hadzic, Tharam S. Dillon
Digital Ecosystems and Business Intelligence Institute (DEBII), Curtin University of Technology, Australia, e-mail: {m.hadzic,f.hadzic,t.dillon}@curtin.edu.au

tors including various genetic and environmental factors, and relationships between these factors, etc.

Many research teams focus only on one factor and perhaps one aspect of a mental illness . For example, in the paper "Bipolar disorder susceptibility region on Xq24-q27.1 in Finnish families" (PubMed ID: 12082562), the research team Ekholm et al. examined one genetical factor (Xq24-q27.1) for one type of mental illness (bipolar disorder). As mental illness does not follow Mendelian patterns but is caused by a number of genes usually interacting with various environmental factors, all factors for all aspects of the illness need to be considered. Some tools have been proposed to enable processing of sentences and produce a set of logical structures corresponding with the meaning of those sentence such as [17]. However, no tool goes as far as examination and analysis of the different causal factors simultaneously.

Data mining is a set of processes that is based on automated searching for actionable knowledge buried within a huge body of data. Frequent pattern analysis has been a focused theme of study in data mining , and many algorithms and methods have been developed for mining frequent sequential and structural patterns [1, 13, 24]. Data mining algorithms have great potential to expose the patterns in data, facilitate the search for the combinations of genetic and environmental factors involved and provide an indication of influence. Much of the mental health information is not in strictly structured form and the use of traditional data mining techniques developed for relational data is not appropriate in this case. The majority of available mental health information can be meaningfully represented in XML format, which makes the techniques capable of mining semi-structured or tree structured data more applicable.

The main objective of this paper is to demonstrate the potential of the tree mining algorithms to derive useful knowledge patterns in the mental health domain that can help disease prevention, control and treatment. Our preliminary ideas have been published in [11, 12]. We use our previously developed IMB3-Miner [26] algorithm and show how tree mining techniques can be applied on patient data represented in XML format. We used synthetic datasets to illustrate the usefulness of the IMB3-Miner algorithm within mental health domain, but we have demonstrated experimentally that the IMB3-Miner algorithm is well scalable when applied to large datasets consisting of complex structures in our previous works [26]. We discussed the implications of using different mining parameters within the current tree mining framework and demonstrated the potential of extracted patterns in providing useful information.

9.2 Information Use and Management within Mental Health Domain

Human diseases can be described as genetically simple (Mendelian) or complex (multifactorial). Simple diseases are caused by a single genetic factor, most frequently by a mutated gene. An abnormal genetic sequence gets translated into a non-functional protein which results in an abnormality within the flow of biochemical reactions within the body and development of disease. Mental illness does not follow Mendelian patterns but is caused by a number of both genetic and environmental factors [15]. For example, genetic analysis has identified candidate loci on human chromosomes 4, 7, 8, 9, 10, 13, 14 and 17 [3]. There is some evidence that environmental factors such as stress, lifecycle matters, social environment, climate etc. are important [14, 22]. Moreover, different types of a specific mental illness exist such as chronic, postnatal, psychotic depression types of the depression. Identification of the precise patterns of genetic and environmental factors responsible for a specific mental illness type still remains unsolved and is therefore a very active research focus today.

A huge body of information is available within the mental health domain. Additionally, as the research continues, new papers or journals are frequently published and added to the various database. Portions of this data may be related to each other, portions of the information may overlap, and portions of the information may be semi-complementary with one another. Retrieving specific information is very difficult with current search engines as they look for the specific string of letters within the text rather than its meaning. In a search for "genetic causes of bipolar disorder", Google provides 95,500 hits which are a large assortment of well meaning general information sites with few interspersed evidence-based resources. Medline Plus (http://medlineplus.gov/) retrieves 53 articles including all information about bipolar disorder plus information on other types of mental illness. A large number of articles is outside the domain of interest and is on the topic of heart defects, eye and vision research, multiple sclerosis, Huntington's disease, psoriasis etc. PubMed (http://www.ncbi.nlm.nih.gov/pubmed) gives a list of 1946 articles. The user needs to select the relevant articles as some of the retrieved articles are on other types of mental illness such as schizophrenia, autism and obesity. Wilczynski et al. [30]: "General practitioners, mental health practitioners, and researchers wishing to retrieve the best current research evidence in the content area of mental health may have a difficult time when searching large electronic databases such as MEDLINE. When MEDLINE is searched unaided, key articles are often missed while retrieving many articles that are irrelevant to the search."

Wilczynski et al. developed search strategies that can help discriminate the literature with mental health content from articles that do not have mental health content. Our research ideas go beyond this. We have previously developed an approach that uses ontology and multi-agent systems to effectively and efficiently retrieve information about mental illness [5, 6]. In this paper, we go a step further and apply data mining algorithms on the retrieved data. The application of data mining algorithms has the potential to link all the relevant information and expose the hidden knowledge and the patterns within this information, facilitate the search for the combinations of genetic and environmental factors involved and provide an indication of influence.

9.3 Tree Mining - General Considerations

Semi-structured data sources enable more meaningful knowledge representations. For this reason they are being increasingly used to capture and represent knowledge within a variety of knowledge domains such as Bioinformatics, XML Mining, Web applications, as well as the mental health domain. The increasing use of semi-structured data sources has resulted in increased interest in the tree mining techniques. Tree mining algorithms enable effective and efficient analysis of semi-structured data and support structural comparisons and association rule discovery. The problem of frequent subtree mining is: "Given a tree database Tdb and minimum support threshold (s), find all subtrees that occur at least s times in Tdb."

Induced and embedded subtrees are the two most commonly mined types within this framework. The parent-child relationships of each node in the original tree are preserved in an induced subtree. An embedded subtree preserves ancestor-descendant relationships over several levels as it allows a parent in the subtree to be an ancestor in the original tree. The further classification of the subtrees is done on the basis of the siblings ordering. The left-to-right ordering among the sibling nodes is preserved in an ordered subtree while in an unordered subtree this ordering is not preserved. Driven by different application needs a number of tree mining algorithms exist that mine different subtree types. To limit the scope we do not provide an overview of the existing tree mining algorithms in this work. Rather, we refer the interested reader to [28] where an extensive discussion of existing tree mining algorithms is provided including important implementation issues and advantages and disadvantages of the different approaches. A discussion on the use of tree mining for general knowledge analysis highlighting the implications of using different tree mining parameters, can be found in [9].

Our work in the field was driven by the need for a general framework that can be used for the mining of all subtree types under different constraints and support definitions. One of our significant contributions to the tree mining field is the Tree Model Guided (TMG) [25,27] candidate generation approach. In this nonredundant systematic enumeration model, the underlying tree structure of the data is used to generate only valid candidates. These candidates are valid in the sense that they are existent in the tree database and hence conform to the tree structure according to which the data objects are organized. Using the general TMG framework a number of algorithms were developed for mining different subtree types. MB3-Miner [25] mines ordered embedded subtrees while IMB3-Miner can mine induced subtrees as well by utlizing the level of embedding constraint. A theoretical analysis of the worst case complexity of enumerating all possible ordered embedded subtrees was provided in [25,27]. As a further extension of this work, we have developed Razor algorithm [23] for mining of ordered embedded subtrees where the distances of nodes relative to the root of the original tree need to be considered. UNI3 algorithm [7] integrated canonical form ordering for unordered subtrees into the general TMG framework and it extracts unordered induced subtrees. The developed algorithms were applied on large and complex tree structures and these experiments successfully demonstrated the scalability of the developed algorithms [7,26,27]. From the application perspective in [10], we have applied our tree mining algorithm to extract useful pattern structures from the Protein Ontology database for Human Prion proteins [21].

9.4 Basic Tree Mining Concepts

A tree is a special type of graph where no cycles are allowed. It consists of a set of *nodes* (or vertices) that are connected by *edges*. Two nodes are associated with each edge. A *path* is defined as a finite sequence of edges. In a tree, there is a single unique path between any two nodes. The *length of a path p* is the number of edges in the path *p*. The *root* of a tree is the top-most node in this tree that has no incoming edges. A node *u* is a *parent* of node *v*, if there is a directed edge from *u* to *v*. Node *v* is a *child* of node *u*. *Leaf* nodes are nodes with no children while *internal nodes* are nodes which have children. In an *ordered tree* all the children of the internal nodes are ordered from left to right. The *level/depth* of a node is the length of the path from the root to that node. The *height of a tree* is the greatest level of its nodes. A tree can be denoted as $T(V,L,E)$ where: V is the set of *vertices* or *nodes*; L is the set of *labels* of vertices; and $E = \{(x,y)|x,y \in V, x \neq y\}$ is the set of *edges* in the tree.

XML documents can be used to capture and represent mental health information in a more meaningful way. In Figure 9.1, we show a top-level layout of a XML document used to capture information about causes of a specific mental illness type.

```
<?xml version="1.0" ?>
<element name = "Mental illness type">
<element name = "Cause">

   <element name="Genetic">
          <element name="Gene x" />
          <element name="Gene y" />
          <element name="Gene z" />
   </element>

   <element name="Environmental">
          <element name ="Climate" />
          <element name ="Relationships" />
          <element name ="Spiritual beliefs" />
          <element name ="Drugs misuse" />
          <element name ="Family conditions" />
          <element name ="Economic conditions" />
          <element name ="Stress" />
     - </element>

</element>
</element>
```

Fig. 9.1 XML representation of causal data

In Figure 9.2, we represent the information from Figure 9.1 in the form of a tree. We will use this simple example to illustrate use of the tree mining algorithms in the identification of specific patterns of illness-causing factors within the mental health domain. We have represented the genetic factors by Gene x, Gene y and Gene z which means that this specific gene has an abnormal structure (i.e. this gene is mutated) and this gene mutation causes a specific mental illness. The environmental factors cover social, economic, physical and cultural aspects (such as climate, relationships, spiritual beliefs, drugs misuse, family conditions, economic conditions and stress). In our model, all these genetic and environmental factors play a role in the onset of mental illness. Our goal is to identify the precise patterns of genetic and environmental causal factors specific to the mental illness under examination. Some properties of the tree represented in Figure 9.2, are:

- 'Mental illness type' $\in V$ is the *root* of the tree
- L = Mental Illness type, Cause, Genetic, Gene x, Gene y, Gene z, Environmental, Climate, Relationships, Spiritual Beliefs, Drugs misuse, Family conditions, Economic conditions, Stress
- The *parent* of node 'Cause' is node 'Mental illness type', of node 'Genetic' is node 'Cause', of node 'Gene x' is node 'Genetic' etc.
- 'Gene x', 'Gene y', 'Gene z', 'Climate', 'Relationships', 'Spiritual Beliefs', 'Drugs misuse', 'Family conditions', 'Economic conditions' and 'Stress' are *leaf nodes*.
- 'Genetic', 'Environmental' and 'Cause' are *internal nodes*.
- The *height* of the tree is equal to 3.

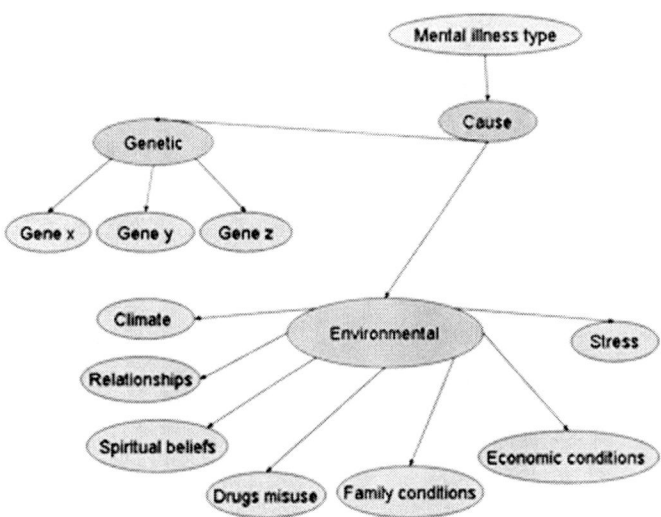

Fig. 9.2 Causal data represented in a tree structure

9 Domain Driven Tree Mining of Mental Health Information

Within the current frequent subtree mining framework, the two most commonly mined types of subtrees are induced and embedded. Given a tree $S = (V_S, L_S, E_S)$ and tree $T = (V_T, L_T, E_T)$, S is an **induced subtree** of T, iff (1) $V_S \subseteq V_T$; (2) $L_S \subseteq L_T$, and $L_S(v) = L_T(v)$; (3) $E_S \subseteq E_T$. S is an **embedded subtree** of T, iff (1) $V_S \subseteq V_T$; (2) $L_S \subseteq L_T$, and $L_S(v) = L_T(v)$; (3) if $(v_1, v_2) \in E_S$ then $parent(v_2) = v_1$ in S and v_1 is ancestor of v_2 in T. Hence, the main difference between an induced and an embedded subtree is that, while an induced subtree keeps the parent-child relationships from the original tree, an embedded subtree allows a parent in the subtree to be an ancestor in the original tree. In addition, the subtrees can be further distinguished with respect to the ordering of siblings. An **ordered subtree** preserves the left-to-right ordering among the sibling nodes in the original tree, while an **unordered subtree** is not affected by the exchange of the order of the siblings (and the subtrees rooted at sibling nodes). Examples of different subtree types are given in Figure 9.3 below. Please note that induced subtrees are also embedded.

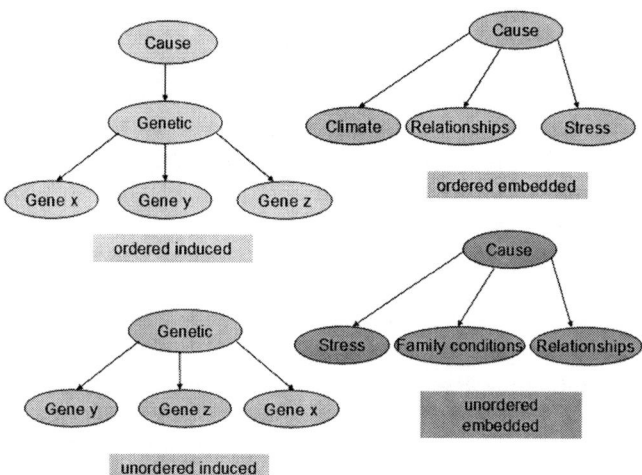

Fig. 9.3 Example of different subtree types

In the previous section, we mentioned the concept of the level of embedding constraint which enabled us to mine both induced and embedded subtrees using the same framework. Furthermore, when the complexity of enumerating all embedded subtrees becomes too high with the large height of a tree, the maximum level of embedding constraint can be used in order to at least extract some subtrees with the limited level of embedding between the nodes. Definitions follow. If $S = (V_S, L_S, E_S)$ is an embedded subtree of tree T, and two vertices $p \in V_S$ and $q \in V_S$ form ancestor-descendant relationship, the **level of embedding** [26], between p and q, denoted by $\delta(p,q)$, is defined as the length of the path between p and q. With this observation a **maximum level of embedding constraint** δ can be imposed on the subtrees extracted from T, such that any two ancestor-descendant nodes present in an embedded subtree of T, will be connected in T by a path that has the maximum length of δ. Hence an induced subtree S_I can be seen as an embedded subtree where the maximum level of embedding that can occur in T is equal to 1, since the level of embedding between two nodes that form a parent-child relationship is equal to 1.

Given a tree $S = (V_S, L_S, E_S)$ and tree $T = (V_T, L_T, E_T)$, S is a ***distance-constrained embedded subtree*** of T, iff (1) $V_S \subseteq V_T$;(2) $L_S \subseteq L_T$, and $L_S(v) = L_T(v)$; (3) if $(v_1, v_2) \in E_S$ then $parent(v_2) = v_1$ in S and v_1 is ancestor of v_2 in T; and (4) $\forall v \in V_S$ an integer is stored indicating the level of embedding between v and the root node of S in the original tree T.

Within the current tree mining framework, the support definitions available are transactional support, occurrence-match, and hybrid support [26]. Formal definitions of transactional support, occurrence-match, and hybrid support are given in [8]. Here we give a more commonsense narrative explanation of these three different levels of support. In tree mining field a transaction corresponds to a part of the database tree whereby an independant instance is described. ***Transactional support*** searches for a specific item within a transaction and counts how many times this item appears within a given set of transactions. The support of an item equals to the number of transactions where the item exists. For example, if we have 100 XML documents capturing the information about causes of postnatal depression and if the item 'Stress' appears in 70 transactions, we say that the support of the item 'Stress' is equal to 70 which is the number of transactions (i.e. XML documents) where this item occurs. ***Occurrence-match*** support also searches for a specific item within a transaction taking into account repetition of items within a transaction but counts the total occurrences in the dataset as a whole. In our example about postnatal depression, the difference between the transactional and occurrence-match support would be noticeable in the datasets which contain transactions where the item 'Stress' appears more than once. Let us say that we have a transaction where the item 'Stress' appears 3 times (for example, stress associated with work, children and finances). Using transactional support definition, as the support of this item would be equal to 1, while with the occurrence-match support definition it would equal to 3. ***Hybrid support*** is a combination of transactional and occurrence-match support and provides extra information about the intra-transactional occurrences of a subtree. If we use hybrid support threshold of x|y, this means that each subtree which occurs in x transactions and at least y times in each of those x transactions is considered to be frequent. Let us say that each transaction is associated with a specific mental illness type. The set of different transactions will contain information specific to different types of mental illness. The aim is to examine the effect of stress on mental health in general. We can then search for the subtree 'Cause' → 'Environmental' → 'Stress' within the different transactions. If we choose to use a hybrid support threshold of 10|7, this will mean that if the 'Cause' → 'Environmental' → 'Stress' subtree occurs at least 7 times in 10 transactions, it will occur in the extracted subtree set and this would lead to a general conclusion that stress negatively affects mental health.

9.5 Tree Mining of Medical Data

The tree mining experiment generally consists of the following steps: (1)Data selection and cleaning, (2)Data formatting, (3)Tree mining, (4)Pattern discovery, and (5)Knowledge testing and evaluation.

Data Selection and Cleaning. The available information source may also contain information which is outside the domain we are interested in. For example, a database may also contain information about other illnesses such as diabetes, arthritis, hypertension, and so on. For this reason, it is important to select the subset of the dataset that contains only the relevant information. It is also possible that the data describing the causal information for mental illness was collected form different sources. Different organizations may find different information relevant to study the causes of mental illness. Additionally, it is required to remove all noise and inconsistent data from the selected dataset. This step needs to be done cautiously as some data may appear to be irrelevant but, in fact, may represent true exceptional cases.

Data Formatting. Simultaneous analysis of the data requires firstly, consistent formatting of all data within the target dataset and secondly, understandability of the chosen format by the data mining algorithm that is used within the application. If only relational type of data is available, the set of features can be grouped according to some criteria, and the relationships between data objects can be represented in a more meaningful way within an XML document. Namely, it may be advantageous to convert the relational data into XML format, when such conversion would lead to a more complete representation of the available information.

Factors to be Considered for the Tree Mining. Firstly, we need to carefully consider what particular subtree type is most suitable for the application at hand. If the data is coming from one organization, it is expected that the format and ordering of the data will be the same. In this case, mining of ordered subtrees will be the most appropriate choice. In the cases where the collected data originates from separate organizations, mining of unordered subtrees will be more appropriate. This is due to the high possibility of different organizations organizing the causal information in different order. The subtrees consisting of the same items that are ordered differently would still be considered as the same candidate. Hence the common characteristics of a particular mental illness type would be correctly counted and found with respect to the set support threshold. Another choice is whether induced or embedded subtrees should be mined. As we extract the patterns where the information about the causal factors has to stay in the context specific to the mental illness in question, the relationship of nodes in the extracted subtrees needs to be limited to parent-child relationships (i.e. induced subtrees). It would be possible to loose some information about the context where particular mental illness characteristic occurred if we allow ancestor-descendant relationships. This is mainly for the reason that some features of the dataset may have a similar set of values and hence it is necessary to indicate which value belonged to which particular feature. On the other hand, when the database contains heterogeneous documents, it may well be possible that same sets of causal information are present in the database but they occur at different levels of the database tree. By mining induced subtrees, these sets of same information will not be considered as same patterns since the level of embedding (see Section 3) between their nodes is different. It is then possible to miss this frequently occurring pattern if the support threshold is not low enough to consider the patterns where the level of embedding between the nodes is equal to one as frequent. There appears to be a trade-off here and in this case the level of embedding may be useful. The difference in levels at which the same information is stored may not be so large while the difference in levels between items with same labels, but used in different contexts, may be much larger. Hence, one could progressively reduce the maximum level of embedding constraint, so that it is sufficiently large to detect the same information that occurs at different levels in the tree and at the same time sufficiently small so that the items with same labels used in different contexts are not incorrectly considered as the same pattern. Generally speaking, allowing large embeddings can result in unnecessary and misleading information, but in other cases, it proves useful as it detects common patterns in spite of the difference in granularity of the information presented. Utilizing the maximum level of embedding constraint can give clues about some general differences among the ways that the mental health causal information is presented by the different organization. It may be necessary to go back to the stage of data formatting so that all the information is presented in a consistent way. If this is achieved then it would be safe to mine induced subtrees since the differences in the levels in which the information is stored has diminished.

In addition, if one is interested in the exact level of embedding between the nodes in the subtree, one could mine distance-constrained embedded subtrees. This would extract all the induced subtrees and the embedded subtrees where the level of embedding between the nodes in the subtree is given. One could do some post-processing on the pattern set to group those subtree patterns together where the level of embedding is acceptable. Another possibility is to mine all the subtree types, induced, embedded and distance-constrained embedded subtrees, and then compare the number of frequent patterns extracted for each subtree type. This would also reveal some information that the user may find useful. For example, if all the subtrees are extracted for a given support threshold and the number of frequent embedded subtrees is much larger than that of distance-constrained embedded subtrees, then it is known that all the additional embedded subtrees occur with the different levels of embedding between the nodes. One could analyze these embedded subtrees in order to check why the difference occurred. If the difference occurred because of the difference in the level that the information is stored then they can be considered as valid patterns. On the other hand, if the difference detected is due to the items with same labels used in different contexts then they should not be considered as valid frequent patterns. These frequent patterns are considered valid with respect to the support threshold.

The XML datasets used in our example will be a collection of rooted labeled ordered subtrees that are organized and ordered in the same way, and our IMB3-Miner algorithm [8] for mining of induced ordered subtrees will be used. This assumption is not necessary for the described method to be applicable, since within our general TMG framework both ordered and unordered induced/embedded subtrees can be

9 Domain Driven Tree Mining of Mental Health Information 137

mined, as well as additional constraints imposed (see Section 3). This assumption was made to allow comparisons with other methods and ease of data generation. Secondly, we need to choose which type of support definition to use. This choice depends on the way the data is organized. Consider the following three scenarios.

Scenario 1: Each patient record is stored as a separate subtree or transaction in the XML document and separate XML documents contain the records describing causal information for different mental illnesses. An illustrative example of this scenario for organizing the mental health data is displayed in Figure 9.4. For ease of illustration, the specific illness-causing factors for each cause are not displayed, but one can assume that they will contain those displayed in Figure 9.2. In this figure and in all subsequent figures, the three dots indicate that the previous structure (or document) can repeat many times. For this scenario, both occurrence match or transaction based support can be used. This is because the aim is to find a frequently occurring pattern for each illness separately, and the records are stored separately for each illness. Hence the total set of occurrences, as well as the existence of a pattern in each transaction, would yield the same result.

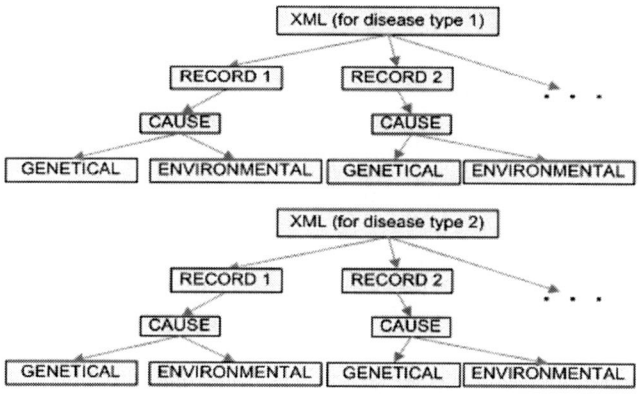

Fig. 9.4 Illustrating Scenario 1 of Data Organization

Scenario 2: The patient records for all types of mental illnesses are stored as separate subtrees or transactions in one XML document, as is illustrated in Figure 9.5. It does not make a difference whether the records for each illness type are occurring together, as is the case in Figure 9.5. In our example, this would mean that there is one XML document where the number of transactions is equal to the number of patient records. Here the transactional support would be more appropriate.

Scenario 3: The XML document is organized in such a way that a collection of patient records for one particular illness is contained in one transaction. This is illustrated in Figure 9.6. In our example, this would mean that there is one XML document where each transaction corresponds to a specific illness type. Each of

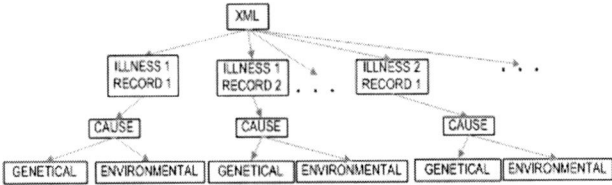

Fig. 9.5 Illustrating Scenario 2 of Data Organization

those transactions would contain records of patients associated with that particular illness type. Hybrid support definition is most suitable in this case.

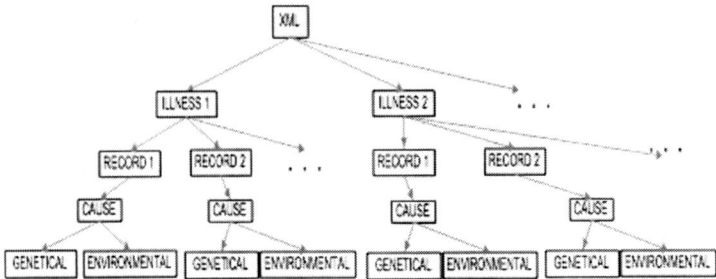

Fig. 9.6 Illustrating Scenario 3 of Data Organization

In order to find the characteristics specific to mental illness under examination, in scenarios 1 and 2 the minimum support threshold should be chosen to be approximately close to the number of patient records (transactions) that the XML dataset contains about a particular illness. Due to noise often being present in data, the minimum support threshold can be set lower. For scenario 3, the number of mental illness types described would be used as the transactional part of the hybrid support, while the approximate number of patient records would be used as the requirement for occurrence of a subtree within each transaction. In the illustrative example we are concerned with the second scenario.

Before data mining takes place, the dataset can be split into two subsets, one for deriving the knowledge model ('source dataset') and one for testing the derived knowledge model ('test dataset'). The 'test dataset' can also come from another organization.

Pattern Discovery Phase. Precise combinations of genetic and environmental illness-causing factors associated with each mental illness type are identified during pattern discovery phase. These kinds of results that reveal the correlation and interdependence between the different genetic and environmental factors are groundbreaking results significantly contributing to the research, control and prevention of mental illnesses.

9 Domain Driven Tree Mining of Mental Health Information

Knowledge Testing and Evaluation. The 'source dataset' is used to derive the hypothesis while the 'test dataset' represents the data 'unseen' by the tree mining algorithm and is used to verify the hypothesis. As it is possible that the chosen data mining parameters affect the nature and granularity of the obtained results, we will also experiment with the data mining parameters and examine their effect on the derived knowledge models.

9.6 Illustration of the Approach

A synthetic dataset was created analogous to the knowledge representation shown in Figure 9.1 and 9.2. The created XML document consisted of 30 patient records (i.e. transactions). The document contains records for three different personality disorder types: antisocial, paranoid and obsessive-compulsive personality disorder. We have applied the IMB3-Miner [26] algorithm on the dataset. We have used transactional support definition with the minimum support threshold equal to 7. We have detected a number of subtree patterns as frequent. The frequent smaller subtree patterns were subsets of the frequent larger subtree patterns (i.e. patterns containing most nodes). We have focused on the larger patterns to derived knowledge models specific to each personality disorder type.

In Figure 9.7, we show three different patterns each associated with a specific personality disorder type. As can be seen the patterns are meaningful in the sense that a specific pattern of causes is associated with a specific personality disorder type. Note that this is only an illustrative example used to clarify the principle behind the data mining application, and by no means indicates the real-world results. The real-world data would be of the analogous nature but much more complex in regard to the level of detail and precision contained within the data. Additionally, the underlying tree structure of the information can vary between different organizations and applications. This issue would pose no limitation for the application of our algorithm since our previous works [26, 27] have demonstrated scalability of the algorithm and its applicability on large and complex data of varying structural characteristics.

9.7 Conclusion and Future Work

We have highlighted a number of important issues present within the mental health domain including increasing number of mentally ill patients, rapid increase of mental health information, lack of tools to systematically analyze and examine this information, and the need for embracing the data mining technology to help address and solve these issues. We have generally explained the high significance of data mining technology for deriving new knowledge that will assist the prevention, diagnosis, treatments and control of mental illness. The appropriate use of tree mining techniques was explained to effectively mine semi-structured mental health data. The choice of the most appropriate tree mining parameters (subtree types and

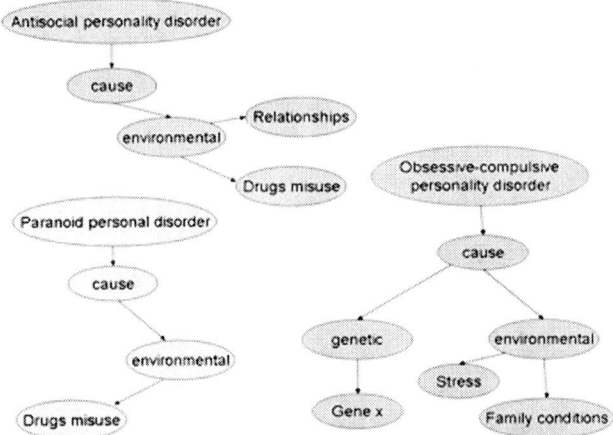

Fig. 9.7 Data patterns derived from artificial dataset specific for a particular type of personality disorder

support definitions) was discussed with respect to the possible representations of mental health data and the aim of the application. In future, we aim to apply the developed tree mining algorithms on real world datasets and benchmark our approach with other related methods. We will work with different XML documents originating from a variety of sources and this will enable us to identify commonalities between the different XML structures. Consequently, we will be able to develop a standardized XML format to capture and represent mental health information uniformly across different organizations and institutions.

References

1. Agrawal R., Srikant R.: Fast algorithms for mining association rules. VLDB, Chile (1994).
2. Asai T., Arimura H., Uno T., Nakano S.: Discovering Frequent Substructures in Large Unordered Trees. Proc. of the Int'l Conf. on Discovery Science, Japan (2003).
3. Craddock N., Jones I.: Molecular genetics of bipolar disorder. The British Journal of Psychiatry, vol. 178, no. 41, pp. 128-133 (2001).
4. Ghoting A., Buehrer G., Parthasarathy S., Kim D., Nguyen A., Chen Y.-K., Dubey P. : Cache-conscious Frequent Pattern Mining on a Modern Processor, VLDB Conf., (2005).
5. Hadzic M., Chang E.: Web Semantics for Intelligent and Dynamic Information Retrieval Illustrated Within the Mental Health Domain, to appear in Advances in Web Semantics: A State-of-the Art, Springer, (2008).
6. Hadzic M., Chang E.: An Integrated Approach for Effective and Efficient Retrieval of the Information about Mental Illnesses', Biomedical Data and Applications, Springer, (2008).
7. Hadzic F., Tan H., Dillon T.S.: UNI3-Efficient Algorithm for Mining Unordered Induced Subtrees Using TMG Candidate Generation. IEEE CIDM Symposium, Hawaii (2007).
8. Hadzic F., Tan H., Dillon T.S., Chang E.: Implications of frequent subtree mining using hybrid support definition, Data Mining and Information Engineering, UK, (2007).

9. Hadzic F., Dillon T.S., Chang E.: Knowledge Analysis with Tree Patterns, HICSS-41, USA, (2008).
10. Hadzic F., Dillon T.S., Sidhu A., Chang E., Tan H.: Mining Substructures in Protein Data, IEEE ICDM DMB Workshop, China (2006).
11. Hadzic M., Hadzic F., Dillon T.: Mining of Health Information from Ontologies, Int'l Conf. on Health Informatics, Portugal, (2008).
12. Hadzic M., Hadzic F., Dillon T.: Tree Mining in Mental Health Domain, HICSS-41, USA, (2008).
13. Han J., Kamber M.: Data Mining: Concepts and Techniques (2nd edition). San Francisco: Morgan Kaufmann (2006).
14. Horvitz-Lennon M., Kilbourne A.M., Pincus H.A.: From Silos To Bridges: Meeting The General Health Care Needs Of Adults With Severe Mental Illnesses. Health Affairs vol. 25, no. 3, pp. 659-669 (2006).
15. Liu J., Juo S.H., Dewan A., Grunn A., Tong X., Brito M., Park N., Loth J.E., Kanyas K., Lerer B., Endicott J., Penchaszadeh G., Knowles J.A., Ott J., Gilliam T.C., Baron M.: Evidence for a putative bipolar disorder locus on 2p13-16 and other potential loci on 4q31, 7q34, 8q13, 9q31, 10q21-24, 13q32, 14q21 and 17q11-12. Mol Psychiatry, vol. 8, no. 3, pp. 333-342 (2003).
16. Lopez A.D., Murray C.C.J.L.: The Global Burden of Disease, 1990-2020. Nature Medicine vol. 4, pp. 1241-1243 (1998).
17. Novichkova S., Egorov S., Daraselia N.: Medscan, a natural language processing engine for Medline abstracts. Bioinformatics, vol. 19, no. 13, pp. 1699-1706, (2003).
18. Onkamo P., Toivonen H.: A survey of data mining methods for linkage disequilibrium mapping. Human genomics, vol. 2, no. 5, pp. 336-340 (2006).
19. Piatetsky-Shapiro G., Tamayo P.: Microarray Data Mining: Facing the Challenges. SIGKDD Explorations, vol. 5, no. 2, pp. 1-6 (2003).
20. Shasha D., Wang J.T.L., Zhang S.: Unordered Tree Mining with Applications to Phylogeny. Int'l Conf. on Data Engineering, USA (2004).
21. Sidhu A.S., Dillon T.S., Sidhu B.S., Setiawan H.: A Unified Representation of Protein Structure Databases. Biotech. Approaches for Sustainable Development, pp. 396-408 (2004).
22. Smith D.G., Ebrahim S., Lewis S., Hansell A.L., Palmer L.J., Burton P.R.: Genetic epidemiology and public health: hope, hype, and future prospects. The Lancet, vol. 366, no. 9495, pp. 1484-1498 (2005).
23. Tan H., Dillon T.S., Hadzic F., Chang E.: Razor: mining distance constrained embedded subtrees. IEEE ICDM 2006 Workshop on Ontology Mining and Knowledge Discovery from Semistructured documents, China (2006).
24. Tan H., Dillon T.S., Hadzic F., Chang E.: SEQUEST: mining frequent subsequences using DMA Strips. Data Mining and Information Engineering, Czech Republic, (2006).
25. Tan H., Dillon T.S., Hadzic F., Chang E., Feng L.: MB3-Miner: mining eMBedded subTREEs using Tree Model Guided candidate generation. MCD workshop, held in conjunction with ICDM05, USA (2005).
26. Tan H., Dillon T.S., Hadzic F., Feng L., Chang E.: IMB3-Miner: Mining Induced/Embedded subtrees by constraining the level of embedding. Proc. of PAKDD, (2006).
27. Tan H., Hadzic F., Dillon T.S., Feng L., Chang E.: Tree Model Guided Candidate Generation for Mining Frequent Subtrees from XML, to appear in ACM Transactions on Knowledge Discovery from Data, (2008).
28. Tan H., Hadzic F., Dillon T.S., Chang E.: State of the art of data mining of tree structured information, CSSE Journal, vol. 23, no 2, (2008).
29. Wang J.T.L., Shan H., Shasha D., Piel W.H.: Treerank: A similarity measure for nearest neighbor searching in phylogenetic databases. Int'l Conf. on Scientific and Statistical Database Management, USA (2003).
30. Wilczynski N.L., Haynes R.B., Hedges T.: Optimal search strategies for identifying mental health content in MEDLINE: an analytic survey. Annals of General Psychiatry, vol. 5, (2006).

Chapter 10
Text Mining for Real-time Ontology Evolution

Jackei H.K. Wong, Tharam S. Dillon, Allan K.Y. Wong, and Wilfred W.K. Lin

Abstract In this paper we propose the novel technique, On-line Contin-uous Ontological Evolution (OCOE) approach, which applies text mining to automate TCM (Traditional Chinese Medicine) telemedicine ontology evolution. The first step of the automation process is opening up a closed skeletal TCM ontology core (TCM onto-core) for continuous evolution and absorption of new scientific knowledge. The test-bed for the OCOE verification was the production TCM telemedicine system of the Nong's Company Limited; Nong's is a subsidiary of the PuraPharm Group in the Hong Kong SAR, which is dedicated to TCM telemedicine system development. At Nong's the skeletal TCM onto-core for clinical practice is closed (does not automatically evolve). When the OCOE is combined with the Nong's enterprise TCM onto-core it: i) invokes its text miner by default to search for new scientific findings over the open web incessantly; and ii) selectively prunes and stores useful new findings in special OCOE data structures. These data structures can be appended to the skeletal TCM onto-core logically to catalyze the evolution of the overall system TCM ontology, which is the logical combination: "original skeletal TCM onto-core plus contents of the special OCOE data structures". The evolutionary process works on the contents of the OCOE data structures only and does not alter any skeletal TCM onto-core knowledge. This onto-core evolution approach is called the "logical-knowledge-add-on" technique. OCOE deactivation will nullify the pointers and cut the association between the OCOE data structures and skeletal TCM onto-core, thus immediately reverting the clinical practice back to the original skeletal TCM onto-core basis.

Jackei H.K. Wong, Allan K.Y. Wong, Wilfred W.K. Lin
Department of Computing, Hong Kong Polytechnic University, Hong Kong SAR, e-mail: jwong@purapharm.com, {csalwong, cswklin}@comp.polyu.edu.hk

Tharam S. Dillon
Digital Ecosystems and Business Intelligence Institute, Curtin University of Technology, Perth, Western Australia, e-mail: tharam.dillon@cbs.curtin.edu.au

10.1 Introduction

In this paper we propose the novel technique, <u>O</u>n-line <u>C</u>ontinuous <u>O</u>ntological <u>E</u>volution (OCOE) technique, which, with text mining support, opens up a closed skeletal TCM ontology core (TCM onto-core) for continuous and automatic evolution. In the OCOE context a closed TCM onto-core does not evolve automatically. The OCOE technique has the following salient features:

a. Master aliases table (MAT): This directory manages all the referential contexts (illnesses) or RC. Every RC has three associating special OCOE data structures to catalyze the on-line TCM onto-core evolution: i) contextual attributes vector (CAV) to record all relevant attributes (e.g. symptoms) to the RC; ii) contextual aliases table (CAT) to record all the illnesses of similar symptoms; and iii) relevance indices table (RIT) to record the degree of similarity of every alias (illness) to the RC.

b. Text mining mechanism (miner is WEKA): This automatically (if active) ploughs through the web for new TCM-related scientific findings, which would be pruned and then added to the three catalytic OCOE data structures (CAV, CAT and RIT). This evolution makes the TCM onto-core increasingly smarter. The knowledge contained in the OCOE data structures is also called the "MAT knowledge".

c. TOES (TCM Ontology Engineering System) interface: This lets the user manage the following mechanisms:

   i Semantic TCM Visualizer (STV): This helps the user visualize and debug the parsing mechanism (parser) which is otherwise invisible.
   ii Text mining: The user activates/stops the text miner anytime.
   iii Master Aliases Table (MAT): This can be manipulated anytime. If it is appended to the skeletal TCM onto-core by managing a set of pointers logically, then the overall TCM onto-core of the running telemedicine system immediately becomes open and evolvable: the "skeletal TCM onto-core + MAT contents" combination literally. Disconnecting MAT from this combination turns the system back to the closed skeletal version that was built initially by consensus certification. The MAT contents, which are updated incessantly by the OCOE mechanism, may be formal. However, they have not gone through the same rigorous consensus certification as the original skeletal TCM onto-core, such as the one used in the Nong's enterprise. "Closed" and "skeletal" are interchangeably used hereafter to describe the same onto-core nature.

To summarize, the TCM onto-core evolution in the OCOE model is based on the "logical-knowledge-add-on" technique, which works only on the contents of the special OCOE data structures. The overall open and evolvable TCM onto-core of a running telemedicine system is the combination, "original skeletal TCM onto core + contents of all the special OCOE data structures (i.e. MAT contents)". Thus, the evolutionary process does not alter any knowledge in the original skeletal TCM onto-core. A skeletal TCM onto-core (e.g. the Nong's enterprise version) contains formal knowledge extracted from the TCM classics by consensus certification [5]. Useful concepts, attributes, and their associations form the subsumption hierarchy of the skeletal onto-core [4, 6]. The interpretations of inter-subontology relationships in this hierarchy are axiomatically constrained. Architecturally an operational TCM telemedicine system has three layers: i) bottom ontology layer (i.e. TCM onto-core); ii) middle semantic layer that should be the exact logical representation of the onto-core subsumption hierarchy; and iii) top syntactical layer for human understanding and query formulations. A query from the top layer is processed by the parser that traces out the logical conclusion (semantics) from the semantic net. The TCM onto-core is functionally the knowledge base; the semantic net (also called DOM (document object model) tree in our research) is the onto-core's machine-processable form; and the syntactical layer is the high-level semantic net representation for human understanding and query formulations.

10.2 Related Text Mining Work

Text mining is a branch of data mining and discovers knowledge patterns in textual sources [7–10, 14]. Various inference techniques can be exploited for effective text mining [11–13], including case-based reasoning, artificial neural network (ANN), statistical approach, fuzzy logic, and algorithmic approaches. Table 10.1 is our survey of text mining techniques and tools in the field. And, our in-house experience indicates that WEKA is the most effective text mining approach [14] for our particular application. For this reason it was adopted and adapted to support the OCOE operation. WEKA gauges how important a term/word is in a document/corpus (a bundle of documents) in a statistical manner.

The following WEKA parameters are useful for the OCOE approach:

a. *Term frequency* (or simply *tf*): This weighs how important the ith term t_i is, by its occurrence frequency in the jth document (or d_j) in the corpus of K documents. If the frequency of t_i is tf_i and $tf_{l,j}$ is the frequency of any other term t_l in d_j, then the relative importance of t_i in d_j is $tf_{i,j}$

10.3 Terminology and Multi-representations

In TCM, terms can be exactly synonymous or similar only to a varying degree. The term or word used as the reference is the context (or contextual reference) in the OCOE model. If *Ter₁* and *Ter₂* are two terms/attributes, they are logically the same (i.e. synonymous connotation) for *Ter₁*=*Ter₂*. They are only OCOE aliases if

Table 10.1 Strengths and weaknesses of the different text mining tools/techniques

Tools	Strengths	Weaknesses
Clementine	• Visual interface • Algorithm breadth	• Scalability
Darwin	• Efficient client-server • Intuitive interface options	• No unsupervised algorithm • Limited visualization
Data Cruncher	• Ease of use	• Single Algorithm
Enterprise Miner	• Depth of algorithms • Visual interface	• Harder to use • New product issues
Gain Smarts	• Data transformations • Built on SAS • Algorithm option depth	• No supervised algorithm • No automation
Intelligent Miner	• Algorithm breadth • Graphical tree/cluster output	• Few algorithms • No model export
Mine Set	• Data visualization	• Few algorithms • No model export
Model 1	• Ease of use • Automated model discovery	• Really a vertical tool
Model Quest	• Breadth of algorithms	• Some non-intuitive interface options
PRW	• Extensive algorithms • Automated model selection	• Limited visualization
CART	• Depth of tree options	• Difficult file I/O • Limited visualization
Scenario	• Ease of use	• Narrow analysis path
Neuroo Shell	• Multiple neural network architectures	• Unorthodox interface • Only neural networks
OLPARS	• Multiple statistical algorithms • Class-based visualization	• Date interface • Difficult file I/O
See5	• Depth of tree options	• Limited visualization • Few data options
S-Plus	• Depth of algorithms • Visualization • Programmable or extendable	• Limited inductive methods • Steep learning curve
Wiz Why	• Ease of use • Ease of model understanding	• Limited Visuallization
WEKA	• Ease of use • Ease of understanding • Depth of algorithms • Visualization • Programmable or extendable	• Not visible

their semantics do not exactly match. As an example, if $P(Ter_1 \cup Ter_2) = P(Ter_1) + P(Ter_2) - P(Ter_1 \cap Ter_2)$ holds logically ($\cup$ and $\cap$ connote union and intersection respectively), then $Ter_1 \neq Ter_2$ implies that Ter_1 and Ter_2 are aliases, not logical synonyms. The $P(Ter_1 \cap Ter_2)$ probability is for the alias part, and $P(Ter_1)$ and $P(Ter_2)$ are probabilities for the multi-representations (other meanings). For example, the

English word "errand" has two meanings (i.e. multi-representations): "a short journey", and "purpose of a journey". Semantic aliasing (defining aliases) is not part of the Nong's enterprise skeletal TCM onto-core; it belongs to the OCOE domain.

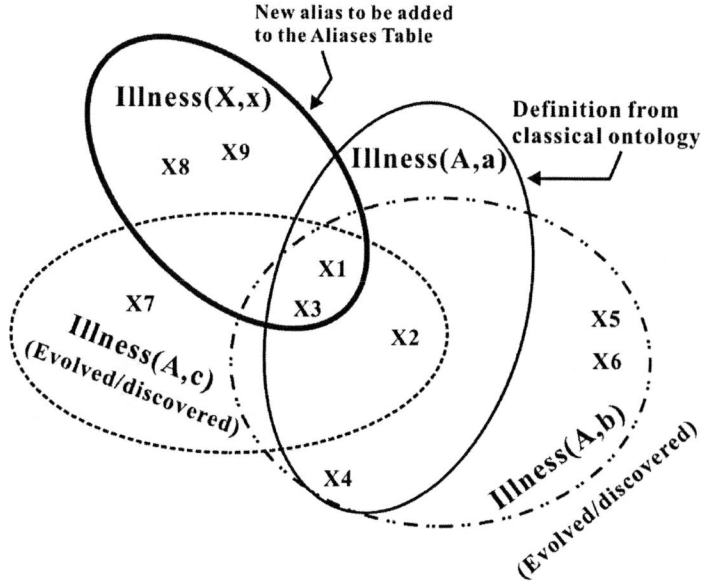

Fig. 10.1 Evolution of a canonical term

Figure 10.1 shows how a canonical term such as **Illness (A, a)** (i.e. illness **A** for geographical location **a** of epidemiological significance) could evolve with geography and time. For example, if this illness, defined by the set of *primary* attributes (PA): x_1, x_2, x_3 and x_4 was enshrined in the referential TCM classics, it is canonical. There might be a need to redefine this canonical term, however, to best suit the clinical usage in geographical region **b** where the same illness is further defined by two more local *secondary* attributes: x_5 and x_6. These additional attributes are secondary in nature if they were enshrined in TCM classics of much later. The addition x_5 and x_6 created the newer term **Illness (A, b)**. Similarly for geographic location **c** the secondary attribute x_7 created another term **Illness (A, c)** of regional epidemiology significance. Medically, **Illness (A, a)**, **Illness (A, b)** and **Illness (A, c)** in Figure 10.1 may be regarded as members of the same illness family; they are aliases to one another. But, in the extant Nong's enterprise skeletal TCM onto-core no concept of aliases was incorporated. That is, the system indicates no degree of similarity between any illnesses (e.g. Illness (A, a), **Illness (A, b)**). This leads to ambiguity problems between the "*key-in*" and "*handwritten*" diagnosis/prescription (D/P) approaches for the Nong's telemedicine system. The *key-in* approach by a physician is "*standard and formal*" in regard to referential TCM classics that provided the Nong's skeletal TCM onto-core the basis for sound clinical practice. While the handwritten approach is traditional practiced since the dawn of TCM, it is "*non-standard and informal*" because it involves terms that are local and have not been enshrined in the classical TCM vocabulary. Worst of all, for telemedicine systems (e.g. the Nong's) [1, 16] the use of informal terms in the handwritten D/P approach prevents direct clinical experience feedback to enrich the TCM onto-core as new "logical-knowledge-add-on" cases immediately. The OCOE approach resolves this problem as a fringe benefit by automating alias definitions (i.e. semantic aliasing), which are mandatory because the aliases and their degrees of similarity to RC must be recorded.

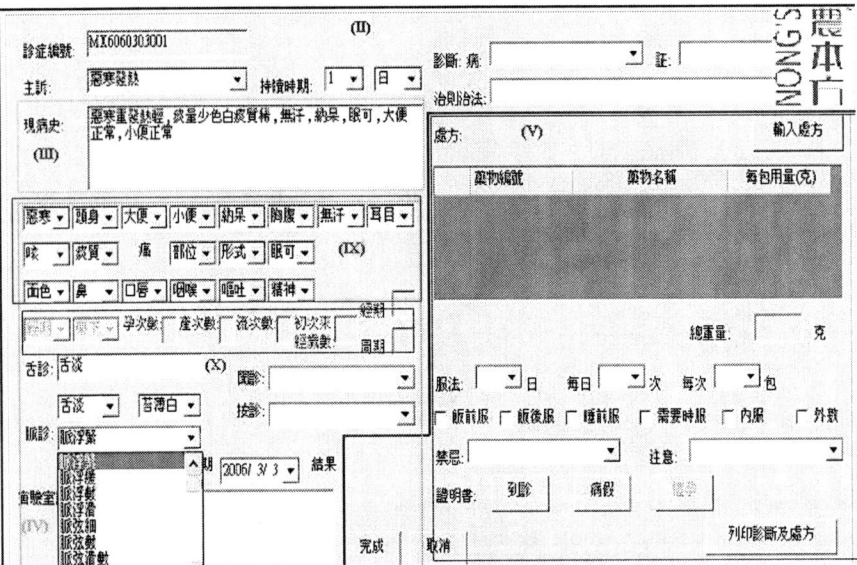

Fig. 10.2 Chinese D/P interface used by a TCM physician to treat a patient (partial view)

Figure 10.2 is the D/P interface of the robusy Nong's telemedicine system [2] in which the service roundtrip time is successfully harnessed [3]. It is the syntactical layer through which the TCM physician creates implicit queries by the standard key-in approach in a computer-aided manner. With respect to the patient's complaint ("主訴") (in section (II) of Figure 10.2), the physician follows the diagnostic procedure of four constituent sessions: look ("望"), listen & smell ("聞"), question ("問"), and pulse-diagnosis ("切") [15]. In a basic "*question & answer & key-in*" fashion, the physician obtains the symptoms and selectively keys them in section (IX). These keyed-in symptoms (e.g. 怕冷) are the parameters of the "implicit query" for the D/P system parser to conclude the illness and curative prescription from the TCM semantic net by inference. The keyed-in symptoms will be echoed immediately in the "Symptoms (現病史)" window in section (III). TThis response is immediate and accurate as these terms are standard in the Nong's enterprise skeletal TCM onto-core. As a decision support, the illness (病) and curative prescription

(處方) concluded by the D/P parser are displayed respectively in the windows in sections (II) and (V). Finally, the physician decides, based on personal experience, if the parser's logical conclusion was acceptable. The above events were stimuli and response in the key-in based D/P process, which normally does not accommodate the common but nonstandard traditional handwritten D/P approach.

In theory, the information in the final decision by the physician can be used as feedback to enrich the open TCM onto-core as new cases in a user transparent fashion. Since all the terms came from the standard TCM vocabulary, *semantic transitivity* (ST) exists among the terms in the three fields: (II), (III) and (V). By the ST connotation, given one field, the parser should conclude the other two unambiguously and consistently. This kind of feedback is impossible for the traditional handwritten D/P approach due to the ambiguity caused by the regional nonstandard TCM terminology. The OCOE approach associates aliases and contextual references automatically (semantic aliasing), and this enables the feedback of handwritten D/P results to enrich the onto-core as a fringe benefit.

Figure 10.3 shows how the *Semantic TCM Visualizer* (STV), which is part of the Nong's telemedicine system, helps the physician visualize the parser's inference of the logical D/P conclusion from an implicit query by the key-in operation. The parser works with the semantic net, which is the logical representation of the subsumption hierarchy for the TCM onto-core; the Influenza (感冒) sub-ontology is part of this hierarchy. For the standard key-in operation *semantic transitivity* should exist in the semantic net. The right side of the STV screen in Figure 10.3 corresponds to the key-in operations in section (IX) of the D/P interface in Figure 10.2. When attributes or symptoms were selected and keyed-in, the STV highlighted them in the scrollable partial subsumption hierarchy (Influenza in this case) on the left side of Figure 10.3. This hierarchy in the OCOE context is the DOM (document object model) tree, which pictorially depicts the relevant portion of the semantic net. The highlighted attributes were the parameters in the implicit query, which guided the parser to infer the unique semantic path as the logical conclusion. In this example, the semantic path indicates Influenza (感冒) as the illness logically.

10.4 Master Aliases Table and OCOE Data Structures

The master aliases table (MAT) virtually pries the skeletal TCM onto-core open for continuous evolution with aid from text mining. The catalytic data structures in MAT disambiguate handwritten information input to the D/P interface (Figure 10.2) by semantic aliasing. This is achieved because semantic aliasing defines possible semantic transitivity for the information. Practically the MAT is the directory for managing the three special OCOE data structures: CAV (contextual attributes vector), CAT (contextual aliases table) and RIT (relevance indices table). Every illness is a referential context (RC) with three unique associating data structures: i) CAV to list all the RC attributes; ii) CAT to list all the RC aliases; and iii) RIT to record the degree of similarity of an alias to the RC. The CAV, CAT and RIT contents grow

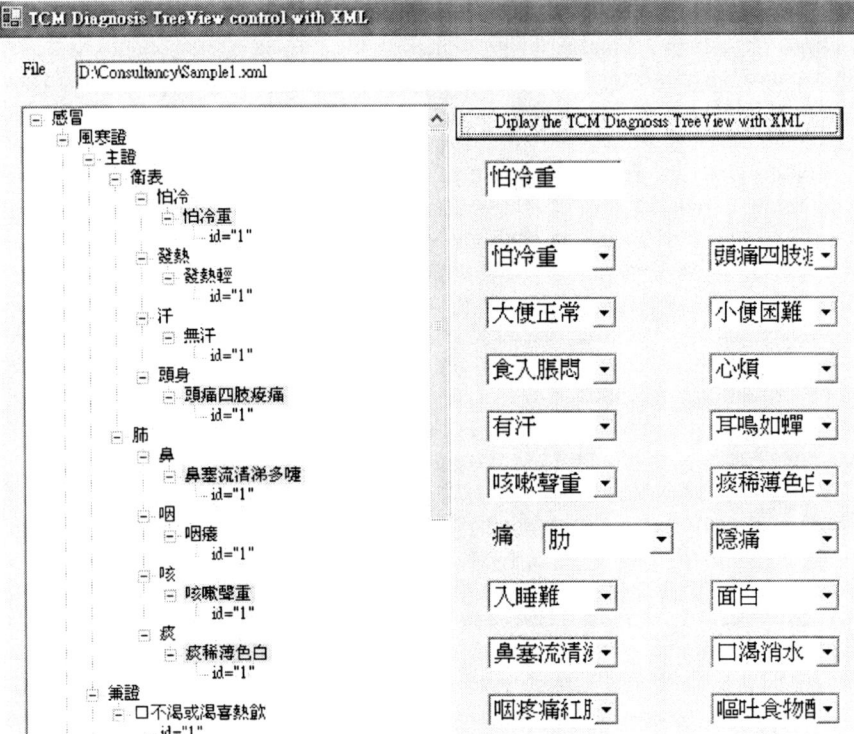

Fig. 10.3 STV - partial TCM Influenza ("感冒") sub-ontology in Chinese (left) (partial view)

with time due to incessant text mining by WEKA in the background. Aliases are weighted by their respective relevance indices (RI) in RIT. If Pneumonia was the ith alias to Common Cold, its RI value is $RI_i=WA_i=\frac{SAC}{\{MAS\in[V]\}}$. The WA_i weight/ratio is defined as "size of the attribute corpus (SAC) of the Common Cold context over the mined attribute set (MAS) for Pneumonia that overlaps". This is the semantic aliasing basis, which is adapted from the aforesaid WEKA parameters: term frequency, and inverse document frequency.

Assuming **Illness (A, a)** in Figure 10.1 is Common Cold defined by four standard primary attributes, x_1, x_2, x_3 and x_4, and **Illness (A, c)** is Pneumonia with primary attributes, x_1, x_2 and x_3, the RI for Pneumonia would be $RI_i=WA_i=\frac{4}{3}$ (or its inverse $\frac{3}{4}$ can be used for the normalization sake). The CAV, CAT, and RIT contents were initialized first with information from the skeletal TCM onto-core of the telemedicine system. Then, the list of aliases for a RC lengthens as time passes. For example the addition of **Illness (X, x)** in Figure 10.1 as a new alias to **Illness (A, a)** expands the CAV, CAT, and RIT of the latter. This expansion represents the OCOE incremental nature.

Table 10.2 RI score calculation example based on the concept in Figure 10.1

Illness/context reference	Attributes, corpus or alias's set	Attribute Classes (with respect to reference)	RI Calculation, 70% PA, 20% SA, 10% TA, 0% NKA	Remarks
Illness (A, a); Common Cold	*corpus*: $\{x_1,x_2,x_3,x_4,x_8,x_9\}$	PA - $\{x_1,x_2\}$, SA - $\{x_3\}$, TA - $\{x_4\}$, NKA - $\{x_8,x_9\}$	$RI=\frac{0.7}{2}(1+1)+0.2(1)$ $+0.1(1)+\frac{0}{2}(1+1)=1$	Referential context (RC)
Illness (A, b)	*alias's set*: $\{x_1,x_2,x_3,x_4,x_5,x_6\}$	PA - $\{x_1,x_2\}$, SA - $\{x_3\}$, TA - $\{x_4\}$, NKA - $\{x_5,x_6\}$	$RI=\frac{0.7}{2}(1+1)+0.2(1)$ $+0.1(1)+\frac{0}{2}(1+1)=1$	Alias of 100% relevance
Illness (A, c)	*alias's set*: $\{x_1,x_2,x_3,x_7\}$	PA - $\{x_1,x_2\}$, SA - $\{x_3\}$, NKA - $\{x_7\}$	$RI=\frac{0.7}{2}(1+1)+0.2(1)$ $+0.1(0)+\frac{0}{2}(1)=0.9$	Alias of 90% relevance
Illness (X, x)	*alias's set*: $\{x_1,x_2,x_3,x_7\}$	PA - $\{x_1,x_2\}$, NKA - $\{x_8,x_9\}$	$RI=\frac{0.7}{2}(1)+0.2(0)$ $+0.1(0)+\frac{0}{2}(1+1)=0.35$	Alias of 35% relevance

MAT construction involves the following steps:

a. *Consensus certification*: TCM domain experts should exhaustively identify all the contexts and their attributes from canonical material. Since this means searching a very large database (VLDB) of textual data, it is innately an operation that can be made effective by text mining. In the next step the experts have to categorize the "mined" attributes. For example, these attributes can be divided into primary, secondary, tertiary and nice-to-know classes.

b. *Weight assignments*: The categorized attributes should be weighted with respect to their significance by domain experts to facilitate RI calculations. For example, the distribution of weights in a context may be 70% for primary attributes, 20% for secondary ones, 10% for tertiary ones, and 0% for nice-to-knows. Assuming: i) **Illness (A, a)** in Figure 10.1 is Common Cold, and ii) its primary attributes (PA) as: $\{x_1,x_2\}$, its secondary attribute (SA) as: $\{x_3\}$, its tertiary attribute (TA) as: $\{x_4\}$, and its nice-to-know attributes (NKA) as: $\{x_{10},x_{11}\}$, then Table 10.2 (based on Figure 10.1), can be constructed to show the RI calculations. With respect to the referential context Common Cold or **Illness (A, a)** different aliases should have unique relevance indices. The RI score has significant clinical meaning; for example, an RI of 35% for the alias **Illness (X, x)**, from a clinically represents 35% confidence that prescriptions for **Illness (A, a)** would be effective for treating **Illness (X, x)**. This is useful for field clinical decision support.

c. *Contextual attributes vector (CAV)*: In a consensus certification process all the referential contexts and their attributes should be identified from classical TCM texts, treatises, and case histories. Attributes for every referential context are then categorized and weighted by experts as shown in Table 10.2. In the OCOE domain this process is CAV construction. It is the key to building a telemedicine system successfully because it provides the basis for achieving automatic aliasing and the semantic transitivity required for standardizing the handwritten D/P approach.

d. *Knowledge expansion in the catalytic data structures*: This real-time process is supported by a few elements that together pry open the skeletal TCM onto-core for continuous, online and automatic evolution and knowledge acquisition. These elements include: i) text mining to plough through the open web looking for new scientific findings that could be pruned and added to enrich the catalytic OCOE data structures; ii) intelligence to execute semantic aliasing and manage MAT, CAV, CAT, and RIT efficiently; and iii) TOES to aid dissociating/appending the MAT contents from/to the skeletal TCM onto-core manually. The contents in the catalytic data structures will grow over time as the conceivable evidence of non-stop onto-core evolution and knowledge acquisition.

10.5 Experimental Results

Many experiments were conducted to verify if text mining (i.e. WEKA) in the OCOE approach indeed contributes to achieving the following objectives:

a. *CAV construction*: This verifies if text mining helps construct CAV effectively. All the mined terms should be verified against the list of classical terms in the enterprise vocabulary [V], which is mandatory for the OCOE model.
b. *Semantic aliasing*: The attributes in the CAV should be weighted by domain experts so that meaningful RI scores can be computed with clinical significance in mind. These RI scores are the core to semantic aliasing by ranking: i) the degree of similarity of an alias to the given referential context; ii) the efficacy of the alias's prescriptions in percentages (i.e. indicative of the confidence level) for treating the referential context; and iii) relevance/usefulness of a text/article retrieved from the web from the angle of the referential context. Ranking of texts and articles quickens useful information retrieval for clinical reference purposes.
c. *CAV expansion*: The OCOE mechanism invokes WEKA by default to text-mine new free-format articles over the web for the purpose of knowledge acquisition. The text miner finds and adds new NKA to the CAV of the referential context. The new NKA would shed light on new and useful TCM information and serve as a means for possible TCM discovery.

The experimental results presented in this section show the essence of how the above objectives were achieved with the help of the text miner WEKA. The experiments were conducted through the TOES (TCM Ontology Engineering System) module, which is an essential element in the OCOE approach. Through TOES the user can command the telemedicine system to work either with the original skeletal TCM onto-core or the open evolvable version, which is the combination: "original skeletal TCM onto-core + MAT contents". In either case, the text miner is updating the catalytic OCOE data structures in the MAT domain continuously, independent of whether MAT is conceptually appended to the skeletal TCM onto-core or not.

10.5.1 CAV Construction and Information Ranking

MAT_f CAV construction is initialization of the OCOE data structures. In the process: i) all the referential contexts and their enshrined attributes (e.g. symptoms) in the skeletal TCM onto-core should be listed/verified manually by domain experts; and ii) the text miner is invoked to help retrieve every RC so that its three associating data structures, emphatically CAV, are correctly filled. Once the CAV construction has finished, the MAT (called foundation MAT_f) should possess the same level of skeletal knowledge as the original TCM onto-core. In the subsequent evolution new knowledge text-mined by WEKA would be pruned and added to the special OCOE data structures in MAT. This logical-knowledge-add-on evolution, a knowledge acquisition process, enriches MAT_t to become the MAT with contemporary knowledge at time (i.e. MAT_t). Logically MAT_t is richer than MAT_f. The newly acquired knowledge or "$MAT_n = MAT_t - MAT_f$", though formal, had not gone through the same rigorous consensus certification process as MAT_f. OCOE activation means telemedicine with MAT_t, which conceptually means appending the new MAT knowledge MAT_n to the skeletal TCM onto-core. The experimental result in Figure 10.4 was produced with MAT_f; this required prior successful CAV construction. In the semantic aliasing experiments the CAV of Flu RC had four types of attributes: A (primary), B (secondary), C (tertiary), and D (NKA/O).

Information ranking has two basic meanings in the OCOE operation: i) semantic aliasing in which the RI of an alias is computed to show its degree of similarity to the given referential context (RC); and ii) ranking the importance of an article/text by its RI value with respect to the given RC. The RI computation is based on the philosophy of Figure 10.1 and the approach shown by Table 10.2. Ranking TCM texts is useful because their RI help practitioners retrieve useful texts quickly. This experiment in Figure 10.4 used Flu as the RC and RI values were computed to rank the 50 TCM classics given in the file C:\Workspace\FluTestset. The WEKA retrieved the text one by one so that the context in the text as well as the relevant attributes could be identified for RI computation. The assigned weights for the attributes were: 0.7 for A or primary attribute (PA); 0.2 for B or secondary attribute (SA); 0.1 for C or tertiary attribute (TA); 0 for D or nice-to-know attribute/one (NKA/O) (similar to Table 10.2).

The experimental result in Figure 10.4 shows: i) the two aliases (to Flu), namely, Common Cold and Pneumonia; ii) the text/article numbers in FluTestset that correspond to each alias; iii) RI scores for ranking the aliases; and iv) texts/articles (e.g. 12, 1, and 16 in the "Related Article Number" window) having the maximum (Max) RI scores.

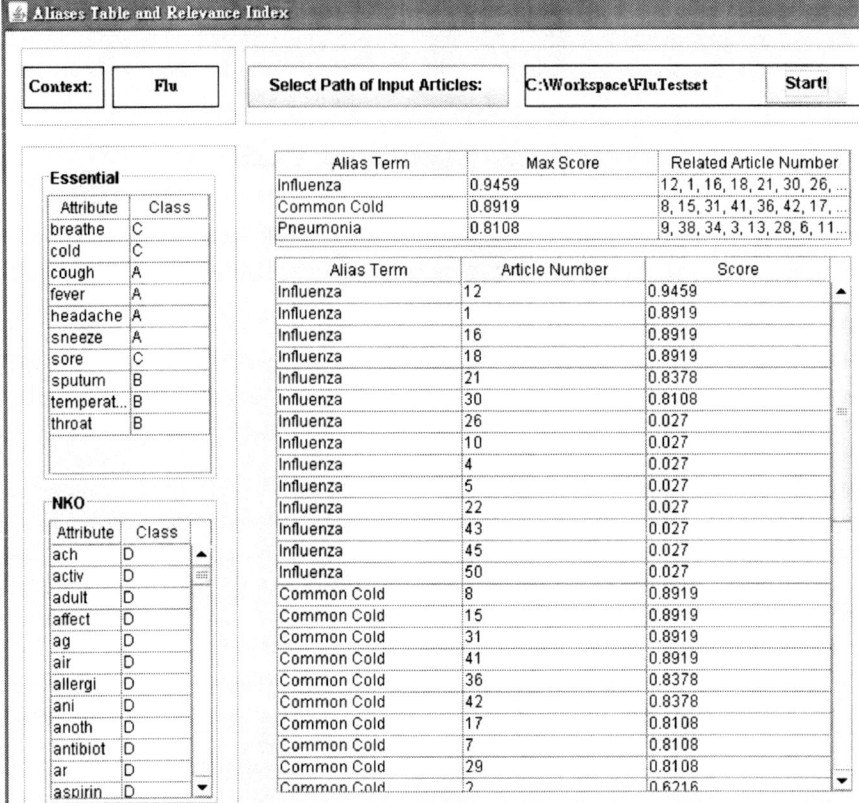

Fig. 10.4 Ranking 50 TCM canonical texts with respect to the Flu RC (partial view)

10.5.2 Real-Time CAV Expansion Supported by Text Mining

The real-time CAV expansion operation is similar to but more complex than the information ranking operation (Figure 10.4) for the following reasons:

a. The article(s) to be processed by WEKA is free-format (FF) and this suggests the following possibilities: i) context in the FF article is not clearly stated; ii) context may not match any canonized form; and iii) names of attributes (e.g. symptoms such as "$PA - (x_1,x_2)$" in Table 10.2 may not be canonical.
b. Intelligence is needed for the OCOE process to verify if a context, which was identified in a free-format article, was indeed valid to become a "new" alias of a canonical referential context. This can be achieved by checking against the enterprise vocabulary [V]. In the present OCOE prototype (written in VB.net and Java together) for verification experiments, this intelligence is basically algorithmic and therefore requires large error tolerance. For intelligence improve-

ment, soft computing techniques (e.g. fuzzy logic) should be explored in the near future.

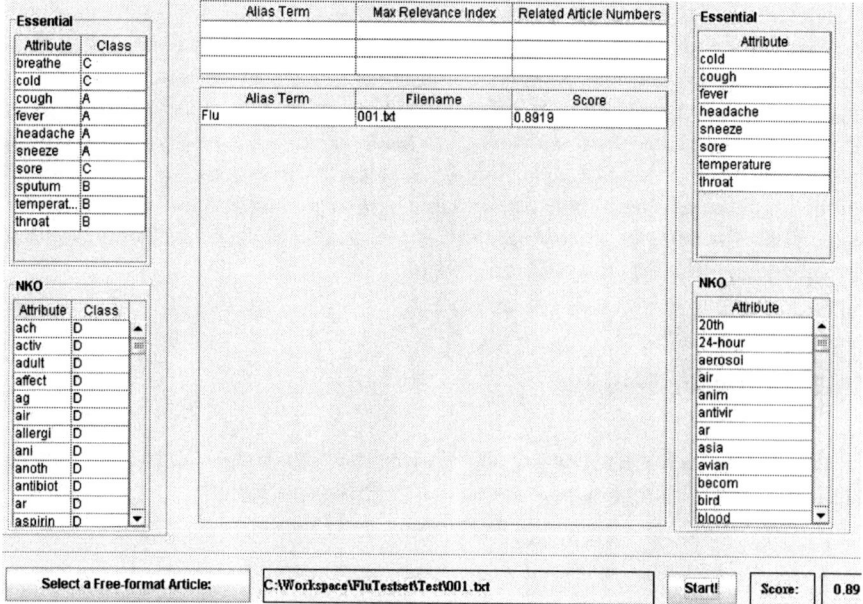

Fig. 10.5 A real-time CAV expansion example (partial view)

Figure 10.5 is the result for a real-time CAV expansion experiment. It shows: i) evaluation of a FF TCM article using "C:\Workspace\FluTestset \Test\001.txt" as the input; and ii) the evaluation result (on the right-hand side of the screen capture). The input article was evaluated by the OCOE mechanism against the Flu RC. The attributes text-mined by WEKA (right side of the screen) were first constrained by the standard terms (attributes) on the left side of the screen. The types of the constraining attributes are marked (i.e. A, B, C, and D). The RI score computed for the article with the essential attributes was 0.8919. That is, for the Nong's production telemedicine system platform on which all the verification experiments with the OCOE prototype were carried out, the chance (confidence level) was 89.19% that the input FF text addressed Flu issues.

10.6 Conclusion

In this paper we have proposed and verified the OCOE approach, which applies text mining to automate the TCM onto-core evolution in a running telemedicine system. The evolutionary process is a real-time, automatic, and incessant manner.

Text mining helps: i) building the MAT_t and MAT_f bases for clinical practice over the web and on-line evolution at the same time; and ii) support continuous TCM onto-core evolution by ploughing through the web for new scientific findings which could be pruned to enrich MAT_t. Evolution of the open TCM onto-core (i.e. MAT_t) is reflected in the non-stop expansion of the CAV, CAT and RIT data structures. Without the support of incessant and automatic text mining MAT expansion would be impossible. Semantic aliasing is natural for MAT_t and it yields semantic transitivity to disambiguate handwritten D/P results. In effect, this experience "becomes standardized" and therefore can be used as feedback to enrich MAT_t. The next step in the research is to explore other inference methods for improving the OCOE intelligence to achieve more effective semantic aliasing. Identifying a context and the associating attributes in a free-format TCM text accurately and quickly is essential for enhancing the MAT_f in a real-time fashion.

10.7 Acknowledgement

The authors thank the Hong Kong Polytechnic University and the PuraPharm Group for the research grants, A-PA9H and ZW93 respectively.

References

1. A. Lacroix, L. Lareng, G. Rossignol, D. Padeken, M. Bracale, Y. Ogushi, R. Wootton, J. Sanders, S. Preost, and I. McDonald, G-7 Global Healthcare Applications Sub-project 4, Telemedicine Journal, March 1999
2. Wilfred W. K. Lin, Jackei H.K. Wong and Allan K.Y. Wong, Applying Dynamic Buffer Tuning to Help Pervasive Medical Consultation Succeed, the 1st International Workshop on Pervasive Digital Healthcare (PerCare) in the 6th IEEE International Conference on Pervasive Computing and Communications (Percom2008), Hong Kong March, 2008, Hong Kong
3. Allan K.Y. Wong, Tharam S. Dillon and Wilfred W.K. Lin, Harnessing the Service Roundtrip Time Over the Internet to Support Time-Critical Applications - Concepts, Techniques and Cases, Nova Science Publishers, New York, 2008
4. R. Rifaieh and A. Benharkat, From Ontology Phobia to Contextual Ontology Use in Enterprise Information System, in Web Semantics & Ontology, ed. D. Taniar and J. Rahayu, Idea Group Inc., 2006
5. T. R. Gruber, A Translation Approach to Portable Ontology Specifications. Knowledge Acquisition, 5(2), 1993, 199-220
6. N. Guarino, P. Giaretta, Ontologies and Knowledge Bases: Towards a Terminological Clarification. In Towards very large knowledge bases: Knowledge building and knowledge sharing, 1995, Amsterdam, The Netherlands: ISO Press, 25-32
7. L.E. Holzman, T.A. Fisher, L.M. Galitsky, A. Kontostathis, W.M. Pottenger, A Software Infrastructure for Research in Textual Data Mining, The International Journal on Artificial Intelligence Tools, 14(4), 2004, 829-849
8. S. Bloehdorn, P. Cimiano, A. Hotho and S. Staab, An Ontology-based Framework for Text Mining. LDV Forum - GLDV Journal for Computational Linguistics and Language Technology, 20(1), 2005, 87-112

9. M.S. Chen, S.P. Jong and P.S. Yu, Data Mining for Path Traversal Patterns in a Web Environment, Proc. of the 16th International Conference on Distributed Computing Systems, May 1996, Hong Kong, 385-392
10. U.M. Fayyad, G.Piatesky-Shapiro, P. Smyth and R. Uthurusamy, Advances in Knowledge Discovery and Data Mining, AAAI Press/MIT Press, 1996
11. W. Pedrycz and F. Gomide, An Introduction to Fuzzy Sets: Analysis and Design, MIT Press, 1998
12. C.M. van der Walt and E. Barnard, Data Characteristics that Determine Classifier Performance, Proc. of the 16th Annual Symposium of the Pattern Recognition Association of South Africa, 2006, 160-165
13. R. Agrawal and R. Srikant, Fast Algorithms for Mining Association Rules, Proc. of the 20th Conference on Very Large Databases, Santiago, Chile, September 1994
14. F. Yu, Collaborative Web Information Retrieval and Extraction - Implementation of an Intelligent System "LightCamel", BAC Final Year Project, Department of Computing, Hong Kong Polytechnic University, Hong Kong SAR, 2006
15. M. Ou, Chinese-English Dictionary of Traditional Chinese Medicine, C & C Joint Printing Co. (H.K.) Ltd., ISBN 962-04-0207-3, 1988
16. http://umls.nlm.nih.gov/

Chapter 11
Microarray Data Mining: Selecting Trustworthy Genes with Gene Feature Ranking

Franco A. Ubaudi, Paul J. Kennedy, Daniel R. Catchpoole, Dachuan Guo, and Simeon J. Simoff

Abstract Gene expression datasets used in biomedical data mining frequently have two characteristics: they have many thousand attributes but only relatively few sample points and the measurements are noisy. In other words, individual expression measurements may be untrustworthy. Gene Feature Ranking (GFR) is a feature selection methodology that addresses these domain specific characteristics by selecting features (i.e. genes) based on two criteria: (i) how well the gene can discriminate between classes of patient and (ii) the trustworthiness of the microarray data associated with the gene. An example from the pediatric cancer domain demonstrates the use of GFR and compares its performance with a feature selection method that does not explicitly address the trustworthiness of the underlying data.

11.1 Introduction

The ability to measure gene activity on a large scale, through the use of microarray technology [1, 2], provides enormous potential within the disciplines of cellular biology and medical science. Microarrays consist of large collections of cloned molecules of deoxyribonucleic acid (DNA) or DNA derivatives distributed and bound in an ordered fashion onto a solid support. Each individual DNA clone represents a particular gene. Several kinds of microarray technology are available to researchers to measure levels of gene expression: cDNA, Affymetrix and Illu-

---

Franco A. Ubaudi, Paul J. Kennedy
Faculty of IT, University of Technology, Sydney, e-mail: {faubaudi,paulk}@it.uts.edu.au

Daniel R. Catchpoole, Dachuan Guo
Tumour Bank, The Childrens Hospital at Westmead, e-mail: {DanielC,dachuang}@chw.edu.au

Simeon J. Simoff
University of Western Sydney, e-mail: S.Simoff@uws.edu.au

mina microarrays. Each of these platforms can have over 30,000 individual DNA features "spotted" in a cm^2 area. These anchored DNA molecules can then, through the process of "hybridization", capture complementary nucleic acid sequences isolated from a biological sample and applied to the microarray. The isolated nucleic acids are labeled with special fluorescent dyes which, when hybridized to their complementary spot on the microarray will fluoresce when excited with a laser–based scanner. The level of fluorescence is directly proportional to the amount of nucleic acid captured by the anchored DNA molecules. In the case of gene expression microarrays, the nucleic acid isolated is messenger ribonucleic acid (mRNA) which results when a stretch of DNA containing a gene is "transcribed" or is "expressed".

This chapter focuses on two color spotted cDNA microarrays, an approach which allows for the direct comparison of two samples, usually a control and a test sample. Data is derived from fluorescent images of hybridized microarray "chips" and usually comprises statistical measures of populations of pixel intensities of each gene feature. For example, mean, median and standard deviation of pixel intensities for two dyes and for the local background are generated for each spot on a cDNA microarray.

All microarray platforms involve assessing images of fluorescent features with the level of fluorescence giving a measure of expression. All platforms are beset by the same data analysis issues of feature selection, noise, high dimensionality, non–specific signal, background, and spatial effects. The approach we take in this chapter is non–platform specific, although we apply it to the "noisiest" of the platforms: glass slide spotted cDNA microarray. Comparing gene expression measurements between different technologies and between measurements on the same technology at different times is a challenge, to some extent addressed by normalization techniques [3]. A major issue in these data is the unreliable variance estimation, complicated by the intensity-dependent technology-specific variance [4]. There is also an agreement that different methods of microarray data normalization have a strong effect on the outcomes of the analysis steps [5]. Another issue is the small number of replicated microarrays because of cost and sample availability, resulting in unreliable variance estimation and thus unreliable statistical hypothesis tests. There has been a broad spectrum of proposed methods for determining the sample size in microarray data collection [6], with the majority being focused on how to deal with multiple testing problems and the calculation of significance level.

Common approaches to dealing with these data issues include visual identification of malformed spots for omission and normalization of gene expression measures [7]. Often arbitrary cut off measures are used to select genes for further assessment in attempts to reduce the data volume and to facilitate selection of "interesting genes". Other researchers (e.g. [8–12]) apply dimensionality reduction to microarray data with the assumption that many genes are redundant or irrelevant to modeling a particular biological problem. Golub [8] calls the redundancy of terms *additive linearity* to signify that many genes add nothing new or different. Feature set reduction of microarray data also helps to deal with the "curse of dimensionality" [13]. The view of John et al [14] is that use of feature selection, prior to model construction, obviates the necessity of relying on learning algorithms to determine which features are relevant and has the additional benefit of avoiding overfitting.

Given the widespread use of feature selection for microarray data and the data quality issues, it is surprising that we were unable to find approaches in the literature specifically addressing data quality [15]. This is also true of more general feature selection literature [16]. Some researchers have endeavored to manage the issues of small sample size and microarray noise in their approaches. Unsupervised methods are used to evaluate the quality of microarray data in [17] and [18] but they are subjective and need manual configuration. Baldi and Hatfield [7] use variance of expression as prior knowledge when training a Bayesian network. Yu et al [19] apply an approach they call "quality" to filter genes, although it only uses expression data.

The feature analysis and ranking method proposed in this chapter addresses the issue of dealing with diverse quality across technologies and the small number of replicated measurements. Our approach explicitly uses quality measures of a spot (such as variance of pixel intensities) to compute a trustworthiness measure for each gene which complements its gene expression measure. Understanding of gene expression can then be based on the quality of the spot as well as its intensity. A "confidence" of the findings based on data quality can be incorporated into models built on the data. Also, training sets with few sample points, as are often found in gene expression may also mean that the assumption that test data has similar quality as training sets is no longer valid. The unsupervised learning we apply to all available data helps to gain an understanding of the quality of gene expression measurements.

11.2 Gene Feature Ranking

Gene Feature Ranking is a feature selection methodology for microarray data which selects features (genes) based on (i) how well the gene discriminates between classes of patient and (ii) the trustworthiness of the data. The motivation behind the first of these is straightforward. However, previous methods have not specifically addressed the issue of the quality of data in feature set reduction. Hence, our emphasis on assessing the trustworthiness of the data. A training subset of data is used to assess classification of patients by genes. All available data is used in an unsupervised learning process to assess the trustworthiness of gene measurements.

Gene Feature Ranking, shown schematically in Fig. 11.1, consists of two consecutive feature selection phases which generate ranked lists of features. Phase 1 ranks genes by the trustworthiness of their data. Genes ranked above a threshold are passed to the second phase. Phase 2 ranks the remaining genes using a traditional feature selection algorithm (such as Gain Ratio [20]). The goal of the approach is to maintain a balance between the trustworthiness of data and its discriminative ability. Following we describe how attributes and data sample points are used in GFR and then we describe the two phases in detail. Although data preprocessing methods which suppress genes that do not respect some measures may be seen as an alternative to GFR, such approaches have the limitation that they are generally ad hoc and usually use expression measurements to assess the spot quality rather than "quality measures". Gene Feature Ranking, on the other hand, is a framework that specifically addresses the quality aspects of the data.

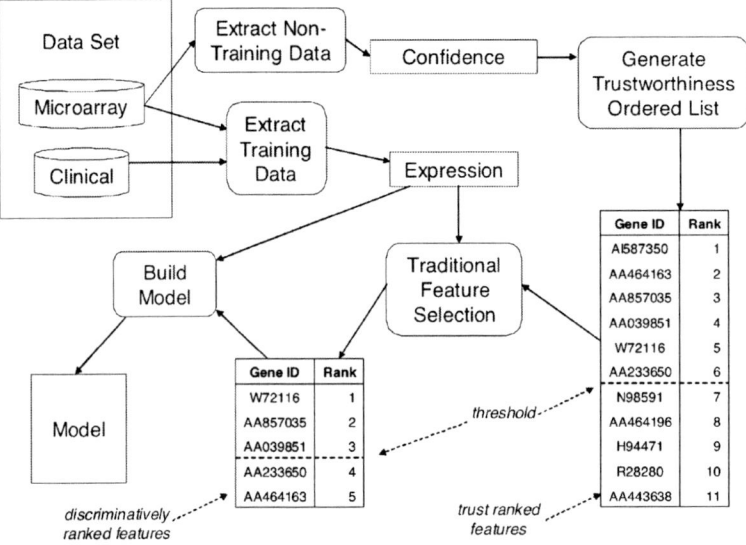

Fig. 11.1 Gene Feature Ranking applies two phases of feature selection to data. Phase 1, labelled "Generate Trustworthiness Ordered List", filters genes based on the trustworthiness of their data, as measured by the "confidence". Phase 2, labelled as "Traditional Feature Selection" filters the trusted genes based on how well they can discriminate classes.

11.2.1 Use of Attributes and Data Samples in Gene Feature Ranking

Attributes associated with individual genes in the microarray data are grouped as either "expression" attributes or "spot quality" attributes, but not both. Expression attributes are considered to measure the level of expression of a gene. Spot quality attributes are statistical measures of likely accuracy of the expression for the gene. An example of an expression attribute is median pixel intensity of the spot and an example of a spot quality attribute is the standard deviation of pixel intensities for a spot. Expression attributes of genes are used to build the model and spot quality attributes are used to assess the trustworthiness (or confidence) of the gene.

Three subsets of available data are recognized in GFR. *Training Data Set* consists of "expression" attributes from the microarray gene expression data together with associated clinical data. The purpose of this data is to build the model and it includes the class of patient. *Test Data Set* is a smaller set of "expression" attributes and clinical data used to evaluate the accuracy of the model built from the training data. *Non–Training Data Set* consists of "spot quality" attributes from all of the microarray gene expression data. This data is used in the first (unsupervised) phase of GFR. As this data is not used directly to build a model it does not contain information about the patient class or any clinical attributes.

11.2.2 Gene Feature Ranking: Feature Selection Phase 1

Phase 1 of GFR determines the trustworthiness of genes by evaluating the confidence of each expression reading (labelled *confidence* in Fig. 11.1) in the non–training dataset. Trustworthiness of a gene is the median value of the N confidence values. The confidence value c for a gene in a specific microarray spot is

$$c = \frac{1}{\log\left(\frac{(\sigma_{control}+\sigma_{test})}{2} + \|\sigma_{control} - \sigma_{test}\|\right)}. \tag{11.1}$$

The spot quality attributes $\sigma_{control}$ and σ_{test} are the standard deviations of pixel intensity for the *control* and *test* channels respectively. High pixel deviation is associated with low confidence in the accuracy of expression measure. As cDNA microarrays have two intensity channels per spot, the definition of c in equation (11.1) comprises an average of the variations of the channels as well as a measure of their difference.

The genes are then ranked by trustworthiness with those below a threshold judged too noisy and then discarded. Gene Feature Ranking does not *a priori* prescribe the threshold. Two issues contribute to setting a threshold: (i) an understanding of the redundancy present in the attributes is needed to avoid setting the threshold too high; and (ii) analysis of the distribution of trustworthiness values for genes is needed to prevent setting the threshold too low. We set the threshold empirically, although an approach based on the calculated trustworthiness values would also be appropriate.

11.2.3 Gene Feature Ranking: Feature Selection Phase 2

Phase 2 of GFR ranks the remaining genes according to their discriminative ability using the training data set (as shown in Fig. 11.1). All genes passed through from phase 1 are considered to have high enough trust for use in phase 2. Genes ranked above a phase 2 threshold are selected to build a classifier. The feature selection method in phase 2 is not prescribed. Here, we use Gain Ratio. The result is a list of genes that are discriminative and trustworthy.

As in the first feature selection phase, choice of the selection threshold is the responsibility of the user. In this chapter, we choose several final feature subset sizes to compare the achieved classification accuracy of models constructed using different feature selection methodologies. Two other possible approaches might use a measure of the minimum benefit provided for each feature with the following metrics: (i) the classification accuracy gained by the model; or (ii) a balance between trustworthiness and discriminative ability.

11.3 Application of Gene Feature Ranking to Acute Lymphoblastic Leukemia data

This section applies GFR to the subtyping of Acute Lymphoblastic Leukemia (ALL) patients. We describe the ALL data then build classifiers using GFR ranked genes and Gain Ratio only ranked genes to classify patients suffering from two subtypes of ALL: B lineage and T lineage.

Biomedical data was provided by the Children's Hospital at Westmead and comprises clinical and microarray datasets linked by sample identifier. Clinical attributes include sample ID, date of diagnosis, date of death and immunophenotype (the class label). The microarray dataset has several attributes for each gene for each array: f635Median, f635Mean, f635SD, f532Median, f532Mean and f532SD. The median and mean measurements are "expression" attributes and the standard deviation attribute is the "spot quality". These relate to median, mean and standard deviation of the pixel intensities for the fluorescent emission wavelength of the two dyes: 635 nm (red) and 532 nm (green). The 635 nm fluorescent dye was associated with leukemia samples and the 532 nm dye represented pooled normal (non–leukemic) samples. Consequently, this study compared on the one array ALL to normal control.

Data was preprocessed. Patients with missing data and genes not present on all microarrays and "non–biological" genes were omitted. Outlier expression measures that were close to the minimum or maximum measurable value were repaired to conform with other data. Microarray data was normalized by adjusting arrays to have the same average intensity. Preprocessing resulted in 120 arrays (47 patients) for B ALL and 44 arrays (14 patients) for T ALL with a total of 9485 genes.

Algorithms other than GFR are implemented in Weka [21]. Gain Ratio [20] was applied in GFR phase 2 to rank genes by class discrimination and alone to compare with GFR. The same parameter settings for Gain Ratio were used in both cases. AdaBoostM1 [22] was used for classification with parameters of a primary classifier of DecisionStump being boosted, 10 iterations of boosting and reweighting with a weight threshold of 100. AdaBoostM1 was chosen to handle the class imbalance. Test error of classifiers was estimated with ten–fold cross validation.

Figure 11.2 explores how trustworthiness differs between genes by graphing gene's trust by rank as calculated in GFR phase 1 with equation (11.1). Data attributes f532SD and f635SD were σ_{test} and $\sigma_{control}$ respectively. From this ranking we empirically set the phase 1 threshold to 7000.

Phase 2 of GFR then ranked the phase 1 selected genes with Gain Ratio using the training dataset. Gain Ratio balances feature trust and class discrimination.

Genes from the training data were also ranked by Gain Ratio without applying the first phase of GFR. Comparison between the GFR ranked list and GainRatio–only

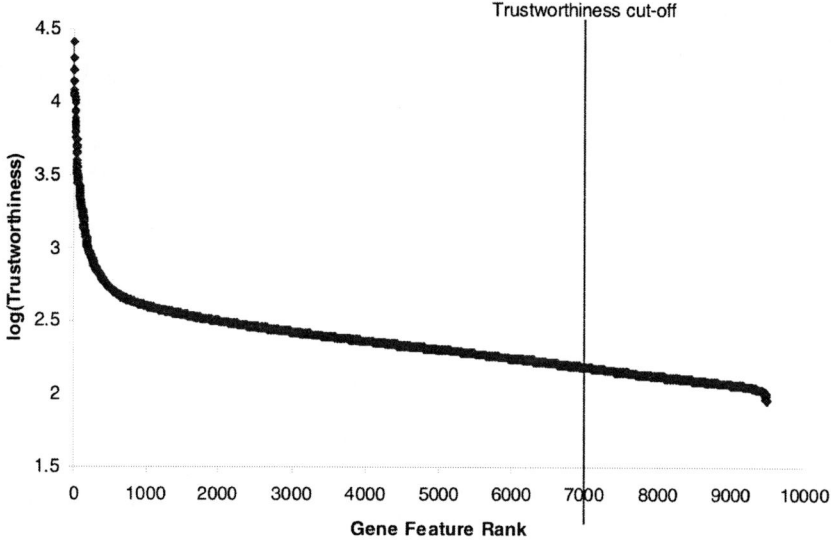

Fig. 11.2 Trustworthiness by gene ranked in phase 1 of GFR. Also shown is the trustworthiness threshold we selected, which is 7,000. All genes ranked after this threshold are removed from further consideration in GFR phase 2.

ranked list show that the top ten ranked genes are the same in both lists. Four of the genes ranked within the top 20 by GainRatio are ranked below the trustworthiness threshold by GFR. Of the top 256 genes, 34 differ between GFR and Gain Ratio.

To further illustrate these changes between GFR and GainRatio–only selected genes, the raw data was converted to a basic expression ratio using

$$Expression = \log_2 \left(\frac{F635 - B635}{F532 - B532} \right) \qquad (11.2)$$

where B is the intensity fluorescence for each channel in the local background region around the feature (F). In eq. (11.2), if the feature is poor quality, faint, misshapen or has a particularly noisy background, often the F − B value in one channel is negative yielding an incalculable logarithm value. The expression value for the 20 genes from each feature selection approach was identified for the 61 ALL patients, subjected to hierarchical clustering and displayed as a "heat map". Expression values which were "null" due to poor quality were represented as a grey square on the heat map. Figure 11.3 indicates that the four genes removed from the top 20 with GFR were all overly represented with null expression values across the patient cohort. Both sets of gene could distinguish B from T–lineage ALL, however GFR leads to the selection of more trustworthy genes.

Classifiers were built using the "immunotype" class attribute and expression attributes ("f635Median" and " f532-Median") for the top 16 ranked genes for GFR and Gain Ratio. The motivation behind choosing this number of features was from considerations of being able to compare models for different classes.

Accuracy of these classifiers derived from 10–fold cross validation is reported in Table 11.1. Features selected with GFR result in a slightly more accurate classifier than with Gain Ratio. We also analyzed the impact of using different sized GFR feature subsets on classification accuracy (see Table 11.2). Classifiers were built using the same parameters as before, with the only difference being the number of features used (16, 1024 or all genes). The best performance arose from using the sixteen features chosen by GFR. Regardless of the number of features used or the feature selection method, all classifiers were accurate. This was expected due to the strong genetic relationship with leukemia cell types.

Table 11.1 Cross validation accuracy of classifiers built using the top 16 ranked features. Column 1 is classification accuracy. Column 2 is the Kappa statistic defined as the agreement beyond chance divided by the amount of possible agreement beyond chance [23] p.115. Column 3 are the total number of classification errors. Column 4 is the feature selection method applied.

Accuracy	Kappa	Total Errors	Feature selection methodology
99.39%	0.9844	1	GFR
98.17%	0.9537	3	Gain Ratio

Table 11.2 Accuracy of classifiers trained using different numbers of the top GFR ranked features. Column 1 is the number of features used. Column 2 is the classification accuracy. Column 3 is the Kappa statistic. Last four columns present counts of true and false positives and negatives.

Features	Accuracy	Kappa	True positive	False negative	False positive	True negative
16	99.39%	0.9844	120	0	1	43
1,024	98.17%	0.9531	119	1	2	42
9,485	98.78%	0.9689	119	1	1	43

11.4 Conclusion

This chapter introduces Gene Feature Ranking, a feature selection approach for microarray data that takes into account the trustworthiness or quality of the measurements by first filtering out genes with low quality measurements before selecting features based on how well they discriminate between different classes of patient. We demonstrate GFR to classify the cancer subtype of patients suffering from ALL. Gene Feature Ranking is compared to Gain Ratio and we show that GFR outperforms Gain Ratio and selects more trustworthy genes for use in classifiers.

References

1. Hardiman, G.: Microarray technologies - an overview. Pharamacogenomics **3** (2002) 293–297
2. Schena, M.: Microarray Biochip Technology. BioTechniques Press, Westborough, MA (2000)
3. Bolstad, B., et al.: A comparison of normalization methods for high density oligonucleotide array data based on variance and bias. Bioinformatics **19** (2003) 185–193
4. Weng, L., Dai, H., Zhan, Y., He, Y., Stepaniants, S.B., Bassett, D.E.: Rosetta error model for gene expression analysis. Bioinformatics **22** (2006) 1111–1121
5. Seo, J., Gordish-Dressman, H., Hoffman, E.P.: An interactive power analysis tool for microarray hypothesis testing and generation. Bioinformatics **22** (2006) 808–814
6. Tsai, C.A., et al.: Sample size for gene expression microarray experiments. Bioinformatics **21** (2005) 1502–1508
7. Baldi, P., Hatfield, G.W.: DNA Microarrays and Gene Expression: from experiments to data analysis and modeling. Cambridge University Press (2002)
8. Golub, T., Slonim, D.K., Tamayo, P., Huard, C., Gaasenbeek, M., Mesirov, J.P., Coller, H., Loh, M.L., Downing, J.R., Caligiuri, M.A., Bloomfield, C.D., Lander, E.S.: Molecular classification of cancer: Class discovery and class prediction by gene expression monitoring. Science **286** (1999) 7
9. Mukherjee, S., Tamayo, P., Slonim, D.K., Verri, A., Golub, T.R., Mesirov, J.P., Poggio, T.: Support vector machine classification of microarray data. AI memo 182. CBCL paper 182. Technical report, MIT (2000) Can be retrieved from ftp://publications.ai.mit.edu.
10. Blum, A.L., Langley, P.: Selection of relevant features and examples in machine learning. Artificial Intelligence **97** (1997) 245–271
11. Yang, J., Hanavar, V.: Feature subset selection using a genetic algorithm. Technical report, Iowa State University (1997+)
12. Efron, B., Tibshirani, R., Goss, V., Chu, G.: Microarrays and their use in a comparative experiment. Technical report, Stanford University (2000)
13. Bellman, R.E.: Adaptive Control Processes. Princeton University Press (1961)
14. John, G.H., Kohavi, R., Pfleger, K.: Irrelevant features and the subset selection problem. In: Eleventh International Conference (Machine Learning), Kaufmann Morgan (1994) 121–129
15. Saeys, Y., Inza, I., et al.: A review of feature selection tecnhiques in bioinformatics. Bioinformatics **23** (2007) 2507–2517
16. Guyon, I., Elisseeff, A.: An introduction to variable and feature selection. Machine Learning Research (2003) 1157–1182
17. Wang, X., Ghosh, S., Guo, S.W.: Quantitative quality control in microarray image processing and data acquisition. Nucleic Acids Research **29** (2001) 8
18. Park, T., Yi, S.G., Lee, S., Lee, J.K.: Diagnostic plots for detecting outlying slides in a cDNA microarray experiment. BioTechniques **38** (2005) 463–471
19. Yu, Y., Khan, J., et al.: Expression profiling identifies the cytoskeletal organizer ezrin and the developmental homeoprotein six- 1 as key metastatic regulators. Nature Medicine **10** (2004) 175–181
20. Quinlan, J.R.: Induction of decision trees. Machine Learning **1** (1986) 81–106
21. Witten, I.H., Frank, E.: Data Mining: Practical machine learning tools and techniques. 2nd edn. Morgan Kaufmann, San Francisco (2005)
22. Freund, Y., Schapire, R.E.: Experiments with a new boosting algorithm. In: International Conference on Machine Learning, San Francisco, Morgan Kaufmann (1996) 148–156
23. Dawson, B., Trapp, R.G.: Basic & Clinical Biostatistics. Third edn. Health Professions. McGraw-Hill Higher Education, Singapore (2001)

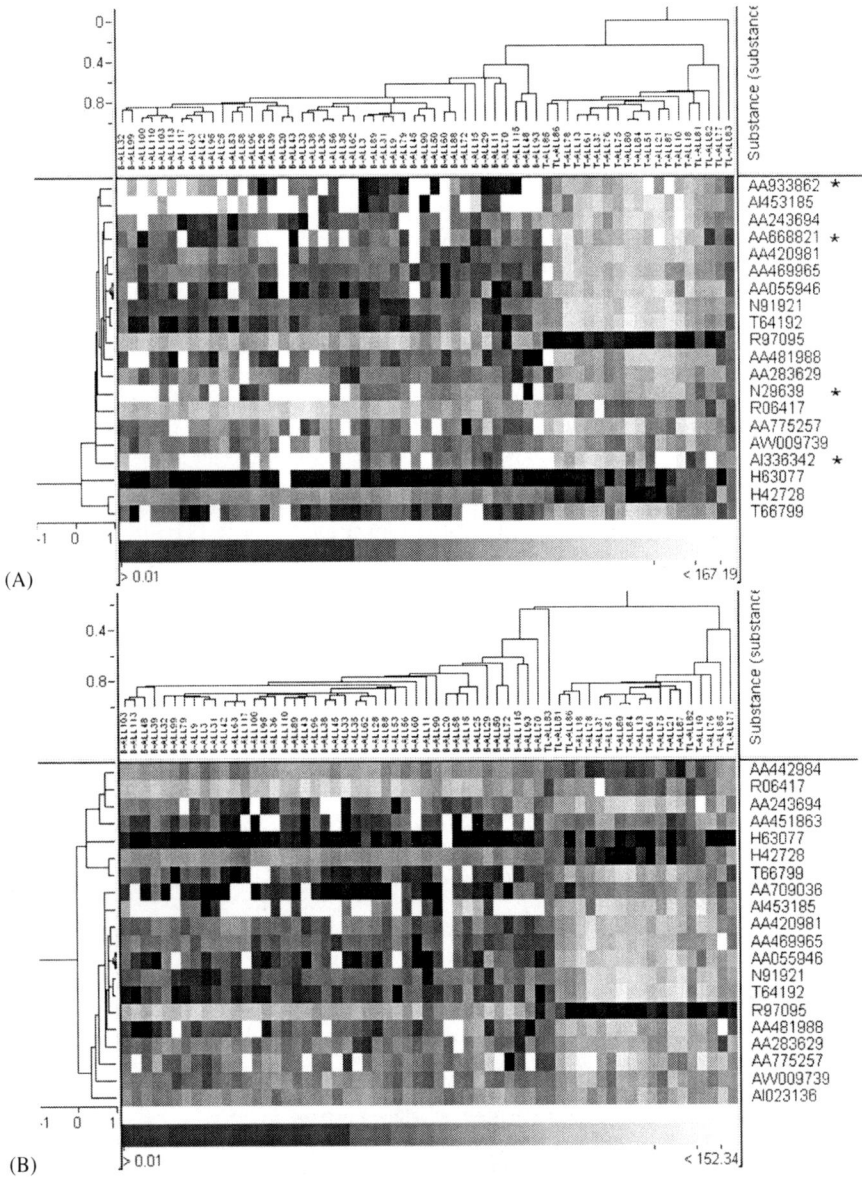

Fig. 11.3 Hierarchical cluster analysis of the top 20 genes selected following (A) Gain Ratio only and (B) GFR feature selection. The dendrogram at the top of each plot identifies the relationship of each patient on the basis of the expression of the 20 genes; B–lineage (on left and marked B–) and T–lineage (on right and marked T– or TL–). Each box on the heat map represents a gray–scale annotation of the expression ratio for each gene (row) in each ALL patient (column), with the gray scale at the base of each plot. Increasing light gray boxes represents increasing expression in ALL patients compared to control samples, whilst dark gray represents decreasing expression. White boxes represent "null" expression value indicative of poor quality features. Genes marked with * were removed by GFR phase 1.

Chapter 12
Blog Data Mining for Cyber Security Threats

Flora S. Tsai and Kap Luk Chan

Abstract Blog data mining is a growing research area that addresses the domain-specific problem of extracting information from blog data. In our work, we analyzed blogs for various categories of cyber threats related to the detection of security threats and cyber crime. We have extended the Author-Topic model based on Latent Dirichlet Allocation for identify patterns of similarities in keywords and dates distributed across blog documents. From this model, we visualized the content and date similarities using the Isomap dimensionality reduction technique. Our findings support the theory that our probabilistic blog model can present the blogosphere in terms of topics with measurable keywords, hence aiding the investigative processes to understand and respond to critical cyber security events and threats.

12.1 Introduction

Organizations and governments are becoming vulnerable to a wide variety of security breaches against their information infrastructure. The severity of this threat is evident from the increasing rate of cyber attacks against computers and critical infrastructure. According to Sophos latest report, one new infected webpage is discovered every 14 seconds, or 6,000 a day [17].

As the number of cyber attacks by persons and malicious software are increasing rapidly, the number of incidents reported in blogs are also on the rise. Blogs are websites where entries are made in a reverse chronological order. Blogs may provide up-to-date information on the prevalence and distribution of various security incidents and threats.

Blogs range in scope from individual diaries to arms of political campaigns, media programs, and corporations. Blogs' explosive growth is generating large vol-

---

Flora S. Tsai, Kap Luk Chan
Nanyang Technological University, Singapore, e-mail: `fst1@columbia.edu,eklchan@ntu.edu.sg`

umes of raw data and is considered by many industry watchers one of the top ten trends. Blogosphere is the collective term encompassing all blogs as a community or social network. Because of the huge volume of existing blog posts and their free format nature, the information in the blogosphere is rather random and chaotic, but immensely valuable in the right context. Blogs can thus potentially contain usable and measurable information related to security threats, such as malware, viruses, cyber blackmail, and other cyber crime.

With the amazing growth of blogs on the web, the blogosphere affects much in the media. Studies on the blogosphere include measuring the influence of the blogosphere [6], analyzing the blog threads for discovering the important bloggers [13], determining the spatiotemporal theme pattern on blogs [12], focusing the topic-centric view of the blogosphere [1], detecting the blogs growing trends [7], tracking the propagation of discussion topics in the blogosphere [8],searching and detecting topics in business blogs [3], and determining latent friends of bloggers [16].

Existing studies have also focused on analyzing forums, news articles, and police databases for cyber threats [14, 21–23], but few have looked at blogs. In this paper, we focus on analyzing security blogs, which are blogs providing commentary or analysis of security threats and incidents.

In this paper, we propose blog data mining techniques for evaluating security threats related to the detection of cyber attacks, cyber crime, and information security. Existing studies on intelligence analysis have focused on analyzing news or forums for security incidents, but few have looked at blogs. We use probabilistic methods based on Latent Dirichlet Allocation to detect keywords from security blogs with respect to certain topics. We then demonstrate how this method can present the blogosphere in terms of topics with measurable keywords, hence tracking popular conversations and topics in the blogosphere. By applying a probabilistic approach, we can improve information retrieval in blog search and keywords detection, and provide an analytical foundation for the future of security intelligence analysis of blogs.

The paper is organized as follows. Section 2 reviews the related work on intelligence analysis and extraction of useful information from blogs. Section 3 defines the attributes of blog documents, and describes the probabilistic techniques based on Latent Dirichlet Allocation [2], Author-Topic model [18], and Isomap algorithm [19] for mining and visualization of blog-related topics. Section 4 presents experimental results, and Section 5 concludes the paper.

12.2 Review of Related Work

This section reviews related work in developing security intelligence analysis and extraction of useful information from blogs.

12.2.1 Intelligence Analysis

Intelligence analysis is the process of producing formal descriptions of situations and entities of strategic importance [20]. Although its practice is found in its purest form inside intelligence agencies, such as the CIA in the United States or MI6 in the UK, its methods are also applicable in fields such as business intelligence or competitive intelligence.

Recent work related to security intelligence analysis include using entity recognizers to extract names of people, organizations, and locations from news articles, and applying probabilistic topic models to learn the latent structure behind the named entities and other words [14]. Another study analyzed the evolution of terror attack incidents from online news articles using techniques related to temporal and event relationship mining [22]. In addition, Support Vector Machines were used for improving document classification for the insider threat problem within the intelligence community by analyzing a collection of documents from the Center for Nonproliferation Studies (CNS) related to weapons of mass destruction [23]. Another study analyzed the criminal incident reporting mainframe system (RAMS) data set used by the police department in Richmond, VA to analyze and predict the spatial behavior of criminals and latent decision makers [21]. These studies illustrate the growing need for security intelligence analysis, and the usage of machine learning and information retrieval techniques to provide such analysis. However, much work has yet to be done in obtaining intelligence information from the vast collection of blogs that exist throughout the world.

12.2.2 Information Extraction from Blogs

Current blog text analysis focuses on extracting useful information from blog entry collections, and determining certain trends in the blogosphere. NLP (Natural Language Processing) algorithms have been used to determine the most important keywords and proper names within a certain time period from thousands of active blogs, which can automatically discover trends across blogs, as well as detect key persons, phrases and paragraphs [7]. A study on the propagation of discussion topics through the social network in the blogosphere developed algorithms to detect the long-term and short-term topics and keywords, which were then validated with real blog entry collections [8]. On evaluating the suitable methods of ranking term significance in an evolving RSS feed corpus, three statistical feature selection methods were implemented: χ^2, Mutual Information (*MI*) and Information Gain (*I*), and the conclusion was that χ^2 method seems to be the best among all, but full human classification exercise would be required to further evaluate such method [15]. A probabilistic approach based on PLSA was proposed in [12] to extract common themes from blogs, and also generate the theme life cycle for each given location and the theme snapshots for each given time period. PLSA has also been previously used for blog search and mining of business blogs [3]. Latent Dirichlet Allocation

(LDA) [2] was used for identifying latent friends, or people who share similar topic distribution in their blogs [16].

12.3 Probabilistic Techniques for Blog Data Mining

This section summarizes the attributes of blog documents that distinguish them from other types of documents such as Web documents. The multiple dimensions of blogs provide a rich medium from which to perform blog data mining. The technique of Latent Dirichlet Allocation (LDA) extended for blog data mining is described. We propose a Date-Topic model based on the Author-Topic model for LDA that was used to analyze the blog dates in our dataset. Visualization is performed with the aid of the Isomap dimensionality reduction technique, which allows the content and date similarities to be easily visualized.

12.3.1 Attributes of Blog Documents

A blog document is structured differently from a typical Web document. Table 12.1 provides a comparison of facets of blog and Web documents. URL stands for the Uniform Resource Locator, that describes the Web address from which a document can be found. A permalink is specific to blogs, and is a URL that points to a specific blog entry after the entry has passed from the front page into the blog archives. Outlinks are documents that are linked from the blog or Web document. Tags are labels that people use to make it easier to find blog posts, photos and videos that are related. One important distinction in blog documents that makes them very different from Web documents are the time and date components.

Table 12.1 Comparison of blog and Web documents

Components	Blog	Web
title	√	√
content	√	√
tags	√	
author	√	
URL	√	√
permalink	√	
outlinks	√	√
time	√	
date	√	

If we consider the different facets of blogs, we can group general blog data analysis into five main attributes (blog content, tags, authors, time, and links), shown in Table 12.2. Each of the attributes itself can be multidimensional.

Table 12.2 Blog attributes

Attributes	Blog Components
Content	title and content
Tags	tags
Author	author or poster
Links	URL, permalink, outlinks
Time	date and time

Another attribute that is not directly present in blogs, but can be extracted from the content or author information, is the blog location, or the geographic location of the blog author. In addition, many blog posts have optional comments for users to add feedback to the blog. Although not part of the original post, comments can provide additional insight into the opinions related to the blog post.

Due to the complexity of analyzing the multidimensional characteristics of blogs, many previous analysis techniques analyze only one or two attributes of the blog data.

12.3.2 Latent Dirichlet Allocation

Latent Dirichlet Allocation (LDA) [2] is a probabilistic technique which models text documents as mixtures of latent topics, where topics correspond to key concepts presented in the corpus. LDA is not as prone to overfitting, and is preferred to traditional methods based on Latent Semantic Analysis (LSA) [5]. For example, in Probabilistic Latent Semantic Analysis (PLSA) [11], the number of parameters grows with the number of training documents, which makes the model susceptible to overfitting.

In LDA, the topic mixture is drawn from a conjugate Dirichlet prior that is the same for all documents, as opposed to PLSA, where the topic mixture is conditioned on each document. In LDA, the steps adapted for blog documents are summarized below:

1. Choose a multinomial distribution ϕ_z for each topic z from a Dirichlet distribution with parameter β.
2. For each blog document b, choose a multinomial distribution θ_b from a Dirichlet distribution with parameter α.
3. For each word token w in blog b, choose a topic t from θ_b.
4. Choose a word w from ϕ_t.

The probability of generating a corpus is thus equivalent to:

$$\iint \prod_{t=1}^{K} P(\phi_t|\beta) \prod_{b=1}^{N} P(\theta_b|\alpha) \left(\prod_{i=1}^{N_b} \sum_{t_i=1}^{K} P(t_i|\theta) P(w_i|t,\phi) \right) d\theta d\phi \qquad (12.1)$$

An extension of LDA to probabilistic Author-Topic (AT) modeling [18] is proposed for the blog author and topic visualization. The AT model is based on Gibbs sampling, a Markov chain Monte Carlo technique, where each author is represented by a probability distribution over topics, and each topic is represented as a probability distribution over terms for that topic [18].

We have extended the AT model for visualization of blog dates. For the Date-Topic (DT) model, each date is represented by a probability distribution over topics, and each topic represented by a probability distribution over terms for that topic.

For the DT model, the probability of generating a blog is given by:

$$\prod_{i=1}^{N_b} \frac{1}{D_b} \sum_d \sum_{t=1}^{K} \phi_{w_i t} \theta_{td} \qquad (12.2)$$

where blog b has D_b dates. The probability is then integrated over ϕ and θ and their Dirichlet distributions and sampled using Markov Chain Monte Carlo methods.

The similarity matrices for dates can then be calculated using the symmetrized Kullback Leibler (KL) distance [10] between topic distributions, which is able to measure the difference between two probability distributions. The symmetric KL distance of two probability distributions P and Q is calculated as:

$$\frac{KL(P,Q) + KL(Q,P)}{2} \qquad (12.3)$$

where KL is the KL distance given by:

$$KL(P,Q) = \sum (P \log(P/Q)); \qquad (12.4)$$

The similarity matrices can be visualized using the Isomap dimensionality reduction technique described in the following section.

12.3.3 Isometric Feature Mapping (Isomap)

Isomap [19] is a nonlinear dimensionality reduction technique that uses Multidimensional Scaling (MDS) [4] techniques with geodesic interpoint distances instead of Euclidean distances. Geodesic distances represent the shortest paths along the curved surface of the manifold. Unlike the linear techniques, Isomap can discover the nonlinear degrees of freedom that underlie complex natural observations [19].

Isomap deals with finite data sets of points in $\mathbb{R}^n$ which are assumed to lie on a smooth submanifold M_d of low dimension $d < n$. The algorithm attempts to recover M given only the data points. Isomap estimates the unknown geodesic distance in M between data points in terms of the graph distance with respect to some graph G constructed on the data points.

Isomap algorithm consists of three basic steps:

1. Find the nearest neighbors on the manifold M, based on the distances between pairs of points in the input space.
2. Approximate the geodesic distances between all pairs of points on the manifold M by computing their shortest path distances in the graph G.
3. Apply MDS to matrix of graph distances, constructing an embedding of the data in a d-dimensional Euclidean space Y that best preserves the manifold's estimated intrinsic geometry [19].

If two points appear on a nonlinear manifold, their Euclidean distance in the high-dimensional input space may not accurately reflect their intrinsic similarity. The geodesic distance along the low-dimensional manifold is thus a better representation for these points. The neighborhood graph G constructed in the first step of allows an estimation of the true geodesic path to be computed efficiently in step two, as the shortest path in G. The two-dimensional embedding recovered by Isomap in step three, which best preserves the shortest path distances in the neighborhood graph. The embedding now represents simpler and cleaner approximations to the true geodesic paths than do the corresponding graph paths [19].

Isomap is a very useful noniterative, polynomial-time algorithm for nonlinear dimensionality reduction if the data is severely nonlinear. Isomap is able to compute a globally optimal solution, and for a certain class of data manifolds (Swiss roll), is guaranteed to converge asymptotically to the true structure [19]. However, Isomap may not easily handle more complex domains such as non-trivial curvature or topology.

12.4 Experiments and Results

We used probabilistic models for blog data mining on our dataset. Dimensionality reduction was performed with Isomap to show the similarity plot of blog content and dates. We extract the most relevant categories and show the topics extracted for each category. Experiments show that the probabilistic model can reveal interesting patterns in the underlying topics for our dataset of security-related blogs.

12.4.1 Data Corpus

For our experiments, we extracted a subset of the Nielson BuzzMetrics blog data corpus[1] that focuses on blogs related to security threats and incidents related to cyber crime and computer viruses. The original dataset consists of 14 million blog posts collected by Nielsen BuzzMetrics for May 2006. Although the blog entries span only a short period of time, they are indicative of the amount and variety of blog posts that exists in different languages throughout the world.

---

[1] http://www.icwsm.org/data.html

Blog entries in the English language related to security threats such as malware, cyber crime, computer virus, encryption, and information security were extracted and stored for use in our analysis. Figure 12.1 shows an excerpt of a blog post related to a security glitch found on voting machines.

Elections officials ... are scrambling to understand and limit the risk from a "dangerous" security hole found in ... touch-screen voting machines. ... Armed with a little basic knowledge of Diebold voting systems ... someone ... could load virtually any software into the machine and disable it, redistribute votes or alter its performance...

Fig. 12.1 Excerpt of blog post related to security glitch in voting machines.

The prevalence of articles and blogs on this matter has led to many proposed legislation reforms regarding electronic voting machines [9]. Thus, security incidents reported in the blogosphere and other online media can greatly effect traditional media and legislation.

There are a total of 2102 entries in our dataset, and each blog entry is saved as a text file for further text preprocessing. For the preprocessing of the blog data, HTML tags were removed, lexical analysis was performed by removing stopwords, stemming, and pruning by the Text to Matrix Generator (TMG) [24] prior to generating the term-document matrix. The total number of terms after pruning and stopword removal is 6169. The term-document matrix was then input to the LDA algorithm.

12.4.2 Results for Blog Topic Analysis

We conducted some experiments using LDA for the blog entries. The parameters used in our experiments are number of topics (10) and number of iterations (1000). We used symmetric Dirichlet priors in the LDA estimation with $\alpha = 50/K$ and $\beta = 0.01$, which are common settings in the literature.

Tables 12.3-12.8 summarizes the keywords found for each of the top six topics.

By looking at the various topics listed, we are able to see that the probabilistic approach is able to list important keywords of each topic in a quantitative fashion. The keywords listed can relate back to the original topics. For example, the keywords detected in the Topic 2 include "malwar", "worm", "threat", and "terror". All of these types are related to the general category of computer malware.

For Topic 5, the keywords such as "vote", "machin", "elect", and "diebold" relate to blog posts about security glitches found on voting machines, as shown in the blog summary from Figure 12.1. The high probability of dates around May 12-13 indicate that many of the blog posts occurred during this period of time. These are examples of events that can trigger conversation in the blogosphere.

Automatic topic detection of security blogs such as those demonstrated above can have significant impact on the investigation and detection of cyber threats in the

Table 12.3 List of terms and dates for Topic 1.

Term	Probability
comput	0.02956
file	0.01451
click	0.01082
search	0.01063
inform	0.00922
page	0.00922
phone	0.00901
track	0.00846
data	0.00813
record	0.00777

Date	Probability
20060513	0.10077
20060512	0.09032
20060503	0.07942
20060505	0.07665
20060516	0.07130
20060502	0.05263
20060507	0.04979
20060514	0.04776
20060523	0.04776
20060504	0.04275

Table 12.4 List of terms and dates for Topic 2.

Term	Probability
browser	0.01660
user	0.01444
secur	0.01277
cyber	0.01194
worm	0.01189
comput	0.01130
instal	0.01097
terror	0.01084
malwar	0.01079
threat	0.01073

Date	Probability
20060523	0.39632
20060522	0.17909
20060524	0.08151
20060521	0.04653
20060519	0.04547
20060512	0.03673
20060505	0.03573
20060513	0.02941
20060504	0.02635
20060515	0.01843

Table 12.5 List of terms and dates for Topic 3

Term	Probability
window	0.01766
encrypt	0.01297
work	0.01182
secur	0.01120
kei	0.01115
network	0.01109
run	0.00955
system	0.00892
server	0.00869
support	0.00751

Date	Probability
20060502	0.10822
20060504	0.08676
20060507	0.08144
20060512	0.07670
20060518	0.07605
20060503	0.07308
20060505	0.07183
20060514	0.06587
20060524	0.06260
20060519	0.04735

Table 12.6 List of terms and dates for Topic 4.

Term	Probability
privaci	0.01601
servic	0.00971
spywar	0.00924
data	0.00896
inform	0.00849
law	0.00847
part	0.00837
time	0.00792
right	0.00774
power	0.00759

Date	Probability
20060519	0.10077
20060521	0.09032
20060513	0.07942
20060505	0.07665
20060512	0.07130
20060518	0.05263
20060522	0.04979
20060524	0.04776
20060523	0.04776
20060504	0.04275

blogosphere. A high incidence of occurrence of a particular topic or keyword can alert the user of potential new threats and security risks, which can then be further

Table 12.7 List of terms and dates for Topic 5.

Term	Probability
vote	0.02172
machin	0.01619
elect	0.01082
state	0.01257
call	0.01155
diebold	0.01106
bush	0.00927
system	0.00912
secur	0.00873
nsa	0.00870

Date	Probability
20060512	0.44553
20060513	0.20167
20060515	0.07230
20060511	0.05536
20060514	0.03944
20060505	0.03734
20060523	0.02488
20060519	0.02310
20060518	0.01288
20060504	0.01280

Table 12.8 List of terms and dates for Topic 6.

Term	Probability
secur	0.02157
softwar	0.01155
peopl	0.01051
compani	0.00991
comput	0.00986
year	0.00982
make	0.00963
technolog	0.00927
system	0.00896
site	0.00861

Date	Probability
20060512	0.10077
20060505	0.09032
20060513	0.07942
20060519	0.07665
20060524	0.07130
20060522	0.05263
20060504	0.04979
20060502	0.04776
20060523	0.04776
20060521	0.04275

analyzed. In addition, the system be altered to detect a higher number of topics; thus, increasing the granularity of cyber threat analysis.

12.4.3 Blog Content Visualization

In order to prepare the dataset, we first created a normalized 6169 × 2102 term-document matrix with term frequency (TF) local term weighting and inverse document frequency (IDF) global term weighting, based on the content of the blogs. From this matrix, we created the 2102 × 2102 document-document cosine similarity matrix, and used this as input to the dimensionality reduction algorithms. The results can be seen in Figure 12.2.

This plot is useful to see the similarities between the blog documents, and can be augmented with metadata such as blog tags or categories to visualize the distinction among blog tags. As our dataset does not contain the tags or labels, the plot is not able to show the distinction of the tags as yet.

12 Blog Data Mining for Cyber Security Threats

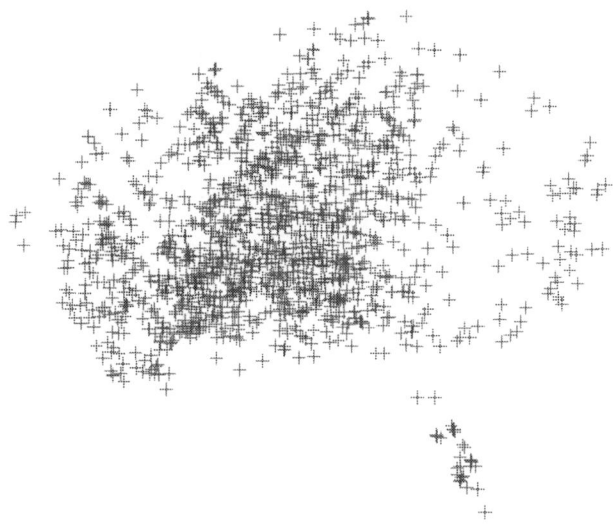

Fig. 12.2 Results on visualization of blog content using Isomap ($k=12$).

12.4.4 Blog Time Visualization

The dataset contained blogs from May 1-24, 2006. The date-document matrix, along with the term-document matrix, were used to compute the date-topic model. In this model, each date is represented by a probability distribution over topics, and each topic is represented as a probability distribution over terms for that topic. The topic-term and date-topic distributions were then learned from the blog data in an unsupervised manner.

For visualizing the date similarities, the symmetrized Kullback Leibler distance between topic distributions was calculated for each date pair. Figure 12.3 shows the 2D plot of the date distributions based on the date-topic distributions. In the plot, the dates were scaled according to the number of blogs in that date. The distances between the dates are proportional to the similarity between dates, based on the topic distributions of the blogs that were posted.

Viewing the date similarities in this way can complement existing analysis such as time-series analysis to provide a more complete picture of the blog time evolution.

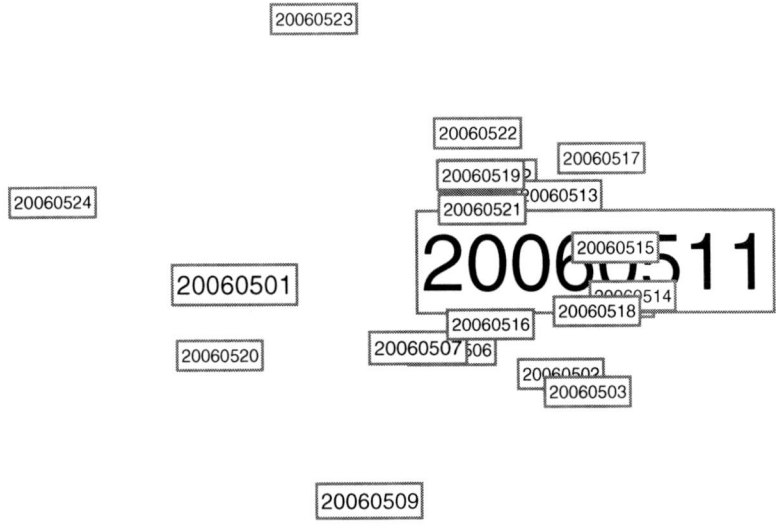

Fig. 12.3 Results on visualization of date similarities using Isomap.

They can also be further subdivided by topic to form a better understanding of topic-time relationships.

12.5 Conclusions

The rapid proliferation of blogs in recent years presents a vast new medium in which to analyze and detect potential cyber security threats in the blogosphere. In this article, we proposed blog data mining techniques for analyzing blog posts for various categories of cyber threats related to the detection of security threats, cyber crime, and information security. The important contribution of this article is the use of probabilistic and dimensionality reduction techniques for identifying and visualizing patterns of similarities in keywords and dates distributed across all the documents in the our dataset of security-related blogs. These techniques can aid the investigative processes to understand and respond to critical cyber security events and threats. Other research contributions include a proposed probabilistic model for topic detection in blogs and the demonstration of our methods for detecting cyber threats in security blogs.

Our experiments on our dataset of blogs demonstrate how our probabilistic blog model can present the blogosphere in terms of topics with measurable keywords, hence tracking popular conversations and topics in the blogosphere. By using probabilistic models, we can improve information mining in blog keywords detection, and provide an analytical foundation for the future of security analysis of blogs.

Future applications of this stream of research may include automatically monitoring and identifying trends in cyber security threats that are present in blogs. The system should be able to achieve real-time detection of potential cyber threats by updating the analysis upon the posting of new blog entries. This can be achieved by applying techniques such as folding-in for automatic updating of new blog documents without recomputing the entire matrix. Thus, the resulting system can become an important tool for government and intelligence agencies in decision making and monitoring of real-time potential international terror threats present in blog conversations and the blogosphere.

References

1. P. Avesani, M. Cova, C. Hayes, P. Massa, Learning Contextualised Weblog Topics, Proceedings of the WWW '05 Workshop on the Weblogging Ecosystem: Aggregation, Analysis and Dynamics, 2005.
2. D.M. Blei, A.Y. Ng, and M.I. Jordan, "Latent dirichlet allocation," *Journal of Machine Learning Research*, vol. 3, pp. 993-1022, 2003.
3. Y. Chen, F.S. Tsai, K.L. Chan, Machine Learning Techniques for Business Blog Search and Mining, *Expert Systems With Applications* 35(3), pp 581-590, 2008.
4. T. Cox and M. Cox, *Multidimensional Scaling*. Second Edition, New York: Chapman & Hall, 2001.
5. S. Deerwester, S. Dumais, T. Landauer, G. Furnas, R. Harshman, Indexing by latent semantic analysis, Journal of the American Society of Information Science 41(6) (1990) 391–407.
6. K.E. Gill, How Can We Measure the Influence of the Blogosphere? Proceedings of the WWW '04 Workshop on the Weblogging Ecosystem: Aggregation, Analysis and Dynamics, 2004.
7. N.S. Glance, M. Hurst, T. Tomokiyo, BlogPulse: Automated Trend Discovery for Weblogs, Proceedings of the WWW '04 Workshop on the Weblogging Ecosystem: Aggregation, Analysis and Dynamics, 2004.
8. D. Gruhl, R. Guha, D. Liben-Nowell, A. Tomkins, Information Diffusion Through Blogspace, Proceedings of the WWW '04 Workshop on the Weblogging Ecosystem: Aggregation, Analysis and Dynamics, 2004.
9. M. Hickins, Congress Lights Fire Under Vote Systems Agency, Business, www.internetnews.com/bus-news/article.php/3655001, 2007.
10. D.H. Johnson and S. Sinanovic, "Symmetrizing the Kullback-Leibler distance," *Technical Report, Rice University.*, 2001.
11. T. Hofmann, Unsupervised Learning by Probabilistic Latent Semantic Analysis, Machine Learning Journal 42(1) (2001) 177–196.
12. Q. Mei, C. Liu, H. Su, C. Zhai, A Probabilistic Approach to Spatiotemporal Theme Pattern Mining on Weblogs, Proceedings of the WWW '06 Workshop on the Weblogging Ecosystem: Aggregation, Analysis and Dynamics, 2006.

13. S. Nakajima, J. Tatemura, Y. Hino, Y. Hara, K. Tanaka, Discovering Important Bloggers based on Analyzing Blog Threads, Proceedings of the WWW '05 Workshop on the Weblogging Ecosystem: Aggregation, Analysis and Dynamics, 2005.
14. D. Newman, C. Chemudugunta, P. Smyth, M. Steyvers, Analyzing Entities and Topics in News Articles Using Statistical Topic Models, Proceedings of the IEEE International Conference on Intelligence and Security Informatics (ISI), 2006.
15. R. Prabowo, M. Thelwall, A Comparison of Feature Selection Methods for an Evolving RSS Feed Corpus, Information Processing and Management 42(6) (2006) 1491–1512.
16. D. Shen, J.-T. Sun, Q. Yang, , Z. Chen, Latent Friend Mining from Blog Data, Proceedings of the IEEE International Conference on Data Mining (ICDM), 2006.
17. Sophos security threat report, http://www.sophos.com/security/whitepapers/, 2008.
18. M. Steyvers, P. Smyth, M. Rosen-Zvi, and T. Griffiths, "Probabilistic Author-Topic Models for Information Discovery," *SIGKDD International Conference on Knowledge Discovery and Data Mining*, 2004.
19. J. Tenenbaum, V. de Silva, and J. Langford, "A Global Geometric Framework for Nonlinear Dimensionality Reduction," *Science*, vol. 290, pp. 2319-2323, Dec. 2000.
20. Wikipedia contributors, Intelligence Analysis, Wikipedia, The Free Encyclopedia, http://en.wikipedia.org/wiki/Intelligence_analysis, 2006.
21. Y. Xue, D.E. Brown, Spatial analysis with preference specification of latent decision makers for criminal event prediction, Decision Support Systems 41(3), (2006) 560–573.
22. C.C. Yang, X. Shi, C.-P. Wei, Tracing the Event Evolution of Terror Attacks from On-Line News, Proceedings of the IEEE International Conference on Intelligence and Security Informatics (ISI), 2006.
23. O. Yilmazel, S. Symonenko, N. Balasubramanian, E.D. Liddy, Leveraging One-Class SVM and Semantic Analysis to Detect Anomalous Content, Proceedings of the IEEE International Conference on Intelligence and Security Informatics (ISI), 2005.
24. D. Zeimpekis, E. Gallopoulos, TMG: A MATLAB Toolbox for generating term-document matrices from text collections, Proceedings of Grouping Multidimensional Data: Recent Advances in Clustering, 2006.

Chapter 13
Blog Data Mining: The Predictive Power of Sentiments

Yang Liu, Xiaohui Yu, Xiangji Huang, and Aijun An

Abstract In this chapter, we study the problem of mining sentiment information from online resources and investigate ways to use such information to predict product sales performance. In particular, we conduct an empirical study on using the sentiment information mined from blogs to predict movie box office performance. We propose Sentiment PLSA (S-PLSA), in which a blog entry is viewed as a document generated by a number of hidden sentiment factors. Training an S-PLSA model on the blog data enables us to obtain a succinct summary of the sentiment information embedded in the blogs. We then present ARSA, an autoregressive sentiment-aware model, to utilize the sentiment information captured by S-PLSA for predicting product sales performance. Extensive experiments were conducted on the movie data set. Experiments confirm the effectiveness and superiority of the proposed approach.

13.1 Introduction

Recent years have seen an emergence of blogs as an important medium for information sharing and dissemination. Since many bloggers choose to express their opinions online, blogs serve as an excellent indicator of public sentiments and opinions.

This chapter is concerned with the predictive power of opinions and sentiments expressed in online resources (blogs in particular). We focus on the blogs that contain reviews on products. Since what the general public thinks of a product can no doubt influence how well it sells, understanding the opinions and sentiments ex-

---

Yang Liu, Aijun An
Department of Computer Science and Engineering, York University, Toronto, ON, Canada M3J 1P3, e-mail: {yliu,ann}@cse.yorku.ca

Xiaohui Yu, Xiangji Huang
School of Information Technology, York University, Toronto, ON, Canada M3J 1P3, e-mail: {xhyu,jhuang}@yorku.ca

pressed in the relevant blogs is of high importance, because these blogs can be a very good indicator of the product's future sales performance. In this chapter, we are concerned with developing models and algorithms that can mine opinions and sentiments from blogs and use them for predicting product sales. Properly utilized, such models and algorithms can be quite helpful in various aspects of business intelligence, ranging from market analysis to product planning and targeted advertising.

As a case study, in this chapter, we investigate how to predict box office revenues using the sentiment information obtained from blog mentions. The choice of using movies rather than other products in our study is mainly due to data availability, in that the daily box office revenue data are all published on the Web and readily available, unlike other product sales data which are often private to their respective companies due to obvious reasons. Also, as discussed by Liu et al. [9], analyzing movie reviews is one of the most challenging tasks in sentiment mining. We expect the models and algorithms developed for box office prediction to be easily adapted to handle other types of products that are subject to online discussions, such as books, music CDs and electronics.

Prior studies on the predictive power of blogs have used the volume of blogs or link structures to predict the trend of product sales [4, 5], failing to consider the effect of the sentiments present in the blogs. It has been reported [4,5] that although there seems to exist strong correlation between the blog mentions and sales spikes, using the volume or the link structures alone do not provide satisfactory prediction performance. Indeed, as we will illustrate with an example, the sentiments expressed in the blogs are more predictive than volumes.

Mining opinions and sentiments from blogs, which is necessary for predicting future product sales, presents unique challenges that can not be easily addressed by conventional text mining methods. Therefore, simply classifying blog reviews as positive and negative, as most current sentiment-mining approaches are designed for, does not provide a comprehensive understanding of the sentiments reflected in the blog reviews. In order to model the multifaceted nature of sentiments, we view the sentiments embedded in blogs as an outcome of the joint contribution of a number of hidden factors, and propose a novel approach to sentiment mining based on Probabilistic Latent Semantic Analysis (PLSA), which we call Sentiment PLSA (S-PLSA). Different from the traditional PLSA [6], S-PLSA focuses on sentiments rather than topics. Therefore, instead of taking a vanilla "bag of words" approach and considering all the words (modulo stop words) present in the blogs, we focus primarily on the words that are sentiment-related. To this end, we adopt in our study the appraisal words extracted from the lexicon constructed by Whitelaw et al. [14]. Despite the seemingly lower word coverage (compared to using "bag of words"), decent performance has been reported when using appraisal words in sentiment classification [14]. In S-PLSA, appraisal words are exploited to compose the feature vectors for blogs, which are then used to infer the hidden sentiment factors.

Aside from the S-PLSA model which extracts the sentiments from blogs for predicting future product sales, we also consider the past sale performance of the same product as another important factor in predicting the product's future sales performance. We capture this effect through the use of an autoregressive (AR) model , which has been widely used in many time series analysis problems, including stock price prediction [3]. Combining this AR model with sentiment information mined from the blogs, we propose a new model for product sales prediction called the Autoregressive Sentiment Aware (ARSA) model. Extensive experiments on the movie dataset has shown that the ARSA model provides superior predication performance compared to using the AR model alone, confirming our expectation that sentiments play an important role in predicting future sales performance.

In summary, we make the following contributions.

- We are the first to model sentiments in blogs as the joint outcome of some hidden factors, answering the call for a model that can handle the complex nature of sentiments. We propose the S-PLSA model, which through the use of appraisal groups, provides a probabilistic framework to analyze sentiments in blogs.
- We propose the Autoregressive Sentiment Aware (ARSA) model for product sales prediction, which reflects the effects of both sentiments and past sales performance on future sales performance. Its effectiveness is confirmed by experiments.

The rest of the chapter is organized as follows. Section 13.2 provides a brief review of related work. In Section 13.3, we discuss the characteristics of online discussions and specifically, blogs, which motivate the proposal of S-PLSA in Section 13.4. In Section 13.5, we propose ARSA, the sentiment-aware model for predicting future product sales. Section 13.6 reports on the experimental results. We conclude the chapter in Section 13.7.

13.2 Related Work

Most existing work on sentiment mining (sometimes also under the umbrella of opinion mining) focuses on determining the semantic orientations of documents. Among them, some of the studies attempt to learn a positive/negative classifier at the document level. Pang et al. [12] employ three machine learning approaches (Naive Bayes, Maximum Entropy, and Support Vector Machine) to label the polarity of IMDB movie reviews. In a follow-up work, they propose to firstly extract the subjective portion of text with a graph min-cut algorithm, and then feed them into the sentiment classifier [10]. Instead of applying the straightforward frequency-based bag-of-words feature selection methods, Whitelaw et al. [14] defined the concept of "adjectival appraisal groups" headed by an appraising adjective and optionally modified by words like "not" or "very". Each appraisal group was further assigned four type of features: attitude, orientation, graduation, and polarity. They report good classification accuracy using the appraisal groups. They also show that when combined with standard "bag-of-words" features, the classification accuracy can be further boosted. We use the same words and phrases from the appraisal words to compute the blogs' feature vectors, as we also believe that such adjectival appraisal words play a vital role in sentiment mining and need to be distinguished from other words. However, as will become evident in Section 13.4, our way of using these appraisal groups is different from that in [14].

There are also studies that work at a finer level and use words as the classification subject. They classify words into two groups, "good" and "bad", and then use certain functions to estimate the overall "goodness" or "badness" score for the documents. Kamps et al. [8] propose to evaluate the semantic distance from a word to good/bad with WordNet. Turney [13] measures the strength of sentiment by the difference of the mutual information (PMI) between the given phrase and "excellent" and the PMI between the given phrase and "poor".

Pushing further from the explicit two-class classification problem, Pang et al. [11] and Zhang [15] attempt to determine the author's opinion with different rating scales (i.e., the number of stars). Liu et al. [9] build a framework to compare consumer opinions of competing products using multiple feature dimensions. After deducting supervised rules from product reviews, the strength and weakness of the product are visualized with an "Opinion Observer".

Our method departs from classic sentiment classification in that we assume that sentiment consists of multiple hidden aspects, and use a probability model to quantitatively measure the relationship between sentiment aspects and blogs, as well as sentiment aspects and words.

13.3 Characteristics of Online Discussions

To make a better understanding of the characteristics of online discussions and their predictive power, we investigate the pattern of blog mentions and its relationship to sales data by examining a real example from the movie sector.

13.3.1 Blog Mentions

Let us look at the following two movies, *The Da Vinci Code* and *Over the Hedge*, which are both released on May 19, 2006. We use the name of each movie as a query to a publicly available blog search engine(http://blog-search.google.com). In addition, as each blog is always associated with a fixed time stamp, we augment the query input with a date for which we would like to collect the data. For each movie, by issuing a separate query for each single day in the period starting from one week before the movie release till three weeks after the release, we chronologically collect a set of blogs appearing in a span of one month. We use the number of returned results for a particular date as a rough estimate of the number of blog mentions published on that day.

In Figure 13.1 (a), we compare the changes in the number of blog mentions of the two movies. Apparently, there exists a spike in the number of blog mentions for the movie *The Da Vinci Code*, which indicates that a large volume of discussions on that movie appeared around its release date. In addition, the number of blog mentions are significantly larger than those for *Over the Hedge* throughout the whole month.

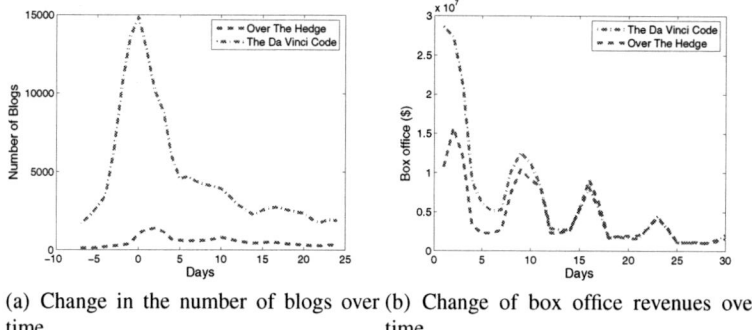

(a) Change in the number of blogs over time
(b) Change of box office revenues over time

Fig. 13.1 An Example

13.3.2 Box Office Data and User Rating

Besides the blogs, we also collect for each movie one month's box office data (daily gross revenue) from the IMDB website (http://www.imdb.com/). The changes in daily gross revenues are depicted in Figure 13.1 (b). Apparently, the daily gross of *The Da Vinci Code* is much greater than *Over the Hedge* on the release date. However, the difference in the gross revenues between the two movies becomes less and less as time goes by, with *Over the Hedge* sometimes even scoring higher towards the end of the one-month period. To shed some light on this phenomenon, we collect the average user ratings of the two movies from the IMDB website. *The Da Vinci Code* and *Over the Hedge* got the rating of 6.5 and 7.1 respectively.

13.3.3 Discussion

It is interesting to observe from Figure 13.1 that although *The Da Vinci Code* has a much higher number of blog mentions than *Over the Hedge*, its box office revenue are on par with that of *Over the Hedge* save the opening week. This implies that the number of blog mentions may not be an accurate indicator of a product's sales performance. A product can attract a lot of attention (thus a large number of blog mentions) due to various reasons, such as aggressive marketing, unique features, or being controversial. This may boost the product's performance for a short period of time. But as time goes by, it is the quality of the product and how people feel about it that dominates. This can partly explain why in the opening week, *The Da Vinci*

Code had a large number of blog mentions and staged an outstanding box office performance, but in the remaining weeks, its box office performance fell to the same level as that for *Over the Hedge*. On the other hand, people's opinions (as reflected by the user ratings) seem to be a good indicator of how the box office performance evolves. Observe that, in our example, the average user rating for *Over the Hedge* is higher than that for *The Da Vinci Code*; at the same time, it enjoys a slower rate of decline in box office revenues than the latter. This suggests that sentiments in the blogs could be a very good indicator of a product's future sales performance.

13.4 S-PLSA: A Probabilistic Approach to Sentiment Mining

In this section, we propose a probabilistic approach to analyzing sentiments in the blogs, which will serve as the basis for predicting sales performance.

13.4.1 Feature Selection

We first consider the problem of feature selection , i.e., how to represent a given blog as an input to the mining algorithms. The traditional way to do this is to compute the (relative) frequencies of various words in a given blog and use the resulting multidimensional feature vector as the representation of the blog. Here we follow the same methodology. But instead of using the frequencies of all the words appearing the blogs, we choose to focus on the set containing 2030 appraisal words extracted from the lexicon constructed by Whitelaw et al. [14], and use their frequencies in a blog as a feature vector. The rationale behind this is that for sentiment analysis, sentiment-oriented words, such as "good" or "bad", are more indicative than other words [14].

13.4.2 Sentiment PLSA

Mining sentiments present unique challenges that cannot be handled easily by traditional text mining algorithms. This is mainly because the sentiments are often expressed in subtle and complex ways. Moreover, sentiments are often multifaceted, and can differ from one another in a variety of ways, including polarity, orientation, graduation, and so on [14]. Therefore, it would be too simplistic to just classify the sentiments expressed in a blog as either positive or negative. For the purpose of sales prediction, a model that can extract the sentiments in a more accurate way is needed.

To this end, we propose a probabilistic model called Sentiment Probabilistic Latent Semantic Analysis (S-PLSA), in which a blog can be considered as being generated under the influence of a number of hidden sentiment factors. The use of hidden factors allows us to accommodate the intricate nature of sentiments, with each hidden factor focusing on one specific aspect of the sentiments. The use of a probabilistic generative model, on the other hand, enables us to deal with sentiment analysis in a principled way. In its traditional form, PLSA [6] assumes that there are a set of hidden semantic factors or *aspects* in the documents, and models the relationship among these factors, documents, and words under a probabilistic framework. With its high flexibility and solid statistical foundations, PLSA has been widely used in many areas, including information retrieval, Web usage mining, and collaborative filtering. Nonetheless, to the best of our knowledge, we are the first to model sentiments and opinions as a mixture of hidden factors and use PLSA for sentiment mining.

We now formally present S-PLSA. Suppose we are given a set of blog entries $\mathscr{B} = \{b_1, \ldots, b_N\}$, and a set of words (appraisal words) from a vocabulary $\mathscr{W} = \{w_1, \ldots, w_M\}$. The blog data can be described as a $N \times M$ matrix $D = (c(b_i, w_j))_{ij}$, where $c(b_i, w_j)$ is the number of times w_i appears in blog entry b_j. Each row in D is then a frequency vector that corresponds to a blog entry.

We consider the blog entries as being generated from a number of hidden sentiment factors, $\mathscr{Z} = \{z_1, \ldots, z_K\}$. We expect that those hidden factors would correspond to blogger's complex sentiments expressed in the blog review. S-PLSA can be considered as the following generative model.

1. Pick a blog document b from $\mathscr{B}$ with probability $P(b)$;
2. Choose a hidden sentiment factor z from $\mathscr{Z}$ with probability $P(z|b)$;
3. Choose a word from the set of appraisal words $\mathscr{W}$ with probability $P(w|z)$.

The end result of this generative process is a blog-word pair (b, w), with z being integrated out. The joint probability can be factored as follows: $P(b, w) = P(b)P(w|b)$, where $P(w|b) = \sum_{z \in \mathscr{Z}} P(w|z)P(z|b)$. Assuming that the blog entry b and the word w are conditionally independent given the hidden sentiment factor z, we can use Bayes rule to transform the joint probability to the following: $P(b, w) = \sum_{z \in \mathscr{Z}} P(z)P(b|z)P(w|z)$.

To explain the observed (b, w) pairs, we need to estimate the model parameters $P(z)$, $P(b|z)$, and $P(w|z)$. To this end, we seek to maximize the following likelihood function: $L(\mathscr{B}, \mathscr{W}) = \sum_{b \in \mathscr{B}} \sum_{w \in \mathscr{W}} c(b, w) \log P(b, w)$, where $c(b, w)$ represents the number of occurrences of a pair (b, w) in the data. This can be done using the Expectation-Maximization (EM) algorithm [2].

13.5 ARSA: A Sentiment-Aware Model

We now present a model to provide product sales predications based on the sentiment information captured from blogs. Due to the complex and dynamic nature of sentiment patterns expressed through on-line chatters, integrating such information is quite challenging. To the best of our knowledge, we are the first to consider using sentiment information for product sales prediction.

We focus on the case of predicting box office revenues to illustrate our methodologies. Our model aims to capture two different factors that can affect the box office revenue of the current day. One factor is the box office revenue of the preceding days. Naturally, the box office revenue of the current day is strongly correlated to those of the preceding days, and how a movie performs in previous days is a very good indicator of how it will perform in the days to come. The second factor we consider is the people's sentiments about the movie. The example in Section 13.3 shows that a movie's box office is closely related to what people think about the movie. Therefore, we would like to incorporate the sentiments mined from the blogs into the prediction model.

13.5.1 The Autoregressive Model

We start with a model that captures only the first factor described above and discuss how to incorporate the second factor into the model in the next subsection.

The temporal relationship between the box office revenues of the preceding days and the current day can be well modeled by an autoregressive (AR) process. Let us denote the box office revenue of the movie of interest at day t by x_t ($t = 1, 2, \ldots, N$ where $t = 1$ corresponds to the release date and $t = N$ corresponds to the last date we are interested in), and we use $\{x_t\}(t = 1, \ldots, N)$ to denote the time series $x_1, x_2, \ldots, x_N$. Our goal is to obtain an AR process that can model the time series $\{x_t\}$. A basic (but not quite appropriate, as discussed below) AR process of order p is as follows: $X_t = \sum_{i=1}^{p} \phi_i X_{t-i} + \varepsilon_t$, where $\phi_1, \phi_2, \ldots, \phi_p$ are the parameters of the model, and ε_t is an error term (white noise with zero mean).

Once this model is learned from training data, at day t, the box office revenue x_t can be predicted by $x_{t-1}, x_{t-2}, \ldots, x_{t-p}$. It is important to note, however, that AR models are only appropriate for time series that are stationary. Apparently, the time series $\{x_t\}$ are not, because there normally exist clear trends and "seasonalities" in the series. For instance, in example 13.3, there is a seemingly negative exponential downward trend for the box office revenues as the time moves further from the release date. "Seasonality" is also present, as within each week, the box office revenues always peak at the weekend and are generally lower during weekdays. Therefore, in order to properly model the time series $\{x_t\}$, some preprocessing steps are required.

The first step is to remove the trend. This is achieved by first transforming the time series $\{x_t\}$ into the logarithmic domain, and then differencing the resulting time series $\{x_t\}$. The new time series obtained is thus $x'_t = \Delta \log x_t = \log x_t - \log x_{t-1}$. We then proceed to remove the seasonality [3]. To this end, we apply the lag operator on $\{x'_t\}$ and obtain a new time series $\{y_t\}$ as follows: $y_t = x'_t - L^7 x't = x'_t - x'_{t-7}$.

By computing the difference between the box office revenue of a particular date and that of 7 days ago, we effectively removed the seasonality factor due to different days of a week. After the preprocessing step, a new AR model can be formed on the resulting time series $\{y_t\}$:

13 Blog Data Mining: The Predictive Power of Sentiments

$$y_t = \sum_{i=1}^{p} \phi_i y_{t-i} + \varepsilon_t. \tag{13.1}$$

It is worth noting that although the AR model developed here is specific for movies, the same methodologies can be applied in other contexts. For example, trends and seasonalities are present in the sales performance of many different products (such as electronics and music CDs). Therefore the preprocessing steps described above to remove them can be adapted and used in the predicting the sales performance.

13.5.2 Incorporating Sentiments

As discussed earlier, the box office revenues might be greatly influenced by people's opinions in the same time period. We modify the model in (13.1) to take this factor into account. Let $\mathscr{B}_t$ denote the set of blogs on the movie of interest that were posted on day t. The average probability of sentiment factor $z = j$ conditional on blogs in $\mathscr{B}_t$, is defined as $\omega_{t,j} = \frac{1}{|\mathscr{B}_t|} \sum_{b \in \mathscr{B}_t} p(z = j|b)$, where $p(z = j|b)(b \in \mathscr{B}_t)$ are obtained based a trained S-PLSA model. Intuitively, $\omega_{t,j}$ represents the average fraction of the sentiment "mass" that can be attributed to the hidden sentiment factor j. Then our new model, which we call the Autoregressive Sentiment-Aware (ARSA) model, can be formulated as follows.

$$y_t = \sum_{i=1}^{p} \phi_i y_{t-i} + \sum_{i=1}^{q} \sum_{j=1}^{K} \rho_{i,j} \omega_{t-i,j} + \varepsilon_t, \tag{13.2}$$

where p, q, and K are user-chosen parameters, while ϕ_i and $\rho_{i,j}$ are parameters whose values are to be estimated using the training data. Parameter q specifies the sentiment information from how many preceding days is taken into account, and K indicates the number of hidden sentiment factors used by S-PLSA to represent the sentiment information.

In summary, the ARSA model mainly comprises two components. The first component, which corresponds to the first term in the right hand side of Equation (13.2), reflects the influence of past box office revenues. The second component, which corresponds to the second term, represents the effect of the sentiments as reflected from the blogs.

Training the ARSA model involves learning the set of parameters $\phi_i (i = 1, \ldots, p)$, and $\rho_{i,j}(i = 1, \ldots, q; j = 1, \ldots, K)$, from the training data that consist of the true box office revenues, and $\omega_{t,j}$ obtained from the blog data. The model can be fitted by least squares regression. Details are omitted due to lack of space.

13.6 Experiments

In this section, we report the results obtained from a set of experiments conducted on a movie data set in order to validate the effectiveness of the proposed model, and compare it against alternative methods.

13.6.1 Experiment Settings

The movie data we used in the experiments consists of two components. The first component is a set of blog documents on movies of interest collected from the Web, and the second component contains the corresponding daily box office revenue data for these movies.

Blog entries were collected for movies released in the United States during the period from May 1, 2006 to August 8, 2006. For each movie, using the movie name and a date as keywords, we composed and submitted queries to Google's blog search engine, and retrieved the blogs entries that were listed in the query results. For a particular movie, we only collected blog entries that had a timestamp ranging from one week before the release to four weeks after, as we assume that most of the reviews might be published close the release date. Through limiting the time span for which we collect the data, we are able to focus on the most interesting period of time around a movie's release, during which the blog discussions are generally the most intense. As a result, the amount of blog entries collected for each movie ranges from 663 (for *Waist Deep*) to 2069 (for *Little Man*). In total, 45046 blog entries that comment on 30 different movies were collected. We then extracted the title, permlink, free text contents, and time stamp from each blog entry, and indexed them using Apache Lucene(http://lucene.apache.org).

We manually collected the gross box office revenue data for the 30 movies from the IMDB websitehttp://www.imdb.com. For each movie, we collected its daily gross revenues in the US starting from the release date till four weeks after the release.

In each run of the experiment, the following procedure was followed:
(1) We randomly choose half of the movies for training, and the other half for testing; the blog entries and box office revenue data are correspondingly partitioned into training and testing data sets. (2) Using the training blog entries, we train an S-PLSA model. For each blog entry b, the sentiments towards a movie are summarized using a vector of the posterior probabilities of the hidden sentiment factors, $P(z|b)$. (3) We feed the probability vectors obtained in step (2), along with the box revenues of the preceding days, into the ARSA model, and obtain estimates of the parameters. (4) We evaluate the prediction performance of the ARSA model by experimenting it with the testing data set.

In this chapter, we use the *mean absolute percentage error* (MAPE) [7] to measure the prediction accuracy: $MAPE = \frac{1}{n}\sum_{i=1}^{N} \frac{|Pred_i - True_i|}{True_i}$, where n is the total

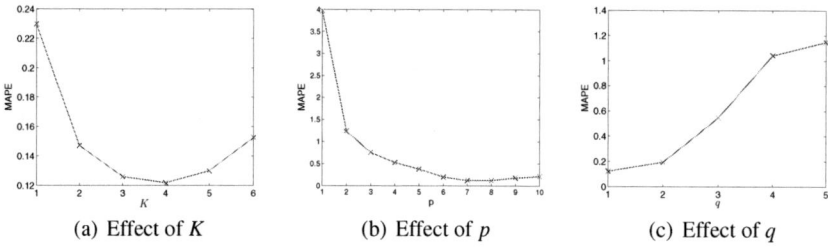

(a) Effect of K (b) Effect of p (c) Effect of q

Fig. 13.2 The effects of parameters on the prediction accuracy

amount of predictions made on the testing data, $Pred_i$ is the predicted value, and $True_i$ represents the true value of the box office revenue. All the accuracy results reported herein are averages of 30 independent runs.

13.6.2 Parameter Selection

In the ARSA model, there are several user-chosen parameters that provide the flexibility to fine tune the model for optimal performance. They include the number of hidden sentiment factors in S-PLSA, K, and the orders of the ARSA model, p and q. We now study how the choice of these parameter values affects the prediction accuracy.

We first vary K, with fixed p and q values ($p = 7$, and $q = 1$). As shown in Figure 13.2 (a), as K increases from 1 to 4, the prediction accuracy improves, and at $K = 4$, ARSA achieves an MAPE of 12.1%. That implies that representing the sentiments with higher dimensional probability vectors allows S-PLSA to more fully capture the sentiment information, which leads to more accurate prediction. On the other hand, as shown in the graph, the prediction accuracy deteriorates once K gets past 4. The explanation here is that a large K may cause the problem of overfitting [1], i.e., the S-PLSA might fit the training data better with a large K, but its generalization capability on the testing data might become poor. Some tempering algorithms have been proposed to solve the overfitting problem [6], but it is out of the scope of our study. Also, if the number of appraisal words used to train the model is M, and the number of blog entries is N, the total number of parameters which must be estimated in the S-PLSA model is $K(M+N+1)$. This number grows linearly with respect to the number of hidden factors K. If K gets too large, it may incur a high training cost in terms of time and space.

We then vary the value of p, with fixed K and q values ($K = 4, q = 1$) to study how the order of the autoregressive model affects the prediction accuracy. We observe from Figure 13.2 (b) that the model achieves it best prediction accuracy when $p = 7$. This suggests that p should be large enough to factor in all the significant influence of the preceding days' box office performance, but not too large to let irrelevant information in the more distant past to affect the prediction accuracy.

Using the optimal values of K and p, we vary q from 1 to 5 to study its effect on the prediction accuracy. As shown in Figure 13.2 (c), the best prediction accuracy is achieved at $q = 1$, which implies that the prediction is most strongly related to the sentiment information captured from blog entries posted on the immediately preceding day.

13.7 Conclusions and Future Work

The proliferation of ways for people to convey personal views and comments online has offered a unique opportunity to understand the general public's sentiments and use this information to advance business intelligence. In this chapter, we have explored the predictive power of sentiments using movies as a case study. A center piece of our work is the proposal of S-PLSA, a generative model for sentiment analysis that helps us move from simple "negative or positive" classification towards a deeper comprehension of the sentiments in blogs. Using S-PLSA as a means of "summarizing" sentiment information from blogs, we develop ARSA, a model for predicting sales performance based on the sentiment information and the product's past sales performance. The accuracy and effectiveness of our model have been confirmed by the experiments on the movie data set. Equipped with the proposed models, companies will be able to better harness the predictive power of blogs and conduct businesses in a more effective way.

References

1. D. Blei, A. Ng, and M. Jordan. Latent dirichlet allocation. *Journal of Machine Learning Research*, 2003.
2. A. P. Dempster, N. M. Laird, and D. B. Rubin. Maximum likelihood from incomplete data via the *em* algorithm. *Journal of Royal Statistical Society*, B(39):1–38, 1977.
3. W. Enders. *Applied Econometric Time Series*. Wiley, New York, 2nd edition, 2004.
4. D. Gruhl, R. Guha, R. Kumar, J. Novak, and A. Tomkins. The predictive power of online chatter. In *KDD '05*, pages 78–87, 2005.
5. D. Gruhl, R. Guha, D. Liben-Nowell, and A. Tomkins. Information diffusion through blogspace. In *WWW '04*, pages 491–501, 2004.
6. T. Hofmann. Probabilistic latent semantic analysis. In *UAI'99*, 1999.
7. W. Jank, G. Shmueli, and S. Wang. Dynamic, real-time forecasting of online auctions via functional models. In *KDD '06*, pages 580–585, 2006.
8. J. Kamps and M. Marx. Words with attitude. In *Proc. of the First International Conference on Global WordNet*, pages 332–341, 2002.
9. B. Liu, M. Hu, and J. Cheng. Opinion observer: analyzing and comparing opinions on the web. In *WWW '05*, pages 342–351, 2005.
10. B. Pang and L. Lee. A sentimental education: Sentiment analysis using subjectivity summarization based on minimum cuts. In *ACL '04*, pages 271–278, 2004.

11. B. Pang and L. Lee. Seeing stars: Exploiting class relationships for sentiment categorization with respect to rating scales. In *ACL '05*, pages 115–124, 2005.
12. B. Pang, L. Lee, and S. Vaithyanathan. Thumbs up? sentiment classification using machine learning techniques. In *Proc. of the 2002 Conference on Empirical Methods in Natural Language Processing (EMNLP)*, 2002.
13. P. D. Turney. Thumbs up or thumbs down?: semantic orientation applied to unsupervised classification of reviews. In *ACL '02*, pages 417–424, 2001.
14. C. Whitelaw, N. Garg, and S. Argamon. Using appraisal groups for sentiment analysis. In *CIKM '05*, pages 625–631, 2005.
15. Z. Zhang and B. Varadarajan. Utility scoring of product reviews. In *CIKM '06*, pages 51–57, 2006.

Chapter 14
Web Mining: Extracting Knowledge from the World Wide Web

Zhongzhi Shi, Huifang Ma, and Qing He

Abstract This chapter addresses existing techniques for Web mining, which is moving the World Wide Web toward a more useful environment in which users can quickly and easily find the information they need. In particular, this chapter introduces the reader to methods of data mining on the Web developed by our laboratory, including uncovering patterns in Web content (semantic processing, classification, clustering), structure (retrieval, classical link analysis method), and event (preprocessing of Web event mining, news dynamic trace, multi-document summarization analysis). This chapter would be an excellent resource for students and researchers who are familiar with the basic principles of data mining and want to learn more about the application of data mining to their problems in Web mining.

14.1 Overview of Web Mining Techniques

The amount of information on the World Wide Web and other information sources such as digital libraries is quickly increasing. These information cover a wide variety of aspects. The huge information space spurs the development of data mining and information retrieval techniques. Web mining, which is moving the World Wide Web toward a more useful environment in which users can quickly and easily find information, can be regarded as the integration of techniques gathered by means of traditional data mining methodologies and its unique techniques.

As many believe, it is Oren Etzioni that first proposed the term of Web mining. He claimed that Web mining is the use of data mining techniques to automatically discover and extract information from World Wide Web documents and services [5]. Web mining is a research area that tries to identify the relevant pieces of infor-

---

Zhongzhi Shi, Huifang Ma, Qing He
Key Laboratory of Intelligent Information Processing, Institute of Computing Technology, Chinese Academy of Sciences, No. 6 Kexueyuan Nanlu, Beijing 100080, People's Republic of China, e-mail: {shizz,mahf,heq}@ics.ict.ac.cn

mation by applying techniques from data mining and machine learning to Web data and documents. In general, Web mining uses document content, hyperlink structure and event organization to assist users in meeting their needed information. Madria et al. claimed the Web involves three types of data [21]: data on the Web, Web log data and Web structure data. Cooley classified the data type as content data, structure data, usage data, and user profile data [18]. M.Spiliopoulou categorized the Web mining into Web usage mining, Web text mining and user modeling mining [15]. Raymond systematically surveyed Web mining, pointed out some confusions regarded the usage of term Web Mining and suggested three Web mining categories [17]. When looked upon in data mining terms, Web mining can be considered to have three operations of interests - clustering (finding natural groupings of information for users), associations (which URLs tend to be more important), and event analysis (organization of information).

Web content mining is the process to discover useful information from the content of a Web page. Since Web data are mainly semi-structured or even unstructured, Web content mining therefore combines available applications of data mining and its own unique approaches.In the following section, we would like to introduce some research results in the field of Web content mining we conclude these years: including semantic text analysis by means of conceptual semantic space; a new way of classification: multi-hierarchy text classification and clustering analysis that is clustering algorithm based on Swarm Intelligence and k-Means.

Web structure mining exploits the graph structure of the World Wide Web. It takes advantage of the hyperlink structure of the Web as an (additional) information source. The Web is viewed as a directed graph whose nodes are the Web pages and the edges are the hyperlinks between them. The primary aim of Web structure mining is to discover the link structure of the hyperlinks at the inter-document level.In the following section, we will analysis Web structure mining through information retrieval's point of view and compare two famous link analysis methods: PageRank vs. HITS.

Web event mining discovers and delivers information and knowledge in a real-time stream of events on the Web. A typical Web event (news in particular) is composed of news title, major reference time, news resource, report time, condition time, portrait and location. We can use a knowledge management model to organize the event.In the following section, these problems will be addressed: preprocessing for Web event, mining news dynamic trace and multi-document summarization.

The remainder of this chapter is organized as follows. Web content mining techniques are explained in Sect. 14.2. Sect. 14.3 deals with Web structure mining. Web event mining is discussed in Sect. 14.4, conclusions and future works are mentioned in the last section.

14.2 Web Content Mining

Web content mining describes the automatic search of information resource available online [21], and involves particularly mining Web content data. It is a combination of novel methods from a wide range of fields including data mining, machine learning, natural language processing, statistics, databases, information retrieval and so on.

Unfortunately, much of the data is unstructured and semi-structured. The Web document usually contains different types of data, such as text, image, audio, video, metadata and hyperlinks. Providing a relational interface to all such databases may be complicated. This unstructured characteristic of Web data forces the Web content mining towards a more complicated approach.

Our lab has implemented a semantic indexing system based on concept space: GHUNT [16] . Some new technologies are integrated in GHUNT. GHUNT can be regarded as an all-sided solution for information retrieval on Internet [34]. In the following, some key technologies concerning Web mining are demonstrated: the way of constructing conceptual semantic space , multi-hierarchy text classification and clustering algorithm based on Swarm Intelligence and k-Means.

14.2.1 Classification: Multi-hierarchy Text Classification

The goal of text classification is to assign one or several proper classes to a document. At present, there are a lot of machine learning approaches and statistics methods used in text classification, including Support Vector Machines (SVM) [26], K-Nearest Neighbor Classification(KNN) [31], Linear Least Square Fit(LLSF) developed by Yang [32], decision trees with boosting by Apte [3], Neural network and Naïve Bayes [24] and so on.

Most of these approaches adopt the classical vector space model (VSM). In this model, the content of a document is formalized as a dot of the multi-dimension space and represented by a vector. The frequently used document representation in VSM is the so-called TF.IDF-vector representation. Lu introduced an improved approach named TF.IDF.IG by combining the information gain from information theory [20].

Our lab has proposed an approach of multi-hierarchy text classification based on VSM [23]. In this approach, all classes are organized as a tree according to some given hierarchical relations, and all the training documents in a class are combined into a class-document. [30].

The basic insight supporting our approach is that classes that are attached to the same node have a lot more in common with each other than other classes. Based on this intuition, our approach divides the classification task into a set of smaller classification problems corresponding to the splits in the classification hierarchy. Each of these subtasks is significantly simpler than the original task, since the classifier at a node of the tree needs only to distinguish between a small number of classes. And this part of classes have a lot more in common with each other, so the models of these classes will be based on a small set of features.

We first construct class models by feature selection after training the documents classified by hand corresponding to the classification hierarchy. In the selection of feature terms, we synthesize two factors, term frequency and term concentration. In the algorithm, all the training documents in one class will be combined into a class-document to perform feature selection. The algorithm CCM (construct class models) is listed as follows:

Input: A tree according to some given hierarchical relations (each node, except the root node, corresponds to a class and all the documents are classified into subclasses corresponding to the leaf nodes in advance)
Output: All the class models, saved as text files

Begin
Judge all the nodes from the bottom layer to the top layer using bottom-up method:

1. If the node V_0 is a leaf node, then analyze the corresponding class-document, including the term frequencies, the number of terms and the sum of all term frequencies.
2. If V_0 is not a leaf node (assume it has t node children from V_1, V_2 to V_t, and there are s terms from T_1, T_2, to T_s in the corresponding class-document), then
   a. Calculate the probability of the class-document d_i corresponding to V_i
   b. Calculate $H(D)$ and $H(D/T_k)$, then get IG_k, where $k = 1,2,...,s$, $H(D)$ is the entropy of the document collection D, $H(D/T_k)$ is the conditional entropy of term T_k
   c. Construct the class model C_i corresponding to V_i, where $i = 1,2,...,t$
      i. Initialize C_i to null
      ii. Calculate term frequency W_ik of T_k, where $k = 1,2,...,s$
      iii. Resort all the terms to a new permutation T_1, T_2,...,T_s according to the descending weights
      iv. Judge the terms from T_1 to T_s individually:

If the number of feature terms in C_i exceeds a certain threshold value NUM_T
Then the construction of C_i ends up
Else
If W_ik exceeds a certain threshold value α, the term frequency of T_k exceeds a certain threshold value β, the term concentration of T_k exceeds a certain threshold value γ and T_k is not in the stop-list given in advance, then T_k will be a feature term and should be added to the class model C_i with its weight.
End

The calculation of term weight in this part considers two factors: term frequency and term position [23]. Then one top-down matching process is hierarchically performed from the root node of the tree until the proper subclass is found corresponding to a leaf node. For more details of the algorithm, you can refer to [23].

14.2.2 Clustering Analysis: Clustering Algorithm Based on Swarm Intelligence and k-Means

Although it is hard to organize the whole Web, it is feasible to organize Web search results of a given query. The standard method for information organization is concept hierarchy and categorization. The popular technique for hierarchy construction is text clustering. Generally, major clustering methods can be classified into five categories: partitioning methods, hierarchical methods, density-based methods, grid-based methods and model-based methods. Many clustering algorithms have been proposed, such as CLARANS [10], DBSCAN [14], STING [27] and so on.

Our lab has proposed a document clustering algorithm based on Swarm Intelligence and K-Means: CSIM [28], which combines Swarm Intelligence with k-Means clustering technique. Firstly, an initial set of clusters is formed by swarm intelligence based clustering method which is derived from a basic model interpreting ant colony organization of cemeteries. Secondly, an iterative partitioning phase is employed to further optimize the results. Self-organizing clusters are formed by this method. The number of clusters is also adaptively acquired. Moreover, it is insensitive to the outliers and the order of input. Actually, the swarm intelligence based clustering method can be applied independently. But by second phase, the outliers which are single points on the ant-work plane are converged on the nearest neighbor clusters and the clusters which are piled too closely to collect correctly on the plane by chance are also split. K-means clustering phase softens the casualness of the swarm intelligence based method which is originated from a probabilistic model. The algorithm can be described as follows:

Input: document vectors to be clustered
Output: documents labeled by clustering number

1. Initialize Swarm similarity coefficient α, ant number maximum iterative times n, slope k, and other parameters;

2. Project the data objects on a plane at random, i.e. randomly give a pair of coordinate (x,y) to each data object;
3. Give each ant initial objects and initial state of each ant is unloaded;
4. for i=1,2... // while not satisfying stop criteria
    a. for $j = 1,2...,ant\_number$;
        i. Compute the Swarm similarity of the data object within a local region with radius r;
        ii. If the ant is unloaded, compute picking-up probability P_p. Compare P_p with a random probability P_r, if $P_p < P_r$, the ant does not pick up this object, another data object is randomly given the ant, else the ant pick up this object, the state of the ant is changed to loaded, a new random pair of coordinate is given the ant;
        iii. If the ant is loaded, compute dropping probability P_d. Compare with a random probability P_r, if $P_d > P_r$, the ant drops the object, the pair of coordinate of the ant is given to the object. the state of the ant is changed to unloaded, another data object is randomly given the ant, else the ant continue moving loaded with the object, a new random pair of coordinate is given the ant.
5. for $i = 1,2,...,pattern\_num$; //for all patterns
    a. if this pattern is an outlier, label it as an outlier;
    b. else label this pattern a cluster serial number; recursively label the same serial number to those patterns whose distance to this pattern is smaller than a short distance *dist*. i.e. collect the patterns belong to a same cluster on the ant-work plane; Serial number *serial_num* + +.
6. Compute the cluster means of the *serial_num* clusters as the initial cluster centers;
7. repeat

a. (re)assign each pattern to the cluster to which the pattern is the most similar, based on the mean value of the patterns in the cluster;
   b. update the cluster means, i.e. calculate the mean value of the patterns for each cluster;
8. until not change.

If you want to know more about the algorithm, see [28].

14.2.3 Semantic Text Analysis: Conceptual Semantic Space

An automatic indexing and concept classification approach to a multilingual (Chinese and English) bibliographic database is presented by H. Chen [8]. A concept space of related descriptors was then generated using a co-occurrence analysis technique. For concept classification and clustering, a variant of a Hopfield neural network was developed to cluster similar concept descriptors and to generate a small number of concept groups to represent (summarize) the subject matter of the database.

A simple way to generate concept semantic space is by using HowNet, which is an on-line common-sense knowledge base unveiling inter-conceptual relations and inter-attribute relations of concepts as connoting in lexicons of the Chinese and their English equivalents [29]. We develop a new way to establish concept semantic space by using clustering algorithm based on Swarm Intelligence and k-Means. Here, the bottom up way is taken. We first classify the Web pages from internet into some domains, and then the Web pages that belong to each domain are clustered. Such would evade subjective warp caused by the departure between document and level in the top down way. Also we can adjust the parameter to make the hierarchy flexible enough.

The concept space can be used to facilitate querying and information retrieval. One of the most important aspects is how to generate the link weights in concept space of specific domain automatically. Before generating the concept space, the concepts of a certain domain must be identified. In the scientific literature domain, the concepts are relatively stable, and there are existing thesauruses that can be adopted. However, in a certain domain, news domain in particular, the concepts are dynamic, so there is no existing thesaurus and it is unrealistic to generate a thesaurus manually. We need to extract the concepts from the document automatically. Using the following formulae we could compute the information gain of each term for classification, which sets a foundation for thesaurus construction.

$$InfGain(F) = P(F)\sum_i P(\psi_i|F)\log\frac{P(\psi_i|F)}{P(\psi_i)} + P(\overline{F})\sum_i P(\psi_i|\overline{F})\log\frac{P(\psi_i|\overline{F})}{P(\psi_i)} \quad (14.1)$$

Where F is a term, $P(F)$ is the probability of that term F occurred, $\overline{F}$ means that term F doesn't occur, $P(\psi_i)$ is the probability of the $i-th$ class value, $P(\psi_i|F)$

is the conditional probability of the *ith* class value given that word F occurred. If $InfGain(F) > \omega$, we choose term F as the concept. Although in this way the thesaurus generated is not as thorough and precise as the way constructed manually in the field of scientific literature, it is acceptable.

After we have recognized the concept of a class, we could generate the concept space of that class automatically. Chen's method that uses co-occurrence analysis and Hopfield net is adopted [9] [25]. By means of using co-occurrence analysis, we compute the term association weight between two terms, and then the asymmetric association between terms is computed; we could activate related terms in response to user's input. This process is accomplished by a single-layered Hopfield network. Each term is treated as a neuron, and the association weight is assigned to the network as the synaptic weight between nodes. After the initialization phase, we repeat the iteration until convergence. For detailed description of establishment of concept semantic space, you can refer to [33].

In this section, some research results we conclude are introduced: including a new way of classification: multi-hierarchy text classification; clustering analysis that is clustering algorithm based on Swarm Intelligence and k-Means and semantic text analysis by means of conceptual semantic space.

14.3 Web Structure Mining: PageRank vs. HITS

Web structure mining is essentially about mining the links on the Web. Web pages are actually instances of semi-structured data, and thus mining their structure is critical to extracting information from them. The structure of a typical Web graph consists of Web pages as nodes and hyperlinks as edges connecting between two related pages. Web structure mining can be regarded as the process of discovering structure information from the Web. In the following, we would like to compare famous link analysis methods: PageRank vs. HITS.

Two most influential hyperlink based search algorithms PageRank and HITS were reported during 1997-1998. Both algorithms exploit the hyperlinks of the Web to rank pages according to their levels of "prestige" or "authority".

PageRank Algorithm is originally formulated by Sergey Brin and Larry Page, PhD students from Stanford University, at Seventh International World Wide Web Conference (WWW) in April, 1998 [22].The algorithm is determined for each page individually according to their authoritativeness.

More specifically, a hyperlink from a page to another page is an implicit conveyance of authority to the target page. The more in-links that a page i receives, the more prestige the page i has. Let the Web as a directed graph $G = (V, E)$ and let the total number of pages be n. The PageRank score of the page i (denoted by $P(i)$) is defined by [2]:

$$P(i) = \sum_{(i,j) \in E} \frac{P(j)}{O_j} \qquad (14.2)$$

O_j is the number of out-link of j.

Unlike PageRank which is a "static" ranking algorithm, HITS is search-query-dependent. HITS was proposed by Jon Kleinberg (Cornel University), at Ninth Annual ACM-SIAM Symposium on Discrete Algorithms, January 1998 [12]. This algorithm is initially developed for ranking documents based on the link information among a set of documents.

More specifically, for each vertex v in a subgraph of interest: $a(v)$ shows the authority of v while $h(v)$ demonstrates the hubness of v. A site is very authoritative if it receives many citations. Citation from important sites weight more than citations from less-important sites. Hubness shows the importance of a site. A good hub is a site that links to many authoritative sites. Authorities and hubs have a mutual reinforcement relationship.

In this section, we briefly introduced prestige link analysis: PageRank and HITS.

14.4 Web Event Mining

Web event mining is the application of data mining techniques to Web event repositories in order to produce results that can be used as the event's cause and effect. Event mining is not a new concept, which has already been used in Petri nets, stochastic modeling, etc. However, there are new opportunities that come from the large amount of data that is stored in various databases.

An event can be defined as related topics in a continuous stream of newswire stories. Concept terms of an event are derived from statistical context analysis between sentences in the news story and stories in the concept database. Detection Methods also includes cluster representation. DeJong uses frame-based objects called "sketchy scripts" [7]. D. Luckham [4] provides a framework for thinking about complex events and for designing systems that use such events.

Our lab has implemented an intelligent event organization and retrieval system [11], which uses machine-learning techniques, and combines specialties of news to organize and retrieve Web news documents. The system consists of a preprocessing process for news documents to get related knowledge, event constructer component to collect the correlative news reports together, cause-effect learning process and event search engine.

There are many attributes for an event such as event ID (ID_{st}), name of the event ($Name_{st}$), time of the event ($Time_{st}$), model of the event ($Model_{st}$), the documents belonging to the event($DocS$). Besides, document knowledge ($Know_d$) is a set of pairs, term and weight. Namely, $Know_d = \{(term_1, weight_1),$
$(term_2, weight_2), ..., (term_s, weight_s)\}$. model knowledge of event ($Know_m$), like $Know_d$, is a set of pairs consists of term and weight. $Know_m(j) = \{(term_1, weight_1, j), ..., (term_n, weight_n, j)\}$.

14.4.1 Preprocessing for Web Event Mining

The preprocessing process consists of two parsing processes: parse HTML files and segment text document. The term segmentation algorithm extracts Chinese terms based on the rule of "long term first" to resolve ambiguity. Meanwhile, the term segmentation program combines Name Entity Recognition Algorithm. That is, human names and place names can be extracted and marked automatically, while segmenting terms.

14.4.1.1 Event Template Learning

An event template represents participants in an event described by the keywords and relations among the participants. The model knowledge of event is the most important reference when constructing event template. We can learn it from a group of training documents that report a same event. The key problem of model knowledge is to compute the support of term to event, denoted as $weight_{t_i,s_t}$, in following expression:

$$weight_{t_i,st} = \sum_{D_j} weight_{t_i,D_j} \tag{14.3}$$

$$weight_{t_i,st} = \frac{weight_{t_i,st}}{\max\{weight_{t_i,D_jst}\}} \tag{14.4}$$

Here, $weight_{t_i,D_j}$ is the term support of t_i to D_j. D_j is a document belongs to the event. $weight_{t_i,st}$ is normalized to the range between 0 and 1.

Due to the diversity of Internet documents, the number of terms in an event is large, and their term support is generally low. The feature selection process at the event level is more sophisticated than that at the document level. From analyzing these feature terms, we find the terms that have bigger weight can be represented as features of the event. For details please refer to [13].

14.4.1.2 Time Information Learning

Time is important factor for a news report, which can reveal time information of the event. According to time information, we can get event time, and organize cause-effect of event in time order. After analyzing many news reports on the Internet, we find there are different kinds of time. The formal format of time is defined as:

$$Time = year - month - day - hour - minute - second$$

Web document developers organize the time in different style according to their habits. According to the character of time, there are 3 sorts of form: Absolute time, Relative time and Fuzzy time. For the algorithm of the document time learner please refer to [11].

14.4.2 Multi-document Summarization: A Way to Demonstrate Event's Cause and Effect

Cause-Effect of the event can represent the complete information coverage of an event, which is organized via listing the titles or abstracts of the documents belonging to the event in some order, such as time. The process of learning the Cause-Effect knowledge can follow the process: *learning time$_d$ → learning summary$_d$ → sorting summary$_d$ or title$_d$ in time order → cause effect of event*.

Another way of producing event's cause and effect is Multi-document summarization, which presents a single summary for a set of related source documents. Multi-document summarizations include extractive and abstractive method. Extractive summarization is based on statistical techniques to identify similarities and differences across documents. It involves assigning salience scores to some units of the documents and then extracting the units with highest scores. While abstraction summarization usually needs information fusion, sentence compression and reformulation.

Researchers from Cornell University used a method of Latent Semantic Indexing to ascertain the topic words and generate summarization [19]. NeATS uses sentence position, term frequency, topic signature and term clustering to select important content [13]. The MEAD system is developed by Columbia University used a method called Maximal Marginal Relevance (MMR) to select sentences for summarization [1]. Newsblaster, a news-tracking tool developed by Columbia University generates summarizations of daily news [6].

In this section, these problems concerning Web event mining are addressed: preprocessing for Web event, mining news dynamic trace and multi-document summarization.

14.5 Conclusions and Future Works

In conclusion, this chapter addresses existing solutions for Web mining, which is moving the World Wide Web toward a more useful environment in which users can quickly and easily find the information they need. In particular, this chapter introduces the reader to methods of data mining on the Web, including uncovering patterns in Web content (semantic processing, classification, clustering), structure (retrieval, classical Link Analysis method), and event (preprocessing of Web event mining, news dynamic trace, multi-document summarization analysis). This chapter demonstrates our implementation of semantic indexing system based on concept space: GHUNT as well as an intelligent Event Organization and Retrieval system.

The approaches described in the chapter represent initial attempts at mining content, structure and event of Web. However, to improve information retrieval and the quality of searches on the Web, a number of research issues still need to be addressed. Research can be focused on Semantic Web Mining, which aims at combining the two fast-developing research areas Semantic Web and Web Mining. The idea is to improve the results of Web Mining by exploiting the new semantic structures in the Web. Furthermore, Web Mining can help to build the Semantic Web. Various conferences are now including panels on Web mining. As Web technology and data mining technology mature, we can expect good tools to be developed to mine the large quantities of data on the Web.

Acknowledgements This work is supported by the National Natural Science Foundation of China (No. 60435010, 60675010), 863 National High-Tech Program (No.2006AA01Z128,2007AA01Z132) and National Basic Research Priorities Programme (No.2007CB311004) .

References

1. Ando R., Kboguraev B., Kbyrd R. J.: Multi-document Summarization by Visualizing Topical Content.ANLP-NAACL 2000 Workshop, Seattle Advanced Summarization Workshop, 2000: 12-19
2. Bing Liu: Web data mining. Springer Verlag, 2007
3. C. Apte, F. Damerau, S. Weiss: Text mining with decision rules and decision trees. In Proceedings of the Conference on Automated Learning and Discovery, Workshop, 1998
4. David C. Luckham, James Vera: An Event-Based Architecture Definition Language. IEEE TRANSANCTION ON Software Engineering, 1995, 21(9): 717-734
5. Etzioni, Oren: World-Wide Web: Quagmire or gold mine. Communications of the ACM, 1996, 39(11): 65-68
6. Evans K., Dklavans J., Lmckeown K. R.: Columbia Newsblaster Multilingual news summarization on the Web.Demonstration Papers at HLT-NAACL, 2004: 1-4
7. G. DeJong: Prediction and substantiation: A new approach to natural language processing. Cognitive Science, 1979: 251-273
8. H. Chen, D. T. Ng.: An algorithmic approach to concept exploration in a large knowledgenetwork (automatic thesaurus consultation): symbolic branch-and-bound vs. connectionist Hopfield net activation. Journal of the American Society for Information Science, 1995, 46(5):348-369
9. H. Chen, J. Martinez, T. D. Ng, B. R. Schatz: A Concept Space Approach to Addressing the Vocabulary Problem in Scientific Information Retrieval: An Experiment on the Worm Community System. Journal of the American Society for Information Science, 1997, 48(1): 17-31
10. J. R. T. Ng, J. Han: Efficient and effective clustering methods for spatial data mining. Proceedings of the 20th VLDB Conference, 1994: 144-155
11. Jia Ziyan, He Qing, Zhang Hai Jun, Li Jiayou, Shi Zhongzhi: A News Event Detection and Tracking Algorithm Based on Dynamic Evolution Model. Journal of Computer Research and Development (in Chinese), 2004, 41(7): 1273-1280
12. Jon M. Kleinberg: Authoritative sources in a hyperlinked environment. Journal of the ACM, 1999, 46(5): 604-632
13. Lin Chin Yew, Hovy Eduard: From Single to Multi-document Summarization: A Prototype System and its Evaluation. In Proceedings of ACL, 2002: 25-34

14. M. Ester, H. P. Kriegel, J. Sander, X. Xu: A Density-Based Algorithm for Discovering Clusters in Large Spatial Databases with Noise. Proceeding of the 2nd Internatioal Conference on Knowledge Discovery and Data Mining, 1996: 226-231
15. M. Spiliopoulou: Data mining for the Web. In Proceedings of Principles of Data Mining and Knowledge Discovery. Third European conference, 1999, 588-589
16. Qing He, Ziyan Jia, Jiayou Li,Haijun Zhang,Qingyong Li, Zhongzhi Shi: GHUNT: A SEMANTIC INDEXING SYSTEM BASED ON CONCEPT SPACE. International Conference on Natural Language Processing and Knowledge Engineering (IEEENLP&KE-2003), 2003: 716-721
17. Raymond Kosala, Hendrik Blockeel: Web mining research: a survey. ACM SIGKDD Explorations Newsletter, 2000, 2(1): 1-15
18. R. Cooley: Web Usage Mining: Discovery and Application of Interesting Patterns from Web data. PhD thesis, Dept. of Computer Science, University of Minnesota. May, 2000
19. Radevr, Jing Hongyan, Budzikowska Malgorzata: Centroid-based summarization of multiple documentsSentence extraction, utility-based evaluationand user studies. ANLP-NAACL 2000 Workshop, 2000: 21-29
20. S. Lu, X. L. Li, S. Bai et al.: An improved approach to weighting terms in text. Journal of Chinese Information Processing (in Chinese), 2000, 14(6): 8-13
21. S. K. Madria, S. S. Rhowmich, W. K. Ng, F. P. Lim: Research issues in Web data mining. Proceedings of Data Warehousing and Knowledge Discovery, First International Conference. 1999: 303-312
22. Sergey Brin, Larry Page: The anatomy of a large-scale hypertextual Web search engine. Proceedings of the Seventh International World Wide Web, 1998, 30(7): 107-117
23. Shaohui Liu, Mingkai Dong, Haijun Zhang, Rong Li, Zhongzhi Shi: An approach of multi-hierarchy text classification. International Conferences on Info-tech and Info-net. 2001, 3: 95-100
24. T. Mitchell: Machine Learning. McGraw: Hill, 1996
25. Teuvo Kohonen, Samuel Kashi: Self-Organization of a Massive Document Collection. IEEE Transactions On Neural Networks, 2000,11(3): 574-585
26. V. Vapnik: The Nature of Statistical Learning Theory. New York. Springer-Verlag, 1995
27. Wei Wang, Jiong Yang, Richard Muntz: STING: A Statistical Information Grid Approach to Spatial Data Mining. Proceedings of the 23rd VLDB Conference, 1997: 186-195
28. Wu Bin, Zheng Yi, Liu Shaohui, Shi Zhongzhi: CSIM: A Document Clustering Algorithm Based On Swarm Intelligence. World Congress on Computational Intelligence, 2002: 477-482
29. www.keenage.com
30. X. L. Li, J. M. Liu, Z. Z. Shi: The concept-reasoning network and its application in text classification. Journal of Computer Research and Development (in Chinese), 2000, 37(9): 1032-1038
31. Y. Yang, C. G. Chute: An example-based mapping method for text categorization and retrieval. ACM Transaction on Information Systems (TOIS), 1994, 12(3): 252-277
32. Y. Yang: Expert Network: Effective and efficient learning from human decisions in text categorization and retrieval. Proceedings of the Fourth Annual Symposium on Document Analysis and Information Retrieval (SIGIR'94), 1994: 13-22
33. Yuan Li, Qing He, Zhongzhi Shi: Association Retrieval based on concept semantic space. (in Chinese) Journal of University of Science and Technology Beijing, 2001, 23(6): 577-580
34. Zhongzhi Shi, Qing He, Ziyan Jia, Jiayou Li: Intelligence Chinese Document Semantic Indexing System. International Journal of Information Technology and Decision Making, 2003, 2(3): 407-424

Chapter 15
DAG Mining for Code Compaction

T. Werth, M. Wörlein, A. Dreweke, I. Fischer, and M. Philippsen

Abstract In order to reduce cost and energy consumption, code–size optimization is an important issue for embedded systems. Traditional instruction saving techniques recognize code duplications only in exactly the same order within the program. As instructions can be reordered with respect to their data dependencies, Procedural Abstraction achieves better results on data flow graphs that reflect these dependencies. Since these graphs are always directed acyclic graphs (DAGs), a special mining algorithm for DAGs is presented in this chapter. Using a new canonical representation that is based on the topological order of the nodes in a DAG, the proposed algorithm is faster and uses less memory than the general graph mining algorithm gSpan. Due to its search lattice expansion strategy, an efficient pruning strategy is applied to the algorithm while using it for Procedural Abstraction. Its search for unconnected graph fragments outperforms traditional approaches for code–size reduction.

15.1 Introduction

We present DAGMA, a new graph mining algorithm for Directed Acyclic Graphs (DAGs). DAG mining is important in general, but our work is inspired by DAGs that appear in code generation, especially in code compaction. Code–size optimization is crucial for embedded systems as cost and energy consumption depend on the size

---

T. Werth, M. Wörlein, A. Dreweke, M. Philippsen
Programming Systems Group, Computer Science Department, University of Erlangen–Nuremberg, Germany, phone: +49 9131 85-28865, e-mail: {werth,woerlein,dreweke, philippsen}@cs.fau.de

I. Fischer
Nycomed Chair for Bioinformatics and Information Mining, University of Konstanz, Germany, phone: +49 7531 88-5016, e-mail: Ingrid.Fischer@inf.uni-konstanz.de

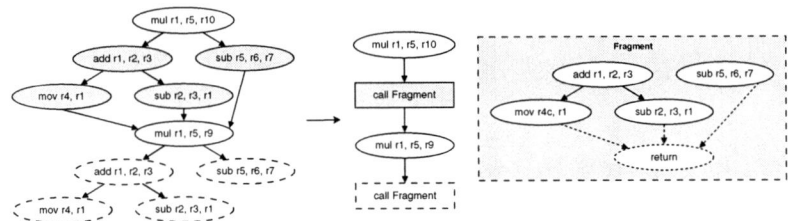

Fig. 15.1 PA example: data flow graph with frequent unconnected fragment (gray and dashed).

of the built–in memory. Moreover, the smaller the code is, the more functionality fits into the memory.

Procedural Abstraction (PA) reduces assembly code size by detecting frequent code fragments, extracting them into new procedures and substituting them with call/jump instructions. Traditionally, text–matching, e.g. suffix trees [6], is used for candidate detection. The obvious disadvantage is that to detect repeated instructions they must occur in the exact same order in the program. Dreweke et al. have shown in [7] that a graph–based PA outperforms the code shrinking results of a purely textual approach. Graph–based PA transforms the instruction sequences of basic blocks into data flow graphs (DFGs). A basic block is a code sequence that has exactly one entry point (i.e. only the first instruction may be the target of a jump instruction) and one exit point (i.e. only the last instruction may be a jump instruction). A DFG is a DAG that represents the instructions as nodes and the data dependencies between these instructions as directed edges. In general, from a single DFG several differently ordered instruction sequences can be generated that have the same semantics but cannot be detected by textual approaches. A graph–based approach, i.e. mining for frequent DFG fragments, can therefore find more opportunities for PA.

This chapter will show that for our domain DAG mining is better than regular graph mining. In addition, DFG–based code compaction has the following domain-specific requirements. While it is hard to extend general purpose miners accordingly, our <u>DAG</u> <u>M</u>ining <u>A</u>lgorithm DAGMA addresses them up front. First, DAGMA can find *unconnected* fragments. This is crucial for PA as shown in Fig. 15.1. In the example DFG an unconnected fragment appears twice and is extracted into a procedure. Existing miners for connected graphs cannot find such unconnected fragments without applying tricks. For example, they use multiple starting points [2] to grow fragments or they add a helper pseudo-root that is connected to all other nodes [16]. Of course DAGMA can also search for connected fragments with one or more roots. Second, in addition to the traditional *graph–based* way to compute support/frequency of a fragment, DAGMA can also calculate it in an *embedding–based* way. Whereas graph–based counting detects that the frequent fragment of Fig. 15.1 appears in one graph (i.e. support = 1), an embedding–based counting distinguishes the two (non-overlapping) embeddings (i.e. support = 2). Since PA can extract both embeddings, PA requires an embedding–based support calculation. To be more exact, PA requires a search for *induced* fragments, because not every *embedded* fragment can be extracted. More details are given in Section 15.3.2.

15 DAG Mining for Code Compaction

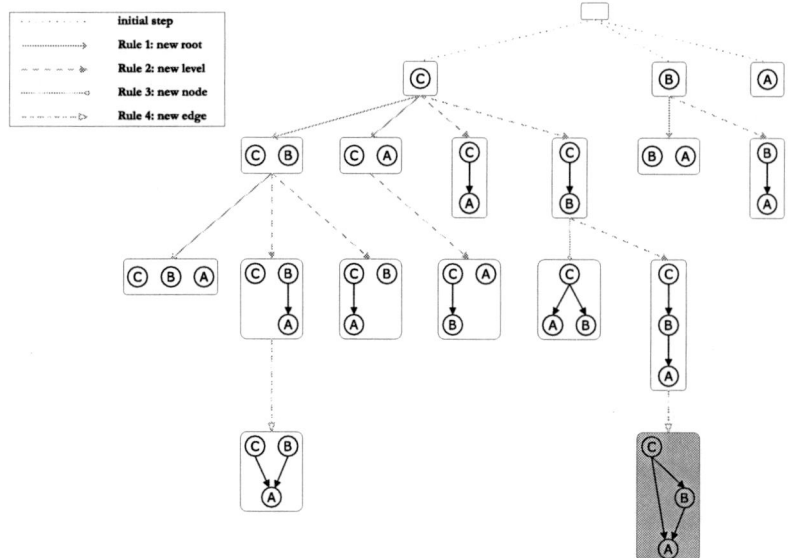

Fig. 15.2 Example search lattice.

15.2 Related Work

Whereas there is a bunch of algorithms for (undirected) mining in trees [4] and graphs [2, 12, 16], the situation is different for DAGs. The only three other DAG miners known to us [3, 14, 17] are not applicable to PA. The first is a miner for gene network data that does not handle induced but just embedded fragments. The others only address single–rooted and connected sub–DAGs and therefore detect fewer extractable fragments. (Note that in research on code compaction, 'template generation', 'clone detection', or 'regularity extraction' are the key words used to denote related forms of DAG mining.) DAGMA is more general and can be used for more than just PA. DAGMA can mine both with embedding–based and with traditional graph–based support and it can also mine for connected fragments (by filtering unconnected ones) or for single–rooted fragments (by using only one root).

After covering some preliminaries, we present DAGMA in detail in Section 15.4. Section 15.5 gives performance results both on synthetic DAGs and on DFGs of ARM assembly programs.

15.3 Graph and DAG Mining Basics

Most graph miners build their *search lattice* by starting from single–node or single–edge graph *fragments* (i.e. common subgraphs of the database graphs) and grow them by adding nodes and/or edges in a stepwise fashion based on a set of expansion rules. Fig. 15.2 holds an example search lattice and shows the relationship between the fragments (nodes) and their expansions (the directed edges) that grow a fragment from another one. The concrete instances of a fragment, i.e. the appearances of its isomorphic subgraphs in the database graphs, are called *embeddings*. A fragment is *frequent* and therefore interesting, if its number of embeddings, counted after each expansion step, exceeds some threshold. This can be a concrete number, denoted by *minimum support*, or a percentage value of the database graphs, denoted by *minimum frequency*. The search process prunes the search lattice at infrequent fragments, since extending such a fragment always leads to other infrequent fragments, so further extensions are worthless (known as *anti–monotone principle*). The main difficulty is to avoid traversing multiple paths to already created fragments, i.e. a fragment should not be generated again if it has already been reached by another sequence of expansion steps. Otherwise such a fragment is processed again, including the detection of embeddings in the database. Since this is costly with respect to space and time, it is essential to check for duplicates efficiently.

15.3.1 Graph–based versus Embedding–based Mining

There are two interpretations of minimum support. Support of a fragment in *graph–based* mining specifies the minimum number of database graphs with one or more embeddings of this fragment. *Embedding–based* mining defines support as the minimum number of non–overlapping embeddings regardless of the database graphs. The number of non–overlapping embeddings is computed by means of a maximum independent set algorithm [13], for PA this process is described in [7].

As mentioned above, the fragment shown in Fig. 15.1 has a graph–based support of 1 but an embedding–based support of 2. In contrast, the example in Fig. 15.3 shows a graph G and a fragment. Although there are two ways to embed the fragment into G, the embedded–based support is just 1, since the two ways of embedding overlap. There are two main reasons for only taking non–overlapping reasons into account. First, PA requires an embedding–based mining because only non-overlapping fragments can be used to shrink the code. An extraction of an embedding replaces all its nodes with a single instruction (call or jump) and therefore afterwards the extraction of an overlapping second fragment is no longer possible. Second, only for edge–disjoint (and therefore disjoint) embeddings the anti–monotone principle can be used to prune the search lattice. If we would also count overlapping embeddings, it is no longer true that the minimum support monotonously decreases with growing fragment sizes.

15 DAG Mining for Code Compaction

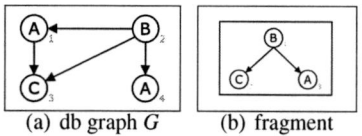

(a) db graph G   (b) fragment

Fig. 15.3 Example of one fragment in two database graphs.

15.3.2 Embedded versus Induced Fragments

Induced fragments are a subset of *embedded fragments*, because of the more strict parent–child relationship in contrast to the more general ancestor–descendant relationship. Induced fragments are real subgraphs of the database graphs, because directly connected nodes also have to be directly connected in the corresponding database graphs. Nodes that are directly connected in an embedded fragment have to be connected in the original graph but the connection may be a path over several nodes and edges. Therefore, a parent node in the embedded fragment has to be an arbitrary ancestor and a child node must be a descendant in the database graph. Only induced fragments are useful for PA, since embedded fragments can skip nodes. For example, if the chain $A \rightarrow B \rightarrow C$ is used as input, an possible embedded but not induced fragment is $A \rightarrow C$ that ignores the dependency of the node B.

15.3.3 DAG Mining Is NP–complete

Whereas the search lattice can be enumerated in polynomial time for trees, general graph mining is *NP*–complete because subgraph isomorphism is *NP*–complete. Graph isomorphism is supposed to be in a complexity class of its own [8]. Unfortunately, sub-DAG isomorphism is in the same complexity class. As a proof, consider the following transformation of a general graph into a DAG: Replace each original edge with a new node (carrying the edge label, if existent) plus two directed edges from the old nodes to the new node. If the source graph is a directed graph, edge labels represent the direction. Obviously, the transformed graph is a DAG since every old node has only outgoing and every new node has only incoming edges. Since this transformation (and the inverse one) can be done in polynomial time and the increase of nodes and edges is polynomial, the transformation is a valid reduction. If two original graphs are isomorphic to each other, they also are isomorphic after the transformation. If a graph contains a subgraph, its transformed graph contains the transformed subgraph. The inverse reduction is obvious, since the DAG can be treated as a general graph. Hence, each (sub-)graph isomorphism problem can be solved by solving the corresponding (sub-)DAG isomorphism problem and vice versa. As a result, DAG isomorphism is in the same complexity class as graph isomorphism and sub-DAG isomorphism is in the same complexity class as subgraph isomorphism and therefore DAG mining is *NP*–complete.

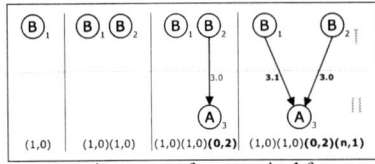
(a) expansion steps of a canonical fragment

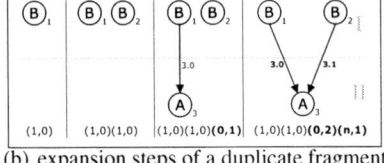
(b) expansion steps of a duplicate fragment

Fig. 15.4 Construction of two isomorphic subgraphs with different canonical forms.

15.4 Algorithmic Details of DAGMA

Because of the *NP*–completeness, one of the challenging problems for DAG mining is to avoid as many costly (sub-)DAG isomorphism tests as possible. The enumeration of the fragments has to quickly detect duplicates, i.e. fragments that are reached through several paths. As other mining algorithms do, DAGMA solves this by encoding fragments in a canonical form that is both simple to construct and more amenable to comparison than costly subgraph isomorphism tests.

15.4.1 A Canonical Form for DAG enumeration

The fundamental idea of DAGMA is a novel canonical form that exploits that DAGs can be sorted topologically (in linear time with respect to the number of nodes and edges [5]). This way each node has a *topological level* based on the length of the longest path from a root node to the node itself. See the two levels in Fig. 15.4, indicated with roman numbers I and II. The main idea is the step–by–step construction of fragments by inserting nodes and edges in topological order. Our canonical form of a DAG contains information about the graph structure, the edge directions, the labels (if any), and the insertion order (when enumerating the search lattice and constructing growing fragments).

In Fig. 15.4(a) fragment expansion starts from a single root B (denoted by index 1). Another node B (index 2) is added in the second step. Step 3 simultaneously inserts the node A (index 3) and the edge from its predecessor B. The edge index 3.0 indicates that the edge was inserted at the same time as node 3. The last step adds an edge to the previously inserted node without adding a new node. This *first* edge targeting node 3 after its insertion is labeled 3.1. The canonical descriptions given below the fragments in Fig. 15.4 consist of tuples of the form (*node label index, predecessor index*). Since edge label indices are irrelevant for PA we omit them to simplify explanation. For efficiency, node labels are sorted according to their frequency (let us assume: $A = 0, B = 1, C = 2, \ldots$). Hence, the first tuple $(1,0)$ states that node B $(= 1)$ has been inserted first with no predecessor (0 as predecessor index). In general, the predecessor index refers to the insertion index of its parent node. For example, the tuple $(0,2)$ indicates that node A $(= 0)$ is inserted with node

```
Data: database with DAGs db, mining parameter s
Result: frequent sub–DAGs res
1 begin
2     res ⟵ ∅
3     n ⟵ frequentNodes(db, s)
4     l ⟵ createLabelFunction(n)
5     while n ≠ ∅ do
6         res ⟵ res ∪ n
7         tmp ⟵ ∅
8         for f ∈ n do
9             tmp ⟵ tmp ∪ insertRoots(f, l)
10            tmp ⟵ tmp ∪ insertLevel(f, l)
11            tmp ⟵ tmp ∪ insertNode(f, l)
12            tmp ⟵ tmp ∪ pruneNonCanonical(insertEdge(f, l))
13        n ⟵ filterInfrequentOrUnExtendibleFragments(tmp, s)
14    res ⟵ filterUnWantedFragments(res, s)
15 end
```

Algorithm 2: DAG mining.

B_2 as its predecessor. If an edge (e.g. 3.1) is inserted without adding a new node at the same time, a special *node label index n* that is bigger than all other label indices is used. Tuple $(n, 1)$ expresses that the edge is connected to the last added node.

A *canonical fragment* is created by the insertion order of nodes and edges with the biggest canonical description. Two canonical descriptions are compared numerically, tuple by tuple and tuple-element by tuple-element. Thus, the fragment in Fig. 15.4(b) with its different edge insertion order is not canonical since $(0, 1)$ is smaller than $(0, 2)$. Hence, it can be pruned during the enumeration of the lattice. The structure of the canonical form can be used to restrict the expansion of fragments and to avoid without explicitly checking the canonical for duplicates in many cases. This will be explained in the next section.

15.4.2 Basic Structure of the DAG Mining Algorithm

The DAG mining algorithm (Algo. 2) computes an initial set of frequent single–node fragments (line 3) that are then expanded in a stepwise fashion. As a consequence of the canonical form, fragments are expanded according to the following rules:

1. insert a new *root node* (at the first topological level, line 9),
2. start a *new topological level* (i.e. insert a new node and a new edge that starts from the current level, line 10),
3. stay at the current topological level and insert a *new node* at that level (and an edge from the previous level, line 11),

```
Data: fragment f, labelIndexFunction l
Result: frequent fragments res
1 begin
2     res ← ∅
3     if containsEdges(f) then
4         return
5     for embedding x ∈ f do
6         for unused node y ∈ database graph of x with l(y) <= l(last ins. node) do
7             tmp ← expand x with y
8             add new embedding tmp to res
9 end
```

Algorithm 3: Expanding Fragments with a *new Root*.

4. insert a *new single edge* to the previously inserted node (whose predecessor has already been inserted, line 12).

The first three rules generate canonical fragments, so just the remaining few duplicates generated by rule 4 must be pruned (line 12). Frequency and other requirements allow more pruning (line 13) before the main loop further expends the lattice.

15.4.3 Expansion Rules

Fig. 15.2 shows a complete search lattice for a database of only one graph (shaded gray) to keep the example simple. Typically, a database contains more than one graph. Different types of edges represent the expansion rules applied. The search lattice nicely demonstrates that in this example none of the fragments (or embeddings) is visited twice, although without a pruning based on the canonical form most fragments could have been reached along several paths. After the initialization, the search lattice holds three single–node fragments.

Rule 1: Because of the topological creation, the first expansion rule (*new root*) can only be applied to fragments with one topological level and without edges (Algo. 3, lines 3–4). This rule completely avoids duplicates because no node with a label index greater than the last one will be inserted (line 7). That is similar to the *candidate item set generation* described in [1]. Consider, for example, the initial fragment B in Fig. 15.2. It is just extended with the root node A and not with C since the fragment (B,C) is already present as (C,B) which has a bigger canonical form $(2,0)(1,0)$ (instead of $(1,0)(2,0)$). More formally: the insertion order of roots is valid and therefore the canonical description $(x_1,0)(x_2,0)...(x_i,0)$ is maximal, if the condition $x_a >= x_b$ holds for every $a < b$.

Rule 2: When a *new topological level* is started by the insertion of a node, the expansion of the current topological level is completed. All duplicates can easily be avoided during this phase by checking *partitions* that reflect the symmetries

```
Data: fragment f, labelIndexFunction l
Result: frequent fragments res
1 begin
2     res ⟵ ∅
3     for node x of f at the current topological level or before the current level do
4         step ⟵ ins. step of x
5         if ¬ samePartition(step, step + 1) then
6             for embedding e ∈ f do
7                 res ⟵ res ∪ expandNewLevel(l,e,x) or expandNewNode(l,e,x)
8 end
```

Algorithm 4: Expanding Fragments with a *new Level* or a *new Node*.

```
Data: labelIndexFunction l, embedding e, node x
Result: frequent fragments res
1 begin
2     res ⟵ ∅
3     superX ⟵ corresponding node to x in supergraph of e
4     for unused edge y to unused node z ∈ supergraph of e do
5         tmp ⟵ expand e with edge y (superX → z)
6         add new embedding tmp to res
7 end
```

Algorithm 5: Subroutine *expandNewLevel*.

in a graph. Partitions are the basis of graph isomorphism tests [10] and can be constructed in polynomial time. Partitions are created by the indegree, outdegree, and node label index of every node and are afterwards iteratively refined based on their neighboring partitions. Regardless of which node of a partition is selected as the predecessor, the resulting graphs are isomorphic. Therefore, a new level can only be started canonically when the last inserted node of a partition (with the highest insertion index) is used as the predecessor (Algo. 4, line 5). Since only neighboring nodes can be in the same partition, the check in line 5 is simple. The subroutine in Algo. 5 finds all unused edges of the supergraph of the current embedding with a used node as startingnode leading to an unused node. Fig. 15.5 shows all possible ways to extend a two–level graph with a new node A, since the predecessor has to be in the last topological level and the last in its partition (the gray boxes).

Rule 3: The insertion of a *new node* at the current level is similar to the previous rule and does not generate duplicates, either. The partition check has to be applied again (Algo. 4, line 5). Since the node label index is the most significant element of each tuple in the canonical form, the next inserted node must have a smaller (or equal) index than its predecessor (Algo. 6, line 4). For equal labels, the new predecessor index also has to be smaller or equal to the previous predecessor index to achieve the maximal canonical description (line 7). In Fig. 15.2 this rule is used once to generate fragment $C \to (A, B)$ with its canonical description

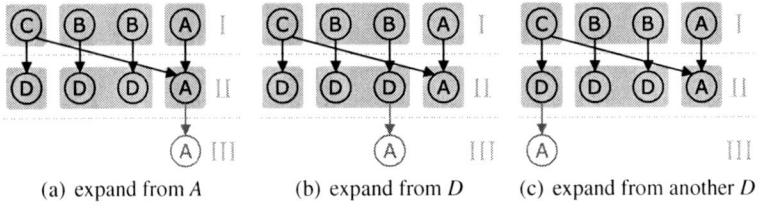

(a) expand from *A* (b) expand from *D* (c) expand from another *D*

Fig. 15.5 Starting a new level III by insertion of a node and an edge.

Data: labelIndexFunction *l*, embedding *e*, node *x*
Result: frequent fragments *res*
1 **begin**
2     $res \longleftarrow \emptyset$
3     $superX \longleftarrow$ corresponding node to *x* in supergraph of *e*
4   **for** *unused edge y to unused node* $z \in$ *supergraph of e with* $l(z) \leq l(\text{last ins. node})$ **do**
5        $stepX \longleftarrow$ ins. step of node *x*
6        $stepLast \longleftarrow$ ins. step of last predecessor
7   **if** $\neg(l(z) = l(\text{last ins. node}) \wedge stepLast < stepX)$ **then**
8           $tmp \longleftarrow$ expand *e* with edge *y* $(superX \rightarrow z)$
9   add new embedding *tmp* to *res*
10 **end**

Algorithm 6: Subroutine *expandNewNode*.

$(2,0)(1,1)(0,1)$. The same fragment could have been reached from the fragment $C \rightarrow A$ with the numerically smaller description $(2,0)(0,1)(1,1)$. Hence it is pruned.

Rule 4:   Probably the most difficult expansion rule is the insertion of a *new single edge* targeting the last inserted node. As before, pruning is based on partitions and predecessor indices (see Algo. 7, line 3 and Algo. 8, line 3). In addition, the set of predecessors of the current and the last inserted node are compared to exclude non-canonical insertion orders (Algo. 8, line 4). This approach can avoid a good portion of potential duplicates but not all of them. Complete avoidance may be possible, but has to be *NP*–complete due to the *NP*–complexity of sub–DAG mining. Hence, there is the usual trade–off: a more complex test is slower but speeds up the search process by more pruning.

Fig. 15.6(a) shows a duplicate of the canonical fragment in Fig. 15.6(b) that is not avoided by the enumeration process and needs to be pruned by an exponential test. We accelerate this test by reusing the partition information computed during expansion. Permuting the insertion order of nodes in the same partition leaves the canonical form unchanged, so only the permutations of partitions at each topological level must be checked. This does not decrease the theoretical complexity compared to permuting all nodes, but speeds up the process considerably.

15 DAG Mining for Code Compaction

Data: fragment f, labelIndexFunction l
Result: frequent fragments res
1 **begin**
2     $res \longleftarrow \emptyset$
3   **if** $samePartition(last\ ins.\ node,\ next\ to\ last\ ins.\ node)$ **then**
4   **return**
5   **for** $embedding\ e \in f$ **do**
6         $superX \longleftarrow$ corresponding node to the last ins. node in supergraph of e
7   **for** $unused\ edge\ y\ from\ used\ node\ z\ to\ last\ ins.\ node \in$ supergraph of e **do**
8             $stepZ \longleftarrow$ ins. step of node z
9             $stepLast \longleftarrow$ ins. step of last edge–adding–node
10   **if** $stepZ < stepLast$ **then**
11                $tmp \longleftarrow expandSingleEdge(l, stepZ, e, y, z)$
12   add new embedding tmp to res
13 **end**

Algorithm 7: Expanding Fragments with a *new Single Edge*.

Data: labelIndexFunction l, ins. step $stepZ$, embedding e, edge y, node z
Result: frequent fragment res
1 **begin**
2     $res \longleftarrow \emptyset$
3   **if** $\neg\ samePartition(stepZ, stepZ + 1) \wedge$
4             $\neg sameLabelAndPredecessors(l,\ last\ ins.\ node,\ next\ to\ last\ ins.\ node)$ **then**
5         $res \longleftarrow$ expand e with edge $y\ (z \rightarrow last\ ins.\ node)$
6 **end**

Algorithm 8: Subroutine *expandSingleEdge*.

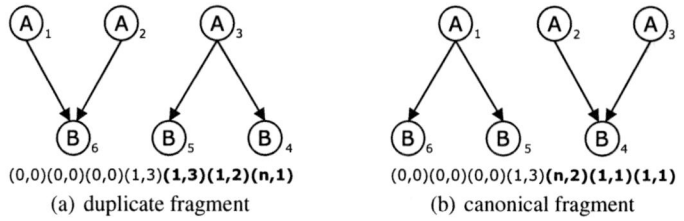

(0,0)(0,0)(0,0)(1,3)**(1,3)(1,2)(n,1)**   (0,0)(0,0)(0,0)(1,3)**(n,2)(1,1)(1,1)**
(a) duplicate fragment   (b) canonical fragment

Fig. 15.6 Duplicate fragment that is not avoided by the enumeration process.

15.4.4 Application to Procedural Abstraction

In general, compilers do not reach minimal code size. PA can reduce code size by extracting duplicate code segments from a program (i.e. the binary code). The instructions of an assembly program do not depend on each other line by line but they can be reordered as long as the data flow dependencies between the instructions are respected. These can be modeled as directed edges in an acyclic graph [11], called *data flow graph*. For PA we minimize the DFGs by removing edges between two nodes if there are alternative chains of dependencies between them. The resulting sparse graphs are faster to mine, but yield the same relevant fragments. Embedding–based DAGMA finds a maximal non-overlapping subset of embeddings of each basic block in the minimized DFGs. Searching for the best maximal non-overlapping subset of *all* embeddings would probably lead to even better results, but its *NP*–complete complexity is too costly for our experiments.

After the mining, we judge the resulting fragments and embeddings with respect to their size, number of occurrences, and type in order to get the maximal code size reduction. Depending on the fragment and extraction type, an extraction by means of a jump my be cheaper than the call–instruction shown in Fig. 15.1. With respect to the compaction profit (if positive), we extract the best fragment by clustering the nodes and edges of the embedding to a new single node and we add new instructions to the graph according to the extraction type (like *return*). Afterwards, DAGMA is applied again and searches for further frequent fragments until no more frequent subgraphs are found or the best compaction profit is below some threshold.

Unfortunately, there are embeddings of induced frequent fragments that cannot be correctly extracted by PA, because they do not respect all original dependencies after such an embedding is extracted. There is a simple way to check if an embedding cannot be extracted: replace all nodes of the embedding with a single new node and redirect edges between in– and outside of the embedding so that they are connected to the new node. If the resulting graph is cyclic, embedding extraction would break dependencies.

As DAGMA expands fragments level by level and node by node some additional pruning is possible. Unregarded dependencies are reflected by cycles in the clustered graph and are the result of missing edges in the embedding. Due to DAGMA's topological expansion, only single–edges towards the last inserted node (or the corresponding instruction) can be added and only cycles that contain this last node can be eliminated by further expansion steps. The expansion of the other nodes is finished at this time and cycles that contain only those finished nodes cannot be included into the embedding. Therefore, those fragments and their expansions can be pruned from the search lattice without affecting the number of instructions PA saves. In our PA experiments, we can prune over 90% of the embeddings that otherwise would be generated.

15.5 Evaluation

To evaluate DAGMA we compared it to gSpan, the most general and flexible graph miner currently available [15]. Since gSpan only addresses connected mining, we extended it with a pseudo–root node that is connected to every other node. This helper node is later removed from the resulting fragments [16]. Both algorithms are implemented in the same Java framework (using Sun JVM version 1.5.0). An AMD Opteron with 2 GHz and 11 GB of main memory has executed our comparisons

15 DAG Mining for Code Compaction

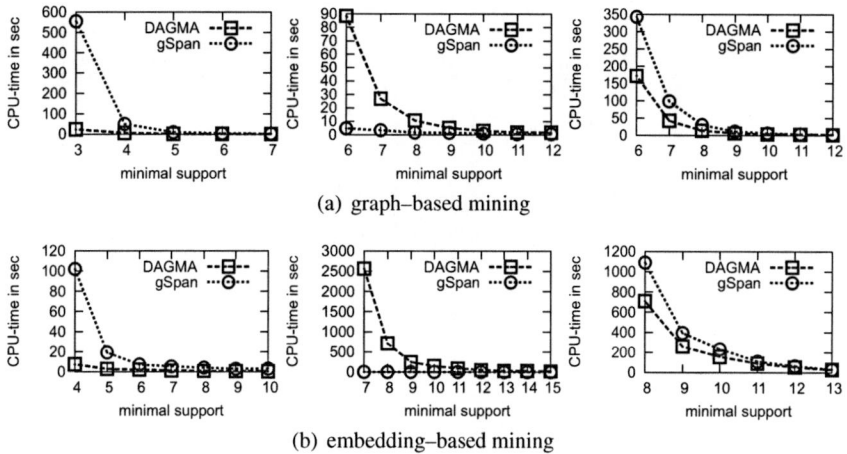

Fig. 15.7 Runtimes for mining single–rooted (left), connected (middle) and unconnected (right) fragments(buttom) in a synthetic DAG database

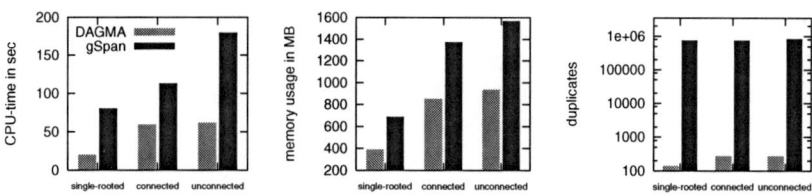

Fig. 15.8 Runtime, memory, and number of duplicates for a fully connected DAG with 7 nodes

on synthetic DAG databases, on a worst case database, and on a database from our application domain (Procedural Abstraction).

Synthetic DAGs were generated as follows to contain similarities: Every node and edge reachable from a randomly selected node in a big random master DAG is copied into the DAG database [3]. Our master DAGs contain 50 nodes, 200 edges, and 10 labels each. We restrict sub–DAGs to 5 topological levels or 25 nodes. Regardless of the random database, we always got almost the same results. Fig. 15.7 compares the runtime for graph– and embedding–based support. For both types, our approach clearly outperforms gSpan, except for connected fragments because of our preprocessing that filters out unconnected fragments. The number of fragments and embeddings is significantly higher when mining unconnected. Since an unconnected fragment can become connected during expansion, no pruning is possible and our approach has to do much unnecessary work. For a decreasing minimal support resp. an increasing number of embeddings the differences between the approaches get more prominent regardless of the fragments' shapes. A simple extension restriction leads to single–rooted mining in DAGMA: After computing the initial set of frequent nodes, no other root is added to the fragments.

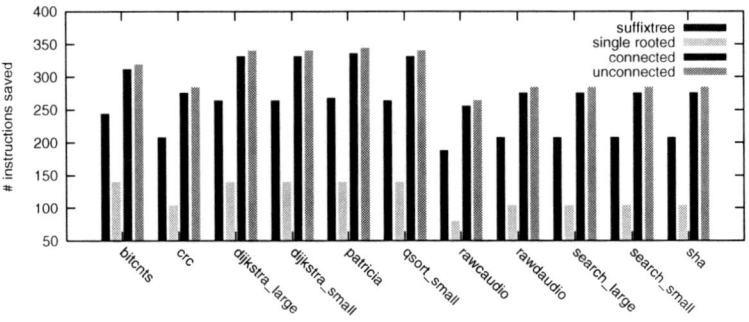

Fig. 15.9 Instruction savings for programs from MiBench

The *worst case* for DAG (and graph) miners is an equally labeled and fully connected DAG that can be created stepwise by inserting a node and connecting it to all previous nodes until the desired number of nodes is reached. Fig. 15.8 shows the results for an embedding–based search with minimal support 1 on such a maximal DAG with seven nodes. In that case 2,895,493 embeddings can be found. Again, DAGMA clearly outperforms gSpan with respect to both runtime and memory consumption in all three mining types. The main advantage and the reason for this behavior become apparent on the right of Fig. 15.8: Due to its DAG–specific canonical form, DAGMA has to handle far less duplicates in costly isomorphism tests than gSpan. The same holds for the synthetic databases.

To evaluate our algorithm for *PA*, we transformed several ARM assembly codes from the MiBench suite [9] into DFGs and mined embedding–based. Fig. 15.9 gives the savings in code size compared to the original code size when mining with suffix trees, mining for single–rooted, connected, and unconnected fragments. Mining for single–rooted DAGs is not as successful as mining with suffix trees. But when searching for connected fragments a lot more instructions can be saved. The search for unconnected fragments leads to the best results, yielding smaller assembly code and therefore higher efficacy.

15.6 Conclusion and Future Work

With DAGMA, we presented a flexible new DAG mining algorithm that is able to search for induced, unconnected or connected, multi– or single–rooted fragments in DAG databases. Since both graph– and (induced) embedding–based mining is possible (the latter is necessary for PA) and since DAGMA can mine both connected and unconnected (necessary for PA), DAGMA can be used for several application scenarios. The novel canonical form and the basic operations of the miner are based on the fact that DAGs have topological levels. The new algorithm faces significantly fewer duplicates in the search space enumeration compared to the general graph miner gSpan. This leads to faster runtime and reduced memory consumption. When applied to Procedural Abstraction, DAGMA achieves more code size reduction than traditional approaches. Procedural Abstraction traditionally searches with a minimal support of 2 embeddings to get the best possible results. For big binaries, the resulting graphs sometimes have been too large to be mined at such a small support. Hence, it seems necessary to study heuristics that guide the mining process. It will probably also require parallel DAG mining.

References

1. Rakesh Agrawal, Tomasz Imielinski, and Arun Swami. Mining association rules between sets of items in large databases. *SIGMOD Record*, 22(2):207–216, May 1993.
2. Christian Borgelt and Michael R. Berthold. Mining Molecular Fragments: Finding Relevant Substructures of Molecules. In *Proc. IEEE Int'l Conf. on Data Mining (ICDM'02)*, pages 51–58, Maebashi City, Japan, December 2002.
3. Y.-L. Chen, H.-P. Kao, and M.-T. Ko. Mining DAG Patterns from DAG Databases. In *Proc. 5th Int'l Conf. on Advances in Web-Age Information Management (WAIM '04)*, volume 3129 of *LNCS*, pages 579–588, Dalian, China, July 2004. Springer.
4. Y. Chi, R. Muntz, S. Nijssen, and J. Kok. Frequent subtree mining – an overview. *Fundamenta Informaticae*, 66(1-2):161–198, 2005.
5. Thomas H. Cormen, Charles E. Leiserson, Ronald L. Rivest, and Clifford Stein. *Introduction to Algorithms, Second Edition*. The MIT Press and McGraw-Hill Book Company, 2001.
6. S.K. Debray, W. Evans, R. Muth, and B. De Sutter. Compiler Techniques for Code Compaction. *ACM Trans. on Programming Languages and Systems*, 22(2):378–415, March 2000.
7. A. Dreweke, M. Wörlein, I. Fischer, D. Schell, T. Meinl, and M. Philippsen. Graph-Based Procedural Abstraction. In *Proc. of the 5th Int'l Symp. on Code Generation and Optimization*, pages 259–270, San Jose, CA, USA, 2007. IEEE.
8. Scott Fortin. The Graph Isomorphism Problem. Technical Report 20, University of Alberta, Edmonton, Canada, July 1996.
9. M. R. Guthaus, J. S. Ringenberg, D. Ernst, T. M. Austin, T. Mudge, and R. B. Brown. MiBench: A free, commercially representative embedded benchmark suite. In *Proc. Int'l Workshop on Workload Characterization (WWC '01)*, pages 3–14, Austin, TX, Dec. 2001.
10. Brendan McKay. Practical Graph Isomorphism. *Congressus Numerantium*, 30:45–87, 1981.
11. Steven S. Muchnick. *Advanced Compiler Design and Implementation*. Morgan Kaufmann Publishers Inc., San Francisco, CA, USA, 1997.
12. Siegfried Nijssen and Joost N. Kok. A Quickstart in Frequent Structure Mining can make a Difference. In *Proc. Tenth ACM SIGKDD Int'l Conf. on Knowledge Discovery and Data Mining (KDD '04)*, pages 647–652, Seattle, WA, USA, August 2004. ACM Press.
13. Robert Endre Tarjan and Anthony E. Trojanowski. Finding a Maximum Independent Set. *SIAM Journal on Computing (SICOMP)*, 6(3):537–546, 1977.
14. A. Termier, T. Washio, T. Higuchi, Y. Tamada, S. Imoto, K. Ohara, and H. Motoda. Mining Closed Frequent DAGs from Gene Network Data with Dryade. In *20th Annual Conf. of the Japanese Society for Artificial Intelligence*, pages 1A2–3, Tokyo, Japan, June 2006.
15. M. Wörlein, T. Meinl, I. Fischer, and M. Philippsen. A quantitative comparison of the subgraph miners MoFa, gSpan, FFSM, and Gaston. In *Proc. Conf. on Knowledge Discovery in Database (PKDD'05)*, volume 3721 of *LNCS*, pages 392–403, Porto, Portugal, October 2005.
16. Xifeng Yan and Jiawei Han. gSpan: Graph-Based Substructure Pattern Mining. In *Proc. IEEE Int'l Conf. on Data Mining (ICDM'02)*, pages 721–724, Maebashi City, Japan, Dec. 2002.
17. David Zaretsky, Gaurav Mittal, Robert P. Dick, and Prith Banerjee. Dynamic Template Generation for Resource Sharing in Control and Data Flow Graphs. In *Proc. 19th Int'l Conf. on VLSI Design*, pages 465–468, Hyderabad, India, January 2006.

Chapter 16
A Framework for Context-Aware Trajectory Data Mining

Vania Bogorny and Monica Wachowicz

Abstract The recent advances in technologies for mobile devices, like GPS and mobile phones, are generating large amounts of a new kind of data: trajectories of moving objects. These data are normally available as sample points, with very little or no semantics. Trajectory data can be used in a variety of applications, but the form as the data are available makes the extraction of meaningful patterns very complex from an application point of view. Several data preprocessing steps are necessary to enrich these data with domain information for data mining. In this chapter, we present a general framework for context-aware trajectory data mining. In this framework we are able to enrich trajectories with additional geographic information that attends the application requirements. We evaluate the proposed framework with experiments on real data for two application domains: traffic management and an outdoor game.

16.1 Introduction

Trajectories left behind moving objects are spatio-temporal data obtained from mobile devices. These data are normally represented as *sample points*, in the form (tid, x, y, t), where *tid* represents the trajectory identifier and x,y correspond to a position in space at a certain time t. Trajectory sample points have very little or no semantics [1], [2], and therefore it becomes very hard to discover interesting patterns

---

Vania Bogorny
Instituto de Informatica, Universidade Federal do Rio Grande do Sul (UFRGS), Av. Bento Gonalves, 9500 - Campus do Vale - Bloco IV, Bairro Agronomia - Porto Alegre - RS -Brasil, CEP 91501-970 Caixa Postal: 15064, e-mail: vbogorny@inf.ufrgs.br

Monica Wachowicz
ETSI Topografia, Geodesia y Cartografa, Universidad Politecnica de Madrid, KM 7,5 de la Autovia de Valencia, E-28031 Madrid - Spain, e-mail: m.wachowicz@topografia.upm.es

from trajectories in different application domains. Figure 16.3(1) shows an example of a trajectory sample.

On the contrary of most conventional data, trajectories can be used in several application domains. For instance, trajectories obtained from GPS devices of car drivers can be used for traffic management, for urban planning, for insurance companies, and so on. In a traffic management application, for instance, to discover causes of traffic jams, different domain spatial information has to be considered, like for instance, streets, traffic lights, roundabouts, objects close to roads that may cause the jams (e.g. schools, shopping centers), the maximal speed, and so on. For car insurance companies, important domain information might be the type of road, high risk places, velocity, etc.

Currently, there are neither tools for automatically adding semantics to trajectories nor for domain-driven data mining. There is a need for new methodologies that allow the user to give the appropriate semantics to trajectories in order to extract interesting patterns in a specific domain.

Since the semantics of trajectory data is application dependent, the extraction of interesting, novel, and useful patterns from trajectories becomes domain dependent. Several data mining methods have been recently proposed for mining trajectories, like for instance [3–10]. In general, these approaches have focused on the geometrical properties of trajectories, without considering an application domain. As a consequence, these methods tend to discover geometric trajectory patterns , which for several applications can be useless and uninteresting. Geometric patterns are normally extracted based on the concept of dense regions or trajectory similarity. Semantic or domain patterns, however, are related to a specific domain, can be independent of x,y coordinates, and may be located in sparse regions without geometric similarity.

Figure 16.1 shows an example of geometric and semantic patterns . In Figure 16.1(left), the trajectories would generate a geometric pattern represented by a dense region B (e.g. 100% of the trajectories cross region *B*). Considering the semantics of the trajectories in the context of a *tourism application*, we can observe in Figure 16.1(right) two domain patterns: (i) a move from Hotel (H) to Restaurant (R) passing by region B; and (ii) a move to Cinema (C), passing by region B.

Trajectory data mining algorithms which are based on density or trajectory similarity, in the example would discover that several trajectories converge to region B or cross region B. From the application domain point of view, this pattern would only be interesting if B is an important place which is relevant for the problem in hand. Otherwise, this pattern will be useless and uninteresting for the user.

Concerning the needs of the user for knowledge discovery from trajectories in real applications, we claim that new techniques have to be developed to extract meaningful, useful, and understandable patterns from trajectories. Recently two different methods for adding semantics to trajectories according to an application domain have been proposed [1], [2]. Based on these methods, one of the authors has developed the first data mining query language to extract multiple-level semantic patterns from trajectories [11].

In this book chapter we present a trajectory data mining framework where the user gives to the data the semantics that is relevant for the application, and therefore the discovered patterns will refer to a specific domain. While in most existing approaches data mining is performed over *a set of trajectories* represented as *sample points*, like for instance, clusters of trajectories located in dense regions [8], a set of trajectories that move between regions in the same time interval [12], trajectories with similar shapes [13], or with similar distances [14], our framework preprocesses *single trajectories* to add semantic information, and then apply data mining methods over *semantic trajectories*.

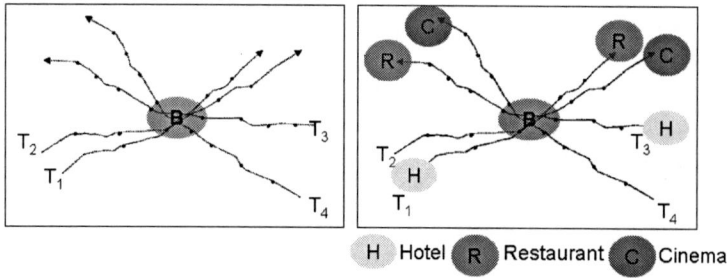

Fig. 16.1 Recurrence geometric pattern X semantic trajectory patterns

The remaining of the chapter is organized as follows: in Section 16.2 we present some basic concepts about geographic data and trajectories. In Section 16.3 we present a framework for context-aware data mining from moving object trajectories. In Section 16.4 we present experiments with real data in two application domains. Finally, in Section 16.5 we conclude the chapter.

16.2 Basic Concepts

Geographic data are real world entities, also called spatial features, which have a location on the Earth's surface [15]. Spatial features (e.g. France, Germany) belong to a feature type (e.g. country), and have both non-spatial attributes (e.g. name, population) and spatial attributes (geographic coordinates x,y). The latter is normally represented as points, lines, polygons, or complex geometries.

In geographic databases every different spatial feature type is normally stored in a different database relation, since most geographic databases follow the relational approach [16]. Figure 16.2 shows an example of how geographic data can be stored in relational databases. There is a different relation for every different geographic object type [17] street, water resource, and gas station, which can also be called as spatial layers.

The spatial attributes of geographic object types, represented by shape in Figure 16.2, have implicitly encoded spatial relationships (e.g. close, far, contains, intersects). Because of these relationships real world entities can affect the behavior of other features in the neighborhood. This makes spatial relationships be the main characteristic of geographic data to be considered for data mining and knowledge discovery, and the main characteristic which differs spatial data mining from non-spatial data mining.

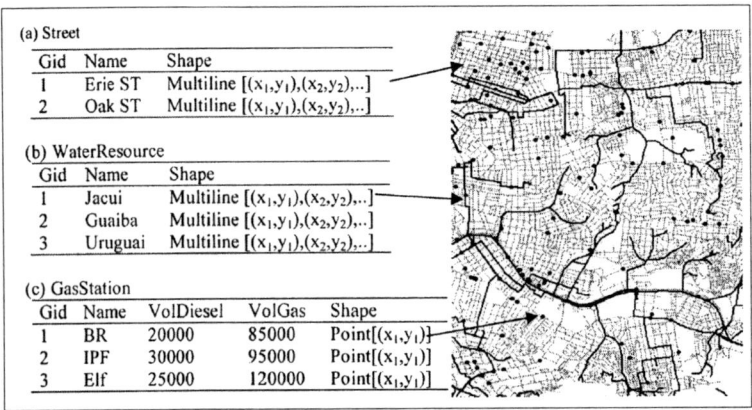

Fig. 16.2 Example of geographic data in relational databases

In order to extract interesting patterns from trajectories, they have to be integrated with geographic information. Trajectory data are normally available as *sample points*, and are in general not related to any geographic information when these data are collected.

Definition 16.1. Trajectory Sample: A trajectory sample is a list of space-time points $\{p_0 = (x_0, y_0, t_0), p_1 = (x_1, y_1, t_1), \ldots, p_N = (x_N, y_N, t_N)\}$, where $x_i, y_i \in \Re$, $t_i \in \Re^+$ for $i = 0, 1, \ldots, N$, and $t_0 < t_1 < t_2 < \cdots < t_N$.

To extract *domain patterns* from trajectories there is a need for annotating trajectories with domain-driven geographic information. For instance, let us analyze Figure 16.3. The first trajectory is a trajectory sample, without any semantic geographic information. The second trajectory is a trajectory sample that has been enriched with geographic information according to a tourism application domain, where the important places are airport, museums, monuments, and hotels. The third trajectory is a sample enriched with information that is relevant for a transportation management application, and the important parts of the trajectory are: crowded places (airport), roundabouts, traffic jams (low velocity), and crossroads.

Trajectory samples integrated with domain geographic information we call *domain trajectories*.

Definition 16.2. Domain Trajectory: a domain trajectory D is a finite sequence $\{I_1, I_2, \ldots, I_n\}$, where I_K is an important place of the trajectory from an application

16 A Framework for Context-Aware Trajectory Data Mining

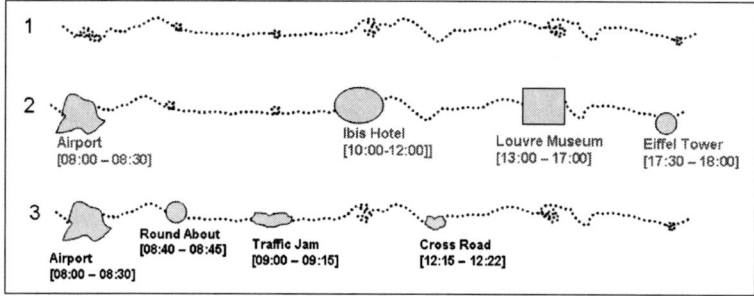

Fig. 16.3 (1) Trajectory Sample, (2)Tourism Domain Trajectory, and (3)Transportation Domain Trajectory

point of view. Every important place is a set $I_m = (G_m, S_m, E_m)$, where G_m represents the geometry of the important place, S_m corresponds to the starting time that the trajectory has entered the place, and E_m is the ending time, when the trajectory has left the important place.

From domain trajectories we are able to compute *domain patterns*. While a geometric pattern could be a move from a region *A* to region *B* passing by region *C*, a domain pattern would be for instance, a move from *Home* to *Work* passing by *ChildrenSchool*.

Definition 16.3. Domain Pattern. Being *A* an application domain and *I* a set of important places, a domain pattern is a subset of *I* that occur in a minimal number of trajectories.

In the next section we present a framework where we go from trajectory sample points to domain-driven semantic trajectories for knowledge discovery.

16.3 A Domain-driven Framework for Trajectory Data Mining

Raw trajectory data are collected from the Earth's surface similarly to any kind of geographic data, as shown in the raw data level in Figure 16.4. It is known that raw geographic data require a lot of work to be transformed into maps, normally stored into shape files (processed data level in Figure 16.4). The *processed geographic data* on the one hand are generated for any application domain, and therefore are *application independent*. They can be used, for instance, to build *geographic databases* for applications of transportation management, tourism, urban planning, etc. *Geographic databases*, on the other hand, are *application dependent*, and therefore, will contain only the entities of processed geographic data that are relevant to the application, as shown in the application level in Figure 16.4.

Data mining and knowledge discovery are on the top level, and are also application dependent. In any data mining task the user is interested in patterns about a specific problem or application. For instance, transportation managers are interested in patterns about traffic jams, crowded roads, accidents, etc, but are not interested in, for instance, patterns of animal migration.

In trajectory pattern mining, in general, mining has been directly performed over raw trajectory data or trajectory sample points, and the background geographic information has not been integrated to the mining process. For data mining and knowledge discovery, trajectories represented as sample points need to be a priori integrated with domain information [11]. Several domain patterns will never emerge in case the meaning is added a posteriori to the patterns, instead of a priori, to the data. This can be observed in the example shown in Figure 16.1. If we apply data mining directly over trajectory samples (Figure 16.1 left), we would only discover that all trajectories pass by region B. By adding semantics to the *patterns* (a posteriori), we would discover that B is a shopping mall. By adding semantics to the *data* (a priori), we would obtain the trajectories shown in Figure 16.1 (right), where the spatial object types (2 hotels, 2 restaurants, and 2 cinemas) have different spatial locations among each other but the same meaning. Considering that stops must appear in 50% of the trajectories to be a pattern, in this example we would discover (1) a move from Hotel to Restaurant passing by Shopping Center and (2) a move from Shopping Center to Cinema. Data mining methods based on density or similarity which consider only space and time, would never semantically group these data to find patterns.

In our framework, shown in Figure 16.4, in the *Application Domain Level*, we propose to generate domain-enriched trajectories, using the model of *stops*, introduced in [18], for finding important places or regions of interest, and generate a semantic trajectory database.

Stops represent the important places of a trajectory where the moving object has stayed for a minimal amount of time. Recently two different methods have been developed for computing stops of trajectories: SMoT [1] and CBSMoT [2]. SMoT is based on the intersection of trajectories with relevant spatial features (places of interest). The important places are defined according to an application, and correspond to different spatial feature types [15] defined in a geographic database (see Section 16.2). For each relevant spatial feature type a minimal amount of time is defined, such that a trajectory should continuously intersect this feature in order to be considered a stop.

In SMoT, a stop of a trajectory T with respect to an application $\mathscr{A}$ is a tuple (G_k, t_j, t_{j+n}) such that a maximal subtrajectory of T $\{(x_i, y_i, t_i) \mid (x_i, y_i)$ *intersects* $G_k\} = \{(x_j, y_j, t_j), (x_{j+1}, y_{j+1}, t_{j+1}), \ldots, (x_{j+n}, y_{j+n}, t_{j+n})\}$, where R_k is the geometry of C_k and $|t_{j+n} - t_j| \geq \Delta_k$.

The method SMoT is interesting for applications in which the velocity of the trajectory is not really important, and the emphasis is on the places where the moving object has stayed for a minimal amount of time. For example, in a tourism application the most important are the places where the tourist has stopped, and not how fast he moved from one place to another. Similarly, in a public transportation application it is important to identify the places where transportation users leave, the intermediate places where they stop to take/change transportation, and the final des-

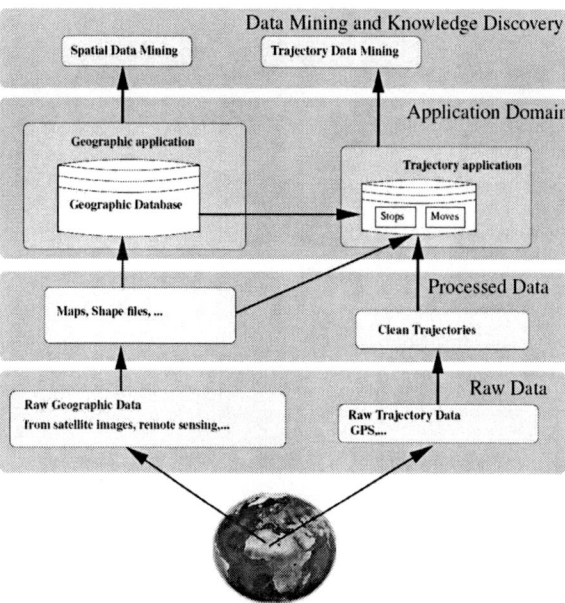

Fig. 16.4 Context-aware trajectory knowledge discovery process

tination. With such information it is possible to the decision maker to create new transportation means which directly connect origin and destination.

CBSMoT is a spatio-temporal clustering-based method where the important places (stops) are the low speed parts that satisfy a minimal time duration threshold. The speed is computed through the distance between two points of the trajectory divided by the time spent in the movement between the two points. This method is very interesting for applications where the speed is the most important, i.e., a stop is generated when the speed is lower than a given threshold. For instance, in traffic management applications the speed of the trajectory is the most important to detect traffic jams (places with low speed). For a car insurance company, the velocity of the trajectory in forbidden/dangerous places as well as the average speed is the most important.

An overview of the architecture of the framework is shown in Figure 16.5. In this framework there are basically three abstraction levels. On the bottom are the data: raw trajectory data or trajectory samples, the geographic information, and the domain-driven trajectories. Domain-driven trajectories are obtained from the integration of trajectory samples and geographic information, as a result of the preprocessing phase.

In the center are the preprocessing tasks, where raw trajectory data are integrated to the domain geographic information, using one of the two methods proposed in [1], [2]. Stops are computed and stored in the data level, as *domain trajectories*. For data mining and knowledge discovery a fundamental preprocessing task is the transformation of the data in different granularity levels [19]. This is essentially

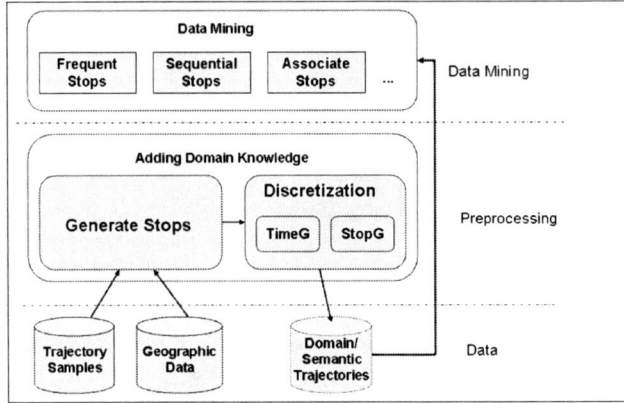

Fig. 16.5 Architecture of a context-aware framework for trajectory data mining

important for trajectory data, where both space and time need to be aggregated in order to allow the discovery of patterns [11]. We provide this task in our framework. In the module *Discretization*, the semantic trajectories or domain trajectories pass through a discretization process and are transformed in different granularity levels. The user may aggregate both space and time in different granularities, like for instance, *morning, afternoon, rushHour* for time and *hotel, 5starhotel, IbisHotel* for space. The modules *TimeG* and *StopG* respectively transform time and space in the granularity specified by the user.

On the top of the framework are the data mining tasks. Once the raw trajectory data are preprocessed and transformed into sequences of stops, different data mining tasks can be applied. The idea is to transform raw trajectories into a sequence of stops, where stops are the important places that are relevant for an application domain. Once we have trajectories represented as a sequence of stops, several classical data mining techniques may be applied, including frequent patterns, association rules, and sequential patterns.

16.4 Case Study

In this section we describe two different applications and present some experiments to show the usability of the framework for trajectory domain-driven data mining. The framework has been implemented as an extension of Weka-GDPM [20] for trajectories. This tool provides a graphical GUI which (1) connects to a Postgres/PostGIS database to read trajectory samples and the relevant geographic information (selected by the user), (2) computes stops with the SMoT and CBSMoT method, and (3) stores the stops back into the database, from where we export their shapefiles to visualize the patterns on a map using the tool ArcExplorer. Our experiments were performed with the sequential pattern mining method [21].

16.4.1 The Selected Mobile Movement-aware Outdoor Game

Paper chase is an old children's game in which an area is explored by means of a set of questions and hints on a sheet of paper. Each team tries to find the locations that are indicated on the questionnaire. Once a place is found, they try to best and quickly answer the questions on the sheet, note down the answer, and proceed to the next point of interest. Based on this game idea, we have used the mobile location-aware game developed by Waag Society, in the Netherlands.

Waag Society developed a 'mobile learning game' pilot together with IVKO, part of the Montessori comprehensive school in Amsterdam [22]. It is a city game using mobile phones and GPS-technology for students in the age of 12-14 (so called HAVO+MAVO basic curriculum). It is a research pilot examining whether it is possible to provide a technology supported educational location-based experience. In the Frequency 1550 mobile game, students are transported to the medieval Amsterdam of 1550 via a medium that's familiar to this age group: the mobile phone. The pilot took place in 2005 from 7 to 9 February.

The game mainly consists of a set of geo-referenced checkpoints associated with multimedia riddles. With the mobile game client a player logs into the game server, receives a historical map with checking points, and the player has to find in real life where the checking point locations are. Each of the check points is geo-referenced by a Gauss Krueger coordinate which is transformed into a screen coordinate and drawn on the map. The player's device includes a GPS receiver which continuously tracks the current position of the device.

Each riddle has associated resources like an image or other additional (media) information that are needed to solve the riddle with the respective interaction(s). The player tries to solve the riddle not only correctly but also as quick as possible, because the time needed to solve all the riddles is accumulated and added to the overall score. The answer to the riddle is communicated to the game server.

In Figure 16.6 we show the background geographic information of the center of Amsterdam (squares - white polygons), in light gray the sample points that correspond to the trajectories of the students, and in black the stops of the students. The students were divided in six groups named with colors: red, purple, green, yellow, blue, and orange. The stops were computed with the method SMoT [1], because the objective is to investigate how long the students have stopped at each place, and not the velocity as they moved from one place to another. In this experiment we have considered 2 minutes as the minimal time duration for a place to be considered a stop.

An analysis over the stops is shown in Table 16.1, which is a summary of the number of stops of each team and the total amount of time that each group has spend at the stops. According to this information we can conclude that the *red* team was

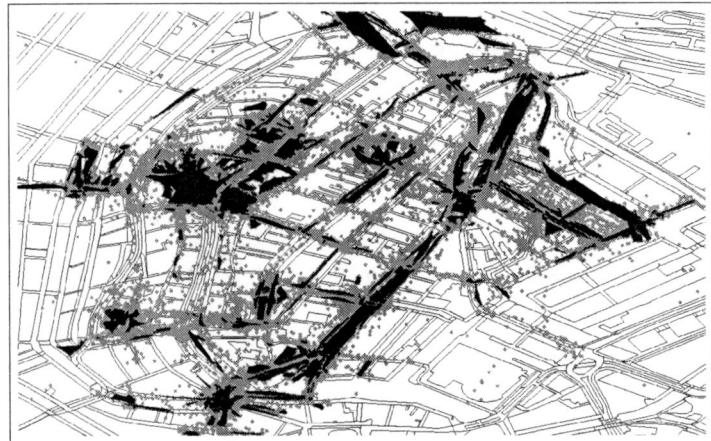

Fig. 16.6 Squares of Amsterdam (white polygons), trajectories (gray dots), and stops (black polygons)

the fastest, and had 21 stops, of at least 2 minutes each, during the game, spending a total of 30 hours. The *yellow* team, although it has the lowest number of stops (19), the group has spend to much time at the stops, summarizing a total of 45 hours. In general, using our framework we can conclude that the red time was the fastest to obtain the information on each historical place, while the yellow time was the slowest.

Table 16.1 Teams, stops, and total time duration

Team	Stops	Duration
Red	21	30:13:09
Green	30	36:20:17
Purple	25	36:21:01
Blue	29	36:33:58
Orange	23	38:56:44
Yellow	19	45:36:09

16.4.2 Transportation Application

A second experiment was performed over trajectory data collected in the city of Rio de Janeiro, Brazil. Each trajectory corresponds to a sensor car equipped with a GPS device. Several cars have been sent to drive in the city of Rio the Janeiro, in several locations, and in different days. The cars play the role of a sensor, and with the collected spatio-temporal data the objective is to identify regions and the time period in which the traffic is slow.

The trajectory dataset has 2,100 trajectories with more than 7 million points. Experiments were performed considering streets and districts of the city of Rio de Janeiro and different time intervals. First, we computed the stops of the trajectories with the method SMoT, considering districts as the relevant spatial features, to identify the sequences of districts that have stops. In this experiment we considered minimum support 10% and minimal stop duration of 120 seconds. Figure 16.7 shows one of the computed patterns and a subset of trajectories. The polygons correspond to the districts, the thick line corresponds to a set of trajectories and the highlighted lines over the trajectories represent stops. In the first pattern (two stops), if there is a stop in the district Sao Conrado at the time interval 17:00-19:00, then there will also be a stop in Joa at this time interaval, in the direction Sao Conrado - Joa.

In the second pattern, shown on the second map in Figure 16.7, a stop occurs at the district Barra da Tijuca and right after in Joa, between 07:00-09:00. In the third pattern, shown on the third map in Figure 16.7, between 07:00-09:00 a stop occurs in Barra da Tijuca, then in Joa, and then in Leblon, in this relative order.

To go deeper into details of the three patterns shown in Figure 16.7, we performed a more refined experiment, considering streets as relevant spatial features. We then discovered that in the districts Sao Conrado and Joa the streets in which stops occur are respectively *Auto Estrada Lagoa Barra* and *Elevada das Bandeiras*. In the districts Barra da Tijuca and Joa, the streets *Elevada das Bandeiras* and two different parts of *Avenida das Americas* have stops. In the third pattern, the streets *Mario Ribeiro* and two parts of *Avenida das Americas* have stops.

A second experiment was performed using the method CBSMoT, which is shown in Figure 16.8. The first map in Figure 16.8 shows an experiment considering a subset of trajectories located in the eastern part of Rio de Janeiro. Stops were computed with minimal time duration of 160 seconds (3 minutes). This pattern is an example of clusters that represent three unknown stops [2].

The second map in Figure 16.8 shows the result of an experiment with a subset of trajectories located in the southern part of Rio de Janeiro, collected in the district of Sao Conrado. In this experiment we considered the time granularity of *weekday* and *weekend*, and only sequential patterns for *weekdays* have been generated, for minimum support 2%, 3% and 4% (with higher support there are no patterns). In this experiment, all generated patterns had the time granularity in the intervals 8:30-9:30 and 16:30-19:30, which characterize rush hours.

The third map, shown in Figure 16.8, presents another experiment, performed with all trajectories, to compare the methods SMoT and CBSMoT. The red lines represent the stops computed by the method SMoT, where each stop is a street that the moving object has intersected for at least 60 seconds. In black are the clusters computed with the method CBSMoT, where the speed of the trajectory is the main threshold. We can observe that the method SMoT generates much more stops than CBSMoT, but the stops generated by SMoT do not necessarily characterize slow movement. On the other hand, the clusters generated by CBSMoT are the regions where the velocity of the trajectories is lower than other parts. This analysis shows that depending on the problem in hand, both methods for adding semantics to trajectories can be useful even for the same application domain.

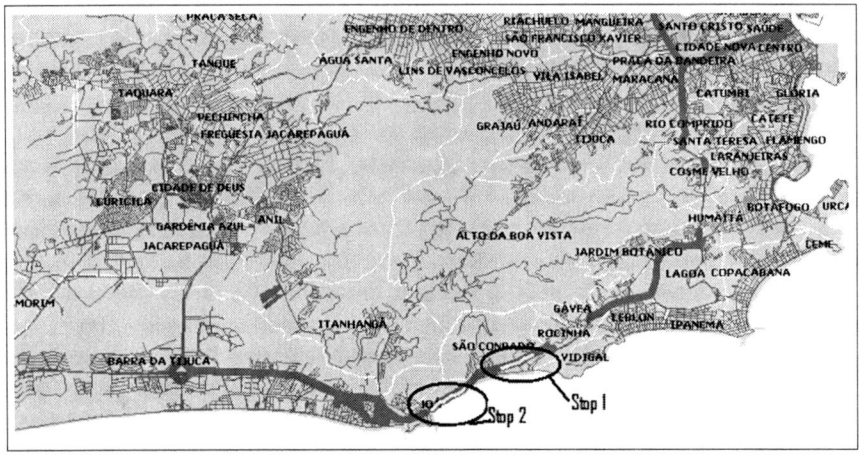
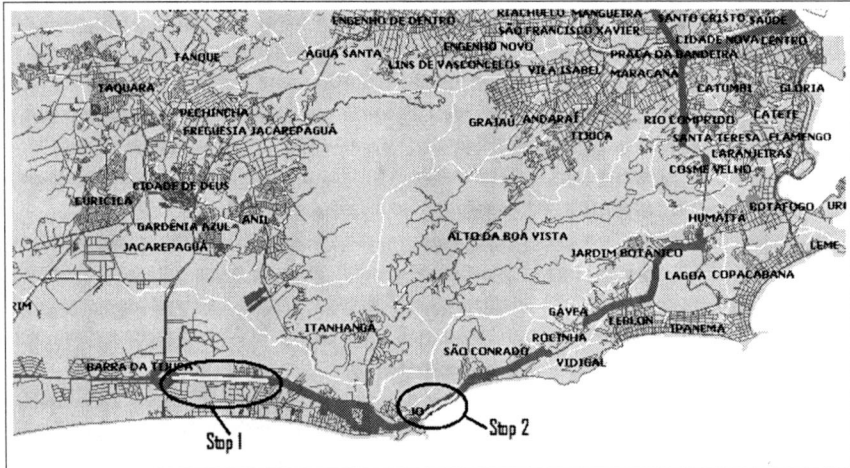
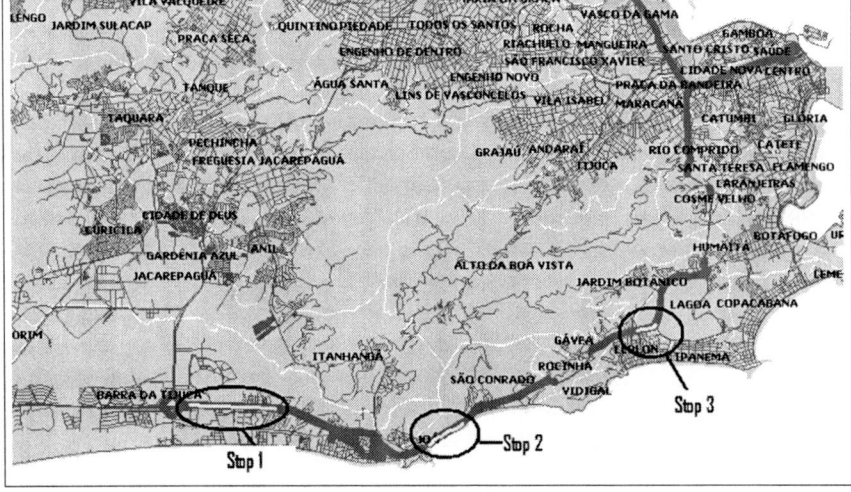

Fig. 16.7 Trajectory patterns with Districts being the relevant spatial features

16 A Framework for Context-Aware Trajectory Data Mining

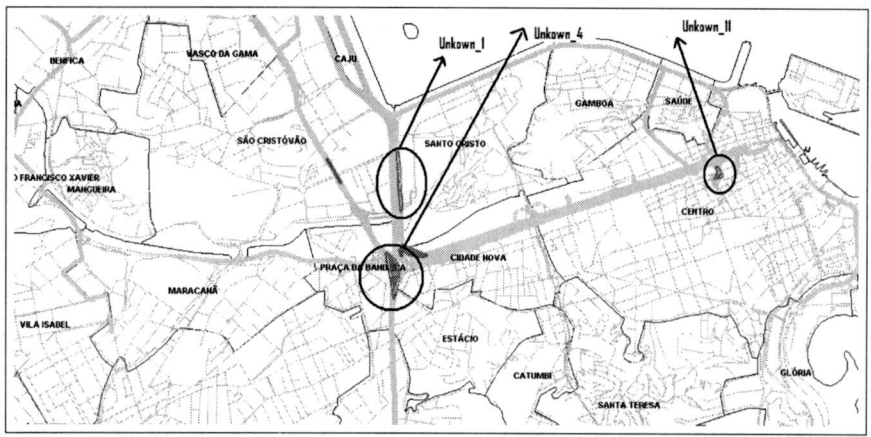

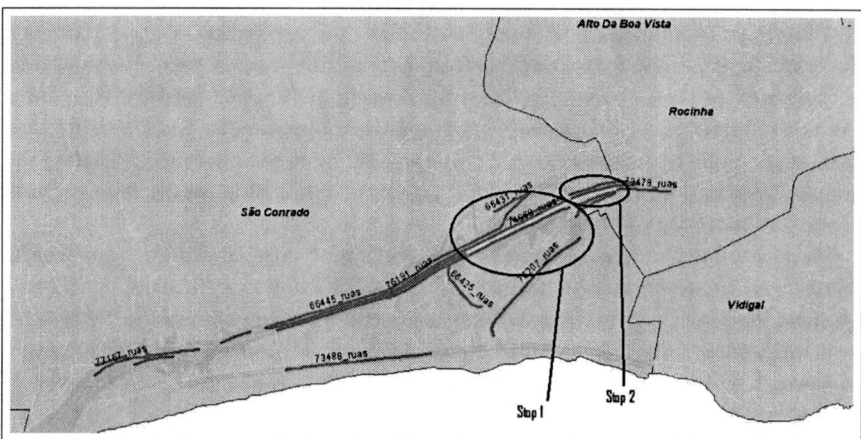

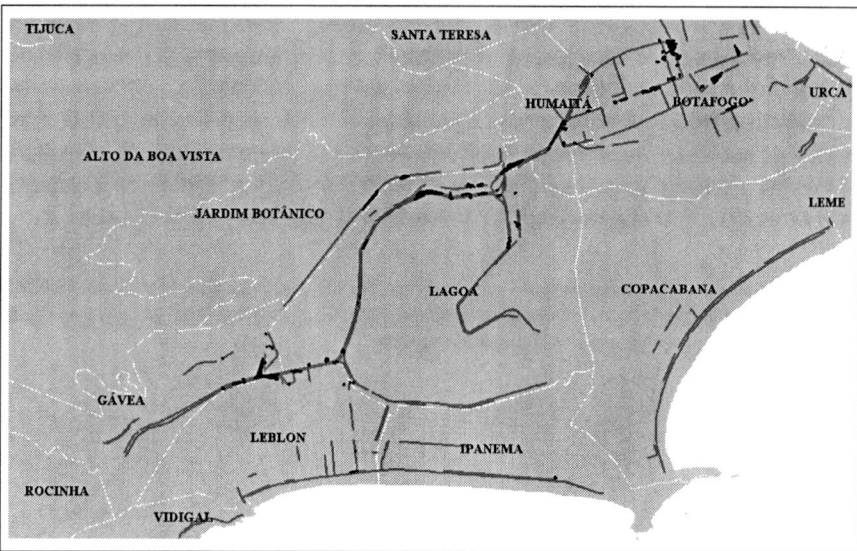

Fig. 16.8 (1) Stops generated with the method CBSMoT (red lines) and Three highlighted unknown stops; (2) Stops (red lines), trajectories (green lines) and 2 sequential stops (highlighted lines); and (3) Stops generated with the methods SMoT (red lines) and CBSMoT (black polygons)

In the transportation application we would like to consider several relevant spatial feature types like roundabouts, semaphores, speed controllers, and other relevant spatial objects that could be related to the stops. However, these data were not available for our experiments, so studies with this kind of data will be done in future works. We will also evaluate the extracted stops and sequential patterns with the user of the application domain.

16.5 Conclusions and Future Trends

Spatio-Temporal data are becoming very common with the advances in technologies for mobile devices. These data are normally available as sample points, with very little or no semantics. This makes their analysis and knowledge extraction very complex from an application point of view. In this chapter we have addressed the problem of mining trajectories from an application point of view. We presented a framework to preprocess trajectories for domain-driven data mining. The objective is to integrate, a priori, domain geographic information that is relevant for data mining and knowledge discovery. With the model of stops the user can specify the domain information that is relevant for a specific application, in order to perform context-aware trajectory data mining.

We have evaluated the framework with real data from two different application domains, what shows that the framework is general enough to be used in different application scenarios. This is possible because the user can choose the domain information that is important for data mining and knowledge discovery. The proposed framework is very simple and easy to use. It was implemented as an extension of Weka-GDPM [20] for trajectory data mining.

Trajectories of moving objects are a new kind of data and a new research field, for which new theories, data models, and data mining techniques have to be developed. Spatio-temporal data generated by mobile devices are raw data that need to be enriched with additional domain information in order to extract interesting patterns.

Domain-driven data mining is an open research field, specially for spatial, temporal, and spatio-temporal data. We believe that in the future new data mining algorithms that consider data semantics and domain information have to be developed in order to extract more meaningful patterns in different application domains.

Acknowledgements Our special thanks to the Waag Society,the Transportation Council of Rio de Janeiro, and Jose Macedo for the real trajectory data. To CAPES (PRODOC Program) and GeoPKDD (www.geopkdd.eu) for the financial support.

References

1. Alvares, L.O., Bogorny, V., Kuijpers, B., de Macedo, J.A.F., Moelans, B., Vaisman, A.: A model for enriching trajectories with semantic geographical information. In: ACM-GIS, New York, NY, USA, ACM Press (2007) 162–169
2. Palma, A.T., Bogorny, V., Kuijpers, B., Alvares, L.O.: A clustering-based approach for discovering interesting places in trajectories. In: ACMSAC, New York, NY, USA, ACM Press (2008) 863–868
3. Cao, H., Mamoulis, N., Cheung, D.W.: Discovery of collocation episodes in spatiotemporal data. In: ICDM, IEEE Computer Society (2006) 823–827
4. Gudmundsson, J., van Kreveld, M.J.: Computing longest duration flocks in trajectory data. [23] 35–42
5. Laube, P., Imfeld, S., Weibel, R.: Discovering relative motion patterns in groups of moving point objects. International Journal of Geographical Information Science **19**(6) (2005) 639–668
6. Lee, J., Han, J., Whang, K.Y.: Trajectory clustering: A partition-and-group framework. In: SCM SIGMOD International Conference on Management Data (SIGMOD'07), Beijing, China (June 11-14 2007)
7. Li, Y., Han, J., Yang, J.: Clustering moving objects. In: KDD '04: Proceedings of the tenth ACM SIGKDD international conference on Knowledge discovery and data mining, New York, NY, USA, ACM Press (2004) 617–622
8. Nanni, M., Pedreschi, D.: Time-focused clustering of trajectories of moving objects. Journal of Intelligent Information Systems **27**(3) (2006) 267–289
9. Tsoukatos, I., Gunopulos, D.: Efficient mining of spatiotemporal patterns. In Jensen, C.S., Schneider, M., Seeger, B., Tsotras, V.J., eds.: SSTD. Volume 2121 of Lecture Notes in Computer Science., Springer (2001) 425–442
10. Verhein, F., Chawla, S.: Mining spatio-temporal association rules, sources, sinks, stationary regions and thoroughfares in object mobility databases. In Lee, M.L., Tan, K.L., Wuwongse, V., eds.: DASFAA. Volume 3882 of Lecture Notes in Computer Science., Springer (2006) 187–201
11. Bogorny, V., Kuijpers, B., Alvares, L.O.: St-dmql: a semantic trajectory data mining query language. International Journal of Geographical Information Science (2009) in Press
12. Giannotti, F., Nanni, M., Pinelli, F., Pedreschi, D.: Trajectory pattern mining. In Berkhin, P., Caruana, R., Wu, X., eds.: KDD, ACM (2007) 330–339
13. Kuijpers, B., Moelans, B., de Weghe, N.V.: Qualitative polyline similarity testing with applications to query-by-sketch, indexing and classification. In de By, R.A., Nittel, S., eds.: 14th ACM International Symposium on Geographic Information Systems, ACM-GIS 2006, November 10-11, 2006, Arlington, Virginia, USA, Proceedings, ACM (2006) 11–18
14. Pelekis, N., Kopanakis, I., Ntoutsi, I., Marketos, G., Theodoridis, Y.: Mining trajectory databases via a suite of distance operators. In: ICDE Workshops, IEEE Computer Society (2007) 575–584
15. OGC: Topic 5, opengis abstract specification - features (version 4) (1999). Available at: http://www.OpenGIS.org/techno/specs.htm. Accessed in August (2005) (1999)
16. Shekhar, S., Chawla, S.: Spatial Databases: A Tour. Prentice Hall (June 2002)
17. Rigaux, P., Scholl, M., Voisard, A.: Spatial Databases: with application to GIS. Morgan Kaufmann
18. Spaccapietra, S., Parent, C., Damiani, M.L., de Macedo, J.A., Porto, F., Vangenot, C.: A conceptual view on trajectories. Data and Knowledge Engineering **65**(1) (2008) 126–146
19. Han, J.: Mining knowledge at multiple concept levels. In: CIKM, ACM (1995) 19–24
20. Bogorny, V., Palma, A.T., Engel, P., Alvares, L.O.: Weka-gdpm: Integrating classical data mining toolkit to geographic information systems. In: WAAMD Workshop, SBC (2006) 9–16
21. Agrawal, R., Srikant, R.: Mining sequential patterns. In Yu, P.S., Chen, A.L.P., eds.: ICDE, IEEE Computer Society (1995) 3–14
22. Society, W.: Frequency 1550. Available at: http://www.waag.org/project/frequentie. Accessed in September (2007) (2005)

Chapter 17
Census Data Mining for Land Use Classification

E. Roma Neto and D. S. Hamburger

Abstract This chapter presents spatial data mining techniques applied to support land use mapping. The area of study is in São Paulo municipality. The methodology is presented in three items: extraction, transformation and first analysis; knowledge discovering and supporting rules evaluation; image classification support. The combined inferences resulted in a good improvement in the digital image classification with the contribution of Census data.

17.1 Content Structure

The intent of this study is to describe the use of spatial data mining of Brazilian Census data as a support to land use mapping using digital image classification. To describe the procedures and evaluate the results obtained **the following items will be described**:

- Land use: what is it and how can it be mapped;
- Remote sensing images as a tool to land use mapping;
- Digital image processing as a technique to classify land use;
- Characteristic of Census data to understand land use distribution;
- Data warehouse and spatial data mining of Census data as a support to land use mapping;
- Integration of data warehouse and spatial data mining and digital image processing to classify land use;
- Results and Discussion;
- Findings and perspectives.

E. Roma Neto, D. S. Hamburger
Av. Eng. Euséio Stevaux, 823 - 04696-000, São Paulo, SP, Brazil, e-mail: elias.rneto@sp.senac.br,diana.hamburger@gmail.com

17.2 Key Research Issues

This chapter presents an application on data warehouse and spatial data mining techniques to support land use mapping through digital image processing.

The availability of satellite data make it easier to achieve information on land use change. The resolution of those images results in difficulties to define the urban area particularly at the urban fringe. This chapter addresses efforts to evaluate the use of census data to improve digital image classification.

The Census data contribution in digital image processing analysis is supported by knowledge (rules) mined through Census data sets, with a proposal on how to extract information from Census data and how to relate it to contribute to land use image classification.

17.3 Land Use and Remote Sensing

The environmental conditions and economic activities result in differentiated spaces. Those elements generate different land uses, a main factor in regional planning and management diagnosis.

Two ways to classify the surface are land use and land cover. The land cover describes the components that are present in the surface, resulting in classes like vegetation, bare soil, etc. The land use refers to the functional activities developed in each area, including agricultural, pasture or urban areas.

Anderson [1] establish a land use and land cover classification system. This system presents four hierarchical levels, each one subdividing the previous. Urban or built land is one class defined in the first level.

Remote sensing data has been used in this process. The need to minimize the time and resources support the use of ancillary data to improve those analysis and procedures.

The following inferences and interpretations are needed to extract urban land use from remote sensing products:

- The understanding on how human activities result in the physical structure in the surface;
- The identification of the elements that compound each land use class; and
- The description on how this distribution is shown in satellite images.

There is an urbanization process and an intensification of the urban centers connection. The land use classes occur because there is a relation between the social and economic behavior and the spatial occupation of the surface in homogeneous zones with spatial and social similarities [12], [8] and [14].

The land use classification system proposed by Anderson el al [1] was developed to make the classification of a large area with an extreme variety of classes possible. The land use classification presents many difficulties resulting in non classified areas as presented in [7] and [3], and [4], such as:

- The present classes can result from a process that occurred in the past;
- Many land use characteristics are not viewed in the urban form;
- The homogeneity of a land use class can be more textural than spectral. There are the same spectral features in many classes, organized in different ways;
- The heterogeneity of the urban environment is not easy to the spectral analysis and classification.

The systematic updated survey through digital image processing, still presents the challenges refered by [3] and [4].

An image is a set of matrices corresponding to an area in the surface. Each matrix correspond to the measurement of spectral radiation according to the satellite bands. Those values expressed in colors or gray tones compose the image.

17.4 Census Data and Land Use Distribution

The land use characteristics can also be understood and inferred from other data sources. Population censuses constitute the source of information about life conditions of the population. The population data and distribution is not directly land use information, but presenting the dwelling information is related to land use.

The data include Brazilian population information and investigates housing conditions. In Brazil, the Demographic Census 2000 presented the results of the survey of an 8 514 215,3 km^2 area, 5 507 municipalities, with a total of 54 265 618 households surveyed. [5].

The survey is proposed to occur every 10 years. In this sense, if the satellite image data is related to the Census data it can be used to update those data. By the other side, Census data can be used to support land use data obtained through digital image processing of remote sensing images.

17.5 Census Data Warehouse and Spatial Data Mining

Data warehouse and data mining technologies have been widely used to support business decisions. Here, both technologies are combined to either support urban analysis or verify data quality. Data mining is mainly used to help finding rules and patterns that may help specialist to classify images this usage is commonly known as knowledge discovery [16].

17.5.1 Concerning about Data Quality

Data quality models can be organized in at least three possible types: establishment of the analysis criteria; frameworks, which can turn it easier to verify desirable characteristics and metric definition. Piattini [13] shows a survey with some of the main proposals presented to help data quality analysis based on frameworks. Muller [11] defines some measures in order to create a metric to evaluate data quality of a given representation. Generally speaking, all these techniques are based on structural elements and are empirically validated.

Focusing on data warehouse environments and due to the fact that spatial databases are not often available as one may need, all patterns and rules mined must be evaluated and the model proposed carries a simple data quality analysis based on three main indicators of quality as presented by Strong [15] and Kimball [9] in a set of most common problems:

- Timeliness - refers to how current the data is at the time of analysis. For this study, a single Brazilian Census was considered for all variables analysed;
- Completeness - analyses whether or not the information contains the whole fact to be showed - not part of it. Here only the areas where the two urban data measures may be calculated are considered;
- Accuracy - defines how well the information reflects the reality. In this case it has been adopted content accuracy. Although it's not that simple to be achieved, gives a higher level result.

Finally, a multidimensional analysis considering land use classes [1] can be defined to help selecting data to the image classification. As can be seen at figure 17.1, these first set of measures helps to choose variables that can support this process in a better way.

Concerning about data quality has, then, conducted to a first data set transformation - removing incomplete and/or inaccurate data.

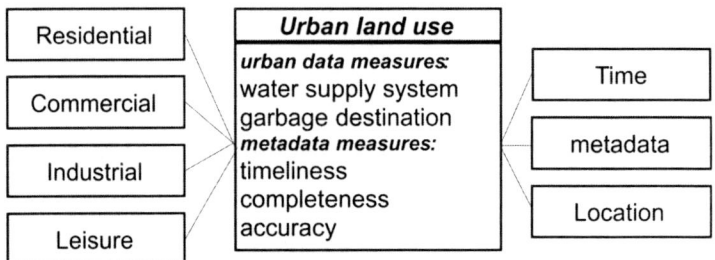

Fig. 17.1 Urban land usage multidimensional model

17.5.2 Concerning about Domain Driven

Besides this data quality analysis, one should also consider that real world needs depend on human and business requirements, that is, pure sets of resulting classes,

17 Census Data Mining for Land Use Classification   245

rules or trees for example are not sufficient to support these real needs. As Cao [2] summarizes, data mining applications must consider business interestingness and constraints. Domain driven in this case was considered as focusing in mapping land use, and two first steps were defined based on this thought: (1) reducing land usages to two types and (2) creating a set of indicators that could help to choose the correct data sets to be inferred by a data mining tool.

1. reducing land usages to two types
   The data sets analyzed were extracted from Brazilian Census [5], which is organized in 4 data groups : household , householder , education and demography . In this application study only the data set demography was not considered. Each data set contains about 14 thousand instances with hundreds of attributes describing San Paulo's Census lowest geographical level for which aggregated data are released. Each level, here called unit, contains, whenever possible, around 200 households, all of them classified as a land use category among 8 (3 urban and 5 rural possibilities defined by the Census). Step one was built in order to analyze how these units were classified among the 8 categories in order to redefine the whole set using only two: urban and rural. Data set instances observation has shown that the number of residents in each category was direct related to one of the two selected categories and this result allowed the land use redefinition desired. Census data considers legal urban limits. In a dynamic metropolitan region, increasing urban areas doesn't follow legal limits. This category simplification aids to adjust census data and supports establishing a relation between legal land use areas and built ones. Concerning domain driven has, then, conducted to a second data set transformation - reclassifying instances between two categories.
2. creating a set of indicators to guide the mining process
   Concerning land use needs and the three resulting data sets, analysis were focused on household and householder information.

A first step on creating a set of indicators was evaluating candidate attributes in both data sets. This evaluating conducted to the 8 indicators presented in the following items:

- Indicator 1: household - houses, apartments and extremely simple houses were studied and a first data set inference has shown that its differences were significantly related to land usage. Houses are present in more than 90% of the rural units, comparing to more than 70% presence of apartments in urban ones;
- Indicator 2: permanent household water supply - distributed system provided by a public company, local system and other types were considered. Less than 60% of rural units are provided by public company distributed system;
- Indicator 3: garbage collecting system - provided by a public company, burned, buried, thrown and other types. Burned garbage are present in 8% of the rural units;
- Indicator 4: drain system - provided by a public company, buried, thrown and other types. Public drain system achieves 90% in urban units comparing to buried types that are present in almost 30% of rural ones;

- Indicator 5: household ownership - own house, rented or other types. Rented houses are 23% of ownership in urban units;
- Indicator 6: sanitary installation (especially rest rooms). Results were not different enough in both urban and rural units to be considered;
- Indicator 7: number of men and women householders. Results were not sufficient different in both urban and rural units to be considered;
- Indicator 8: householder's age. Results were not different enough in both urban and rural units to be considered.

Indicators were created and analyzed in order to reduce the 242 attributes available in two data sets to only 10 candidates.

17.5.3 Applying Machine Learning Tools

A 13257 instances data set was used in the following queries and mining simulations - 10 variables have been selected after these 8 indicators analysis over household and householder full data sets - see table 17.1.

Table 17.1 Variables used in the mining analysis (water supply + garbage system).

Information type	Variables	Description
Water supply	3	Distributed system provided by a public company, local system and other types.
Garbage collecting system	6	Organized collecting system provided by a public company, burned, buried, thrown and other types.
Land use	1	Concerns about how Brazilian census describes each analyzed spatial unit - it was used for verifying purposes only.

Mining was then processed by an implementation of the C4.5 r8, implemented by Weka software [16] - rules are obtained from a partial decision tree as can be seen at table 17.2. J48 is an implementation of C4.5, one of the most famous and traditional classification algorithms, based on decision trees, which are built over a divide and conquer recursive approach.

Table 17.2 Rules obtained (number of residences each unit).

Rule	If clause	Then clause
1	Water supplied by public company > 23 AND other types of water supply ≤ 4 AND other types of collecting garbage ≤ 0	Land use = Urban
2	other types of water supply ≤ 20 AND garbage collected by a public company > 119	Land use = Urban

17.6 Data Integration

17.6.1 Area of Study and Data

The area of study is in Sao Paulo Municipality (Brazil). Sao Paulo had more than 10,000,000 inhabitants, as counted in the Brazilian Census Data [5]. There is a need of updated information to planning and management purposes. The development of this project was possible using the following datasets: the High Resolution CCD Cameras CBERS-2 image (China Brazil Earth Resources Satellite) and the Brazilian census data. Those data will be described bellow. The characteristics of CBERS-2 cameras are presented in Table 17.3.

Table 17.3 CBERS Instruments Characteristics [6]

Instrument characteristics	CCD Camera
Spectral bands	0,51 - 0,73 μm (pan), 0,45 - 0,52 μm (blue), 0,52 - 0,59 μm (green), 0,63 - 0,69 μm (red), 0,77 - 0,89 μm, (near infrared)
Spatial resolution	20 x 20 m
Swath width	113 km
Temporal resolution	26 days nadir view (3 days revisit)

The Brazilian Census [5] present data collected to 13257 census spatial units with around 200 dwellings. The data collected is usually organized in 4 groups: dwellings, responsible, education and demographic. The development of this study considered the dwelling data as the data that can describe the urban features and regions. A preliminary analysis was developed with a Landsat image from the year 2000. The use of CBERS image data from the year 2004 was used considering that the urban areas in 2000 would not be rural in 2004 that was confirmed by the closeness of both results.

17.6.2 Supported Digital Image Processing

The data set was processed under the two rules obtained in order to check the original classification against the resulting one. The proposed approach achieved 85% of the records (the algorithm was not able to classify 15%), in which 97% of the classified records were correct according to the census variable. The urban fringe areas were reclassified according to the results of census data classification defined above. The areas defined as urban fringe were assigned vegetation for the rural areas or residential class for the urban areas, according to the results obtained. This assignment was defined based on the observation of the conditions that characterize the urban fringe. The relation of the image classification with the urban and rural areas as defined in the census data was calculated - see table 17.4.

Table 17.4 Image classes comparison with census data (%).

Classes	Water	Vegetation	Commercial	Residential	Industrial	Urban Fringe
Urban	3.97	10.06	2.46	65.75	14.18	3.58
Rural	0.76	20.23	0.00	1.57	0.46	1.84

The urban area is mostly residential, industrial and vegetated areas and the rural ones are mainly covered by the Sño Paulo water reservoirs. The urban fringe class was reclassified considering that the urban areas should be residential and that the rural ones should be vegetated.

17.6.3 Putting All Steps Together

As presented above this application is based on the steps shown by the next items and presented at figure 17.2, some were simultaneously realized:
Semi-automatic set up:

- Characterization of the land use classes and image classification;
- Data analysis of spatial databases;
- Analysis of a multidimensional schema - available information and its quality;
  - Concerning about data quality - first data set transformation, removing incomplete and/or inaccurate data;
  - Concerning about domain driven - second data set transformation: reducing land usages categories by reclassifying instances between two categories;
  - Creating a set of indicators to guide the mining process - third data set transformation: reducing the 242 attributes available in two data sets to only 10 candidates;

Assisted Analysis:

17 Census Data Mining for Land Use Classification

- Applying Machine Learning - census data set mining analysis: rules based on water supply and garbage collecting systems;
- Spatial databases mining;
- Evaluation of the procedures and results.

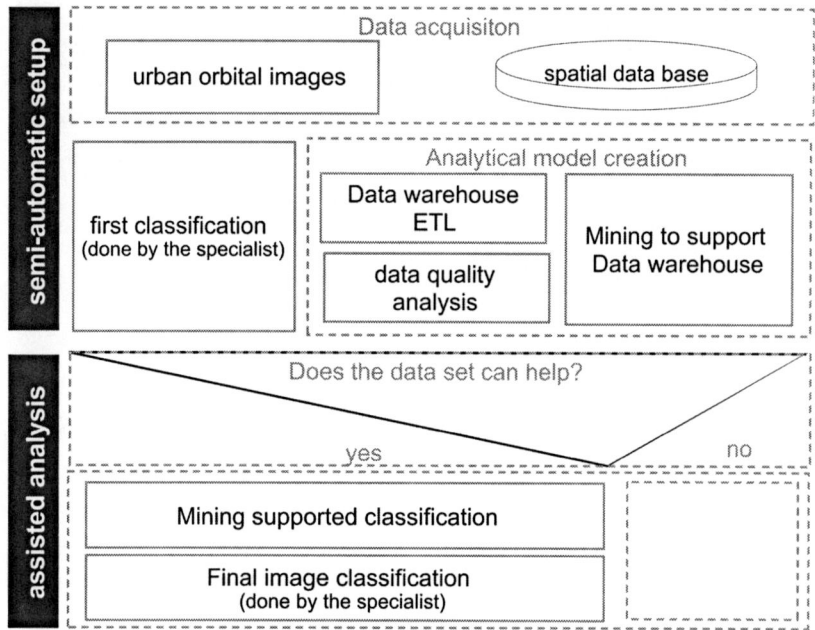

Fig. 17.2 Flow diagram for the data mining supported image classification.

17.7 Results and Analysis

The results were verified by comparing it to two different classifications: Brazilian Census classification, resulting in more than 85% of the untreated areas being correctly supported and, finally, it was compared to a specialist classification, resulting in more than 80% of the untreated areas being correctly supported. The resulting map is shown in the figure 17.3.

The classification of Census data in urban and non urban areas was used to classify the transition areas in CBERS images. Those areas where classified in urban or non urban areas and associated with vegetated, when non urban or residential, when urban. A sample of 30 points was used to check the improvement in the classification with this method. The analysis of the urban border areas supported by census data was possible as far as the rule verification demonstrated to be helpful (Min-

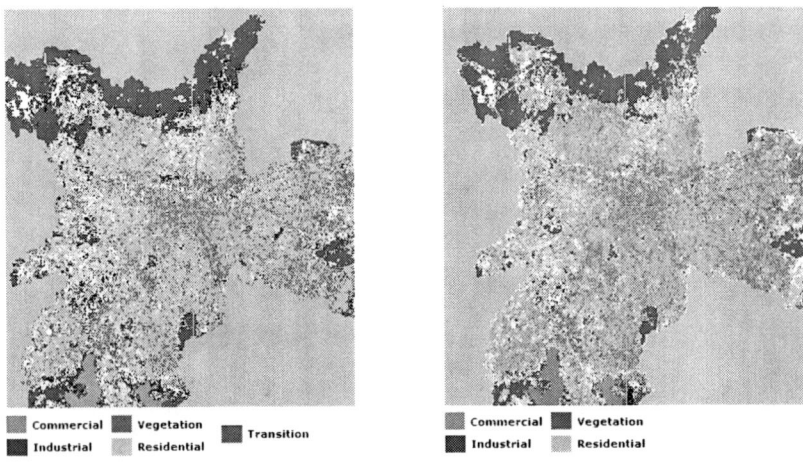

Fig. 17.3 (a) CBERS image classification without support; (b) final image.

ing Support Classification). The final image classification has shown the following results: With the first inference:

- The areas identified as urban through the data mining processes were correctly classified in 68% of the samples;
- The areas identified as rural through the data mining processes were correctly classified in 20% of the samples. With the second inference:
- The areas identified as urban through the data mining processes were correctly classified in 68% of the samples;
- The areas identified as rural through the data mining processes were correctly classified in 40% of the samples.

The urban areas that were classified as rural and associated with vegetation are in the border of the residential area and could be associated to both classes - figure 4a. The urban areas that had some vegetation (like squares and transmission lines, for instance) can be improved by this method - figure 4b. The residential areas classified as vegetation are those located at the urban sprawl area and had their classification compromised by the time between the Census (2000) and the image (2004).

The results show that the areas classified as urban based on census data improved the classification in the urban border. The classification was precise when the areas had already some urban characteristics. Some of the areas classified as rural in the Census (2000), were already changing to urban areas in the 2004 image. To those areas located in the borderline of a vegetated area with a built one, the Census data was not useful. Next step have already been started and some tasks are being defined:

- Consider a few more variables: demography information, in order to create a kind of balanced indicator;

- Apply the method to different data sets;
- Improve the measures and quality analysis;
- Analysis a timeline data set.

References

1. Anderson, J.R.; Hardy, E.E.; Roach, J.T. and Witner, R.E. (1976) "Sistemas de classificação do uso do solo para utilização com dados de sensoriamento remoto", Trad. H.Strang, Rio de Janeiro, IBGE.
2. Cao L. et al (2007), DDDM2007: Domain Driven Data Mining, SIGKDD Explorations Volume 9, Issue 2, pp 84.
3. Forster, B.C. (1984) Combining ancillary and spectral data for urban applications, International archives photogrammetry and remote sensing. V.XXV part A7, Commission 7, INTERNATIONAL SYMPOSIUM ARCHIVES PHOTOGRAMMETRY AND REMOTE SENSING, XVth Congress, Rio de Janeiro 1984. p.207-216.
4. Forster, B.C. (1985) An examination of some problems and solutions in monitoring urban areas from satellite platforms, International journal of remote sensing, 6(1): 139-151.
5. IBGE Brazilian Census 2000.(2005) [On Line] www.ibge.br.
6. INPE, National Spatial Research Institute. (2005) CBERS. [On Line] www.cbers.inpe.br.
7. Jensen, J.R. (1983) "Urban/suburban land use analysis. In: Manual of remote sensing" 2ed. Falls Church, American Society of Photogrammetry. v.2, chapter.30, p.1571-1666.
8. Jim, C.Y. (1989) Tree canopy cover, land use and planning implications in urban Hong Kong. Geoforum, 20(1):57-68.
9. Kimball, R, (1996). The Data Warehouse Toolkit: Practical Techniques for Building Dimensional Data Warehouses (John Wiley & Sons Inc) 416 pp.
10. Liu, S. E Zhu, X. (2004) An Integrated GIS approach to accessibility analysis. Transactions in GIS, 8 (1): 45-62, 2004.
11. Muller, R. J. (1999) "Database design for smarties: using UML for data modeling", San Francisco: Morgan Kaufmann.
12. Mumbower, L.; Donoghue, J. (1967) "Urban poverty study. Photogrammetric engineering", 33(6):610-618.
13. Piattini, M. et al. (2001) "Information and Database Quality", Kluwer Academic Publishers.
14. Roma Neto, E. ; Hamburger, D. S. Data warehouse and spatial data mining as a support to urban land use mapping using digital image classification - A study on Sao Paulo Metropolitan area with CBERS - 2 Data. In: 25th Urban Data Management Symposium, Aalborg, 2006.
15. Strong, D. M. et al. (1997) "Data Quality in Context", Communications of the ACM. New York, vol.40 no 5, p. 103-110, May.
16. Witten, I. H. & Frank, E. (2005) Data Mining: Practical machine learning tools and techniques. 2nd Edition, Morgan Kaufmann, 560 pp.

Chapter 18
Visual Data Mining for Developing Competitive Strategies in Higher Education

Gürdal Ertek

Abstract Information visualization is the growing field of computer science that aims at visually mining data for knowledge discovery. In this paper, a data mining framework and a novel information visualization scheme is developed and applied to the domain of higher education. The presented framework consists of three main types of visual data analysis: Discovering general insights, carrying out competitive benchmarking, and planning for High School Relationship Management (HSRM). In this paper the framework and the square tiles visualization scheme are described and an application at a private university in Turkey with the goal of attracting brightest students is demonstrated.

18.1 Introduction

Every year, more than 1,5 million university candidates in Turkey, including more than half a million fresh high school graduates, take the University Entrance Exam (Öğrenci Seçme Sınavı- ÖSS) to enter into a university. The exam takes place simultaneously in thousands of different sites and the candidates answer multiple-choice questions in the 3-hour exam that will change their life forever. Entering the most popular departments -such as engineering departments- in the reputed universities with full scholarship requires ranking within the top 5,000 in the exam.

In recent years, the establishment of many private universities, mostly backed-up by strong company groups in Turkey, have opened up new opportunities for university candidates. As the students compete against each other for the best universities, the universities also compete to attract the best students. Strategies applied by universities to attract the brightest candidates are almost standard every year: Publishing past years' placement results, promoting success stories in press -especially

---

Gürdal Ertek
Sabancı University, Faculty of Engineering and Natural Sciences, Orhanlı, Tuzla, 34956, Istanbul, Turkey, e-mail: `ertekg@sabanciuniv.edu`

newspapers-, sending high-quality printed and multimedia catalogs to students of selected high schools, arranging site visits to selected high schools around the country with faculty members included in the visiting team, and occasionally spreading bad word-of-mouth for benchmark universities.

Sabancı[1] University was established in 1999 by the Sabancı Group, the second largest company group in Turkey at that time, at the outskirts of Istanbul, the megacity of Turkey with a population of nearly 20 million people. During 2005 and 2006 an innovative framework -based on data mining- was developed at Sabancı University with the collaboration of staff from the Student Resources Unit, who are responsible of promoting the university to high school students, and the author from Faculty of Engineering and Natural Sciences. The ultimate goal was to determine competitive strategies through mining annual ÖSS rankings for attracting the best students to the university. In this paper, this framework and the square tiles visualization scheme devised for data analysis is described.

The developed approach is based on visual data mining through a novel information visualization scheme, namely *square tiles visualization*. The strategies suggested to the managing staff at the university's Student Resources Unit are built on the results of visual data mining. The steps followed in visual data mining include performing competitive benchmarking of universities and departments, and establishment of the *High School Relationship Management* (HSRM) decisions, such as deciding on which high schools should be targeted for site visits, and how site visits to these high schools should be planned.

In the study, information visualization was preferred against other data mining methods, since the end-users of the developed Decision Support System (DSS) would be staff at the university and undergraduate students. In information visualization, patterns such as outliers, gaps and trends can be easily identified without requiring any knowledge of the mathematical/statistical algorithms. Development of a novel visualization scheme was motivated by the difficulties faced by the author in the perception of the area information from irregular tile shapes of existing schemes and software.

In this paper, a hybrid visualization scheme is proposed and implemented to represent data with categorical and numerical attributes. The visualization that is introduced and discussed, namely square tiles, shows each record in a the results of a query as a colored icon, and sizes the icons to fill the screen space. The scheme is introduced in Section 18.2. In Section 18.3 related work is summarized. The mathematical model solved for generating the visualizations is presented in Section 18.4 and the software implementation is discussed. In Section 18.5 the analysis of ÖSS data demonstrated with snapshots of the developed *SquareTiles* software that implements square tiles visualization. In Section 18.6 future work is outlined. Finally in Section 18.7 the paper is summarized and conclusions are presented.

---

[1] pronounced as it Saa-baan-jee

18.2 Square Tiles Visualization

Information visualization is the growing field of computer science that studies ways of visually mining high-dimensional data to identify patterns and derive useful insights. Patterns such as trends, clusters, gaps and outliers can be easily identified by information visualization. Keim [10] presents a taxonomy of information visualization schemes based on the data type to be visualized, the visualization technique used, and the interaction and distortion technique used. Recent reviews of information visualization literature have been carried out by Hoffman & Grinstein [6] and de Oliveira & Levkowitz [3]. Many academic and commercial information visualization tools have been developed within the last two decades, some of which are listed by Eick [4]. Internet sources on information visualization include [7] and [11].

The main differences of information visualization from other data mining methods such as association rule mining and cluster analysis are two-folds: Information visualization takes advantage of the rapid and flexible pattern recognition skills of humans [13], and relies on human intuition as opposed to understanding mathematical/statistical algorithms [10].

In the square tiles visualization scheme (Figure 18.1) each value of a selected categorical attribute (such as high schools in this study) is represented as a distinct box, and the box is filled with strictly-square tiles that represent the records in the database based on the value of the categorical attribute. Colors of the tiles correspond to the values of a selected numerical attribute (ÖSS ranking in this study). In Figure 18.1 icons with darker colors denote students with better rankings in the exam. One can use the names *partitioning attribute* and *coloring attribute* for these attributes, respectively, similar to the naming convention in [9]. The labels in the figure refer to the the variables and parameters in the associated mathematical model, which is described in Section 18.4.

Tile visualization has been widely used before, and has even been implemented in commercial software such as Omniscope [12]. However, existing systems either can not use the screen space efficiently, or display the data with the same tile size through irregularly shaped rectangles. The novelty and the advantage that square tiles visualization brings is the most efficient use of the screen space for displaying data when the tiles are strictly square. The problem of "maximizing the utilization of the screen space with each queried record being represented as a square tile" is formulated as a nonlinear optimization problem, and can be solved to optimality in reasonable time through exhaustive enumeration.

Square tiles can be considered as a two-dimensional extension of the well-known Pareto Charts. A Pareto chart is a two-dimensional chart which plots the cumulative impact on the y-axis against the percentage of elements sorted on the x-axis based on their impact. The cumulative impact is typically a non-linear, concave function of the percentage of the elements: A small percentage of the most important elements are typically observed to account for a great percentage of the impacts. In square tiles visualization, the areas of the most important sets and the distribution of the elements in different sets with respect to the coloring attribute can be compared.

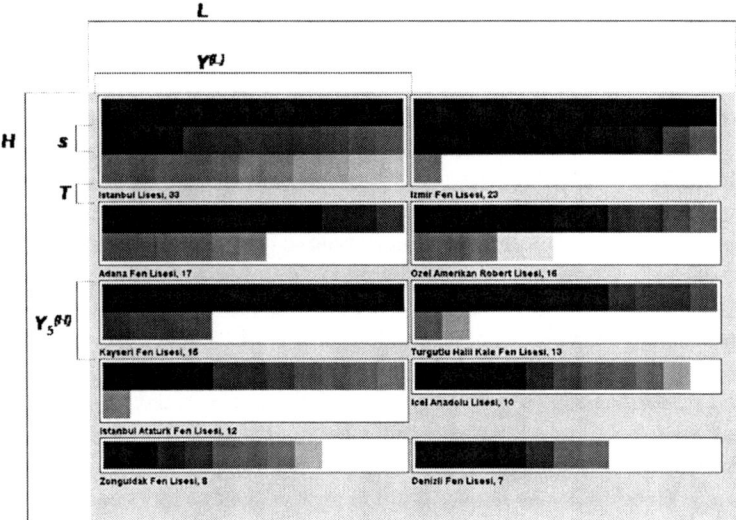

Fig. 18.1 Composition of entrants to a reputed university with respect to high schools (HS_NAME)

The color spectrum used to show the numerical attribute starts from yellow, which shows the largest value, continues to red, and ends at black, which shows the smallest value. This color spectrum allows easy identification of patterns on a grey-scale printout, and has also been selected in [1].

The placement of icons within boxes is carried out from left to right and from top to bottom according to the coloring attribute. The layout of boxes within the screen is carried out again from left to right and from top to bottom based on the number of icons within each box. The PeopleGarden system [16] developed at MIT also considers a similar layout scheme.

18.3 Related Work

The icon-based approach followed in this paper is closest to the approach taken by Sun [13]. The author represents multidimensional production data with colored icons within *smashed tables* (framed boxes). In both [13] and the research here the icons are colored squares which denote elements in a set, with the colors representing values of a numerical attribute. In both papers, *a small multiple design* (Tufte [14], p42, p170, p174) is implemented, where a standard visual design is repeatedly presented side by side for each value of one or more categorical attribute(s).

Space-filling visualizations seek full utilization of the screen space by displaying attributes of data in a manner to occupy the whole screen space. Square tiles visualization in this paper adjusts the sizes of icons and the layout of the icons to achieve this objective, so it can be considered as a space-filling visualization scheme.

One type of space-filling visualization is pixel-based visualization, where a spacefilling algorithm is used to arrange pixels on the screen space at full space-utilization [8], [9]. Pixel-based visualizations are able to depict up to hundreds of thousands of elements on the screen space, since each pixel denotes an element (such as a person). The research presented in here is very similar to pixel-based visualization research, but also shows one important difference: In [8] and [9] each element of a set is denoted by a single pixel. In here, each element is denoted by a square tile. On the other hand, the research here is also similar to [8] and [9] in the sense that in all these studies a mathematical optimization model, with objective function and constraints, that determines the *best* layout is discussed.

18.4 Mathematical Model

Each square tiles visualization is generated based on the optimal solution of the mathematical model presented below. First we will define the sets, and then the parameters and variables, and then the mathematical model, which consists of an objective function to be maximized, and a set of constraints that must be satisfied. Let

$\mathscr{I}$: the set of all boxes to be displayed, with $|\mathscr{I}| = n$
N_i: the number of icons in box i.

Let the parameters be defined as follows:

T: text height
B: space between boxes
P: pixel allowance within each box
m: minimum length of each box
$\bar{S}$: maximum icon size for each element
L: length of the screen area
H: height of the screen area

The most important variables are

s: the size (side length) of each icon, and
$x^{(h)}$: the number of horizontal icons placed in each box.

In the solution algorithm the values of these two variables are changed to find the best set of variable values.

Let the other variables be defined as follows:

$x_i^{(v)}$: number of vertical icons in box i
$y^{(L)}$: length of each box
$y_i^{(H)}$: height of box i
$Y^{(L)}$: total length of each box
$Y_i^{(H)}$: total height of box i
$Z^{(h)}$: number of horizontal boxes
$Z^{(v)}$: number of vertical boxes

It should be noted that $s, x^{(h)}, x^{(v)}, y^{(L)}, y_i^{(H)}, Y^{(L)}, Y_i^{(H)}, Z^{(h)}, Z^{(v)} \in \mathbb{Z}^+$, where $\mathbb{Z}^+$ is the set of positive integers.

The mathematical model is given below:

$$\max \quad \alpha$$

s.t.

$$\alpha = \frac{\sum_{i \in \mathscr{I}} y^{(L)} y_i^{(H)}}{LH} \tag{18.1}$$

$$y^{(L)} = 2P + x^{(h)} s \tag{18.2}$$

$$x_i^{(v)} = \lceil N_i / x^{(h)} \rceil, \forall i \in \mathscr{I} \tag{18.3}$$

$$y_i^{(H)} = 2P + x_i^{(v)} s, \forall i \in \mathscr{I} \tag{18.4}$$

$$Y_i^{(H)} = y_i^{(H)} + B + T, \forall i \in \mathscr{I} \tag{18.5}$$

$$Y^{(L)} = y^{(L)} + B \tag{18.6}$$

$$Z^{(h)} = \lfloor L / Y^{(L)} \rfloor \tag{18.7}$$

$$Z^{(v)} = \left\{ k : \max j \text{ s.t.} \sum_{i=1, Z^{(h)}+1, \ldots, jZ^{(h)}+1} Y_i^{(H)} \leq H \right\} \tag{18.8}$$

$$\alpha \leq 1 \tag{18.9}$$

$$n \leq Z^{(h)} Z^{(v)} \tag{18.10}$$

$$m \leq y^{(L)} \leq L \tag{18.11}$$

$$1 \leq s \leq \overline{S} \tag{18.12}$$

The objective in this model is to maximize α subject to (s.t.) all the listed constraints are satisfied. α is defined in (1) as the ratio of the total area occupied by the boxes to the total screen area available. Thus the objective of the model is to maximize screen space utilization. The length of each box $y^{(L)}$ is calculated in (2) as the summation of the pixel allowances $2P$ within that box and the vertical length $x^{(h)} s$ of the icons in that box. (3) calculates the number of vertical icons of box i, namely $x_i^{(v)}$. Calculation of $y_i^{(H)}$, the height of box i in (4), is similar to the length calculation in (2). Calculations in (5) and (6) take the space between boxes B and the text height T into consideration. The number of horizontal boxes $Z^{(h)}$ is calculated in (7). The number of vertical boxes $Z^{(v)}$ is calculated in (8) by finding the maximum j value such that the total height of the boxes does not exceed H, the height of the screen area. (9) states that α can not exceed 1, since it is denoting utilization. (10) guarantees that all the required boxes are displayed on the screen. (11) puts bounds on the minimum and maximum values of $y^{(L)}$, and thus indirectly s. The last constraint (12) bounds the range of s.

To solve the problem to optimality, the variables s and $x^{(h)}$ are changed within bounds that are calculated based on (11), (6), (2) and (12), and the feasible solution that yields the maximum α value is selected as optimum. For determining feasibility of a $(s,x^{(h)})$ combination, the calculations in (2) through (8) are carried out and the feasibility conditions in (9) and (10) are checked. Once the best $(s,x^{(h)})$ combination is determined, the visualization is generated based on the values of the calculated parameters and variables. In the extreme case, each icon would be a single pixel on the screen, and thus the number of records that can be visualized by the methodology is bounded above by the number of pixels on the screen.

Implementation

The *SquareTiles* software has been developed to create the visualizations, and adopted to the analysis of a particular data set. The software is implemented using Java under Eclipse Integrated Development Environment (IDE) [5]. The data is stored in a Microsoft Access database file, and is queried from within the Java program through ODBC connection. The software developed allows a user without any prior knowledge of a querying language (such as SQL) to create queries that generate visualizations.

Example Layout

Figure 18.1 displays the parameters L, H, T, and the variables $s, Y^{(L)}, Y_5^{(H)}$ on a sample output. These refer to the number of pixels. From the figure, we can also deduce that the number of horizontal and vertical boxes are $Z^{(h)} = 2$ and $Z^{(v)} = 5$, respectively. The number of icons in boxes 1, 2, ... are $N_1 = 33, N_2 = 23$, etc. The number of horizontal and vertical icons in box 1 can be counted as $x^{(h)} = 11$ and $x_1^{(v)} = 3$. In the model $x^{(h)}$ and s are the most important variables, whose values are determined such as to maximize screen space utilization. The values of other variables are calculated based on these two.

18.5 Framework and Case Study

In this section, the framework developed for analyzing and understanding the ÖSS data through square tiles will be presented and demonstrated. The selected data set contains information on the top ranking students in ÖSS for a selected year. The selected ÖSS data set includes 5,965 records, covering students within the top 5,000 with respect to two types of scores. The attributes (dimensions) in the data set include HS_NAME (high school name), HS_TYPE_TEXT (high school type in text format), UNIV_NAME (university name), UNIV_DEPT (university department), RANK_SAY (the student's rank according to score type *sayısal* (science and mathematics based)). All of these attributes are categorical, except the rank attribute, which is numerical.

Sabancı University is a newly established private university which accepts students mostly from the top 1% of the students that take ÖSS. Traditionally (until 2007) the Student Resources Unit at Sabancı University assembled the data on top 5,000 students in the exam and analyzed it using spreadsheet software. However, only basic graphs which provide aggregate summaries were generated using the yearly data sets.

The ÖSS data set provided by the Student Resources Unit had to be cleaned to carry out the analysis with square tiles visualization. The main problems were multiple entries for the same value, and missing attribute values for some records. A taxonomy of dirty data and explanation of the techniques for cleaning it is presented by Kim et al. [2]. According to this taxonomy, the issues faced in here all "require examination and repair by humans with domain expertise".

The *SquareTiles* software allowed a range of analysis to be carried out -by freshmen students with no database experience- and interesting and potentially useful insights to be derived. A report was prepared for the use of Student Resources Unit at Sabancı University that contains competitive benchmarking for 7 selected universities and guidelines for developing strategies in managing relationships with 52 selected high-schools. The study suggested establishment of a new approach for High School Relationship Management (HSRM), where the high schools are profiled through information visualization.

Several suggestions were received from staff within the Student Resources Unit during discussions: One suggestion was the printing of the number of icons in each box (thus, the cardinality of each set). This suggestion was implemented within the software.

The proposed framework consists of three main types of analysis described in the below subsections.

18.5.1 General Insights and Observations

The visualizations can be used to gain insights into the general patterns. One example is given in Figure 18.2, which displays the distribution of top 5,000 students with respect to top 10 high school types.

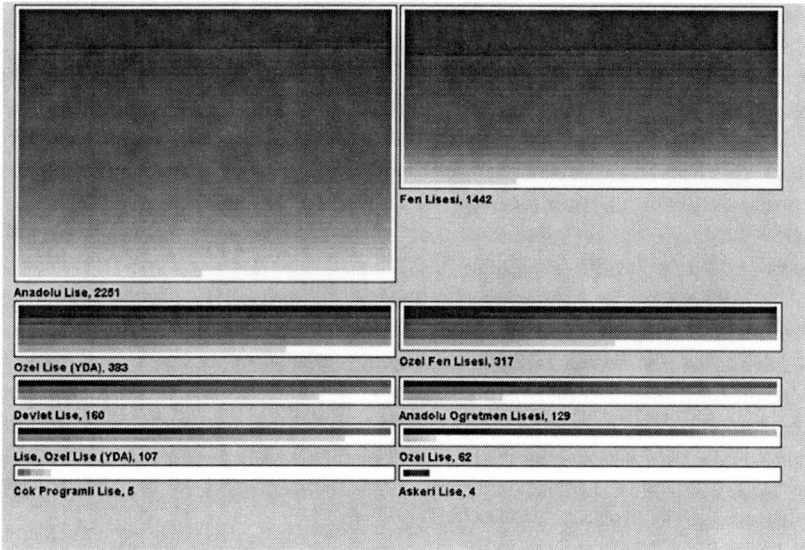

Fig. 18.2 Distribution of top 5,000 students with respect to top 10 high school types

From Figure 18.2 it can be seen that the high school types Anadolu Lise (public Anatolian high schools which teach foreign languages, and are preferred by many families for this reason) and Fen Lisesi (public science high schools) are the most successful. When comparing these two school types, one can observe that the number of darker icons are approximately the same for both. This translates into the fact that science high schools have a greater proportion of their students in the high ranks (with dark colors).

From the figure, once can also observe a pattern that would be observed in a Pareto Chart: The two high school types account for more than half of the screen area; that is, these two high school types impact the Turkish Education System much more significantly than others by accounting for more than half of the top 5,000 students in ÖSS.

Ozel Lise (private high schools) and Ozel Fen Lisesi (private science high schools) follow the first two high school types. The Turkish word *özel* means *private*, and this word is placed in front of the names of both private high school types and private high school names. One pattern to notice is the low success rate of Devlet Lise (regular public high schools). Even though regular public high schools outnumber other types of high schools by far in Turkey, their success rate is very much below the high school types discussed earlier.

18.5.2 Benchmarking

Benchmarking High Schools

Figure 18.1 gives the composition of entrants from within top 5,000 to a reputed university with respect to top 10 high schools. This figure highlights a list of high schools that Sabancı University should focus on. Top performing high schools, such as Istanbul Lisesi and Izmir Fen Lisesi should receive special attention and active promotion should be carried out at these schools. One striking observation in the figure is that almost all of the significant high schools are either Anadolu Lise (public Anatolian high schools) or Fen Lisesi (public science high schools). The only private high school in the top 10 is Ozel Amerikan Robert Lisesi, an American High School that was established in 1863.

Detailed benchmarking analysis of selected universities revealed that there can be significant differences between the universities with respect to the high schools of the entrants. One strategy suggested to the staff of the Student Resources Unit at Sabancı University was to identify high schools that send a great number of students to selected other universities, and carry out a focused publicity campaign geared towards attracting students of these high schools.

Benchmarking Departments

Figure 18.3 gives the distribution of entrants from within 5,000 to top 10 departments of the discussed university. One can visually see and compare the capacities for each department. From the color distributions it can be deducted immediately that the departments Bilgisayar (Computer Engineering), Endustri (Industrial Engineering), and Elektrik-Elektronik (Electrical-Electronics Engineering) are selected by the higher-ranking students in general. Among these three departments, Electrical-Electronical Engineering has the distribution of students with the highest rankings. Makine (Mechanical Engineering) and other departments are selected by lower-ranking students from within the top 5,000. It is worthy to observe that there is one student that entered Iktisat (Economics) with a significantly higher ranking than others who entered the same department. The same situation can be observed in the least populated four departments in the figure: There exist a number of higher ranking students who selected these departments, who probably had these departments as their top choices.

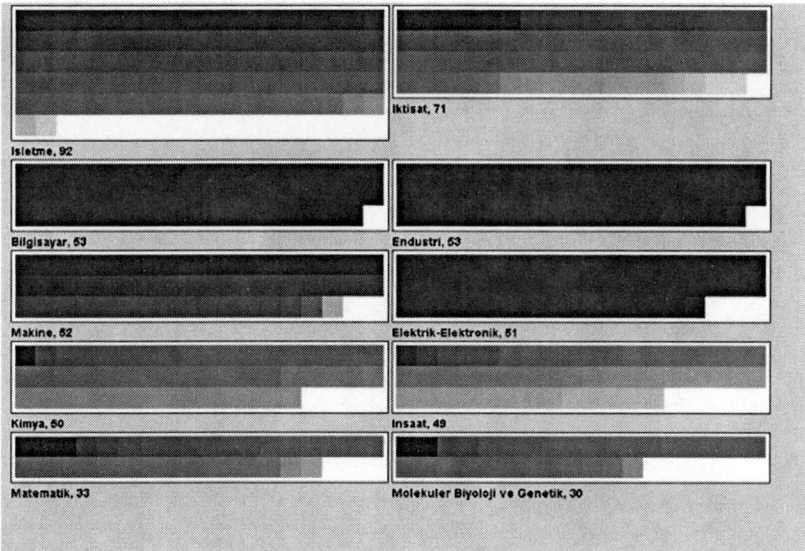

Fig. 18.3 Composition of entrants to a reputed university with respect to top 10 high schools

18.5.3 High School Relationship Management (HSRM)

Figure 18.4 depicts the department preferences of students from a reputed high school within the top 5,000. This distribution is particularly important when planning publicity activities towards this high school. The most popular selections are Endustri (Industrial Engineering) and Bilgisayar (Computer Engineering). The existence of two additional students with high rankings who selected Endustri (Burslu) (Industrial Engineering with scholarship[2]) further indicates the vitality of industrial engineering. So when a visit is planned to this high school, the faculty member selected to speak should be from the industrial engineering department who also has a fundamental understanding of computer engineering. It could also be a good strategy to ask this faculty member to emphasize the relationship between industrial engineering and computer science/engineering. Throughout the analysis of 52 selected high schools, significantly differing departmental preferences have been observed, which suggests that publicity activities should be customized based on the profiles of the schools.

---

[2] the Turkish word *burslu* means *with scholarship*

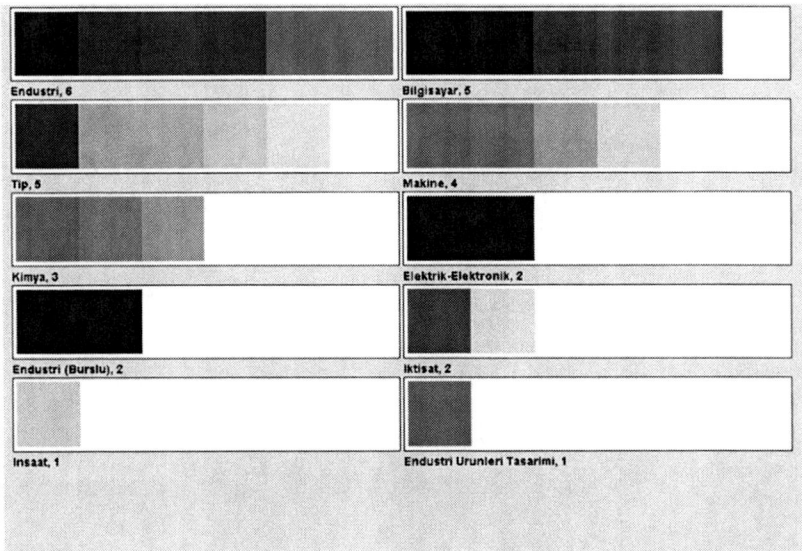

Fig. 18.4 Department (UNIV_DEPT) preferences of students from a reputed high school in Istanbul

18.6 Future Work

The most obvious extension to the research presented here is the innovation or adoption of effective and useful visualization schemes that allow carrying out currently unsupported styles of analysis. One such analysis is the analysis of changes in the queried sets over time. To give an example with the ÖSS data set, one would most probably be interested in visually comparing the distribution of students from a high school to universities in two successive years.

Ward [15] provides a taxonomy of icon (glyph) placement strategies. One area of future research is placing icons and boxes in such a way to derive the most insights.

One weakness of the current implementation is that it does not allow user's interaction with the visualization. The software presented here can be modified and its scope can be broadened to enable visual querying and interaction.

18.7 Conclusions

In this paper, the application of data mining within higher education is illustrated. Meanwhile, the novel visualization scheme used in the study, namely square tiles was introduced and its applicability was illustrated throughout the case study. The selected data contains essential information on top ranking students in the National University Entrance Examination in Turkey (ÖSS) for a selected year. The software implementation of the visualization scheme allows users to gain key insights, carry out benchmarking, and develop strategies for relationship management. As the number of attributes increases, the potential of finding interesting insights has been observed to increase.

The developed *SquareTiles* software that implements the visualization scheme requires no mathematical or database background at all, and was used by two freshmen students to carry out a wide range of analysis and derive actionable insights. A report was prepared for the use of Student Resources Unit at Sabancı University that contained competitive benchmarking analysis for seven of the top universities and guidelines for HSRM for 52 selected high schools.

Detailed benchmarking analysis of selected universities revealed that there exist significant differences between the universities with respect to the high schools of the entrants. One strategy suggested to the staff of the Student Resources Unit at Sabancı University was to identify high schools that send a large number of students to selected other universities, and carry out a focused publicity campaign geared towards attracting students of these high schools. The analysis for HSRM provided the details of managing relations with the 52 selected high schools.

The described framework was implemented at Sabancı University for one year, but was discontinued due to the high costs of retrieving the data from ÖSYM, the state institution that organizes ÖSS, and due to the difficulties in arranging faculty members to participate in HSRM activities. Still, we believe that the study serves as a unique example in the data mining literature, as it reports discussion of practical issues in higher education and derivation of actionable insights through visual data mining, besides development of a generic information visualization scheme motivated by a domain-specific problem.

Acknowledgements The author would like to thank Yücel Saygın and Selim Balcısoy for their suggestions regarding the paper, Mustafa Ünel and Ş. Ilker Birbil for their help with LaTeX. The author would also like to thank Fethi M. Özdöl and Barış Değirmencioğlu for their help with mining the ÖSS data.

References

1. Abello, J., Korn, J.: MGV: A system for visualizing massive multidigraphs. IEEE Transactions on Visualization and Computer Graphics **8**, no.1, 21–38 (2002)
2. Kim, W., Choi, B., Hong E., Kim, S., Lee, D.: A taxonomy of dirty data. Data Mining and Knowledge Discovery. **7**, 81–99 (2003)
3. de Oliveira, M. C. F., Levkowitz, H.: From visual data exploration to visual data mining: a survey. IEEE Transactions on Visualization and Computer Graphics **9**, no.3, 378–394 (2003)
4. Eick, S. G.: Visual discovery and analysis. IEEE Transactions on Visualization and Computer Graphics **6**, no.1, 44–58 (2000)
5. http://www.eclipse.org
6. Hoffman, P. E., Grinstein, G. G.: A survey of visualizations for high-dimensional data mining. In: Fayyad, U., Grinstein, G. G., Wierse, A. (eds.) Information visualization in data mining and knowledge discovery, pp. 47-82 (2002)
7. http://iv.homeunix.org/

8. Keim, D. A., Kriegel, H.: VisDB: database exploration using multidimensional visualization. IEEE Computer Graphics and Applications. September 1994, 40–49 (1994)
9. Keim, D. A., Hao, M. C., Dayal U., Hsu, M.: Pixel bar charts: a visualization technique for very large multi-attribute data sets. Information Visualization. **1** 20–34 (2002)
10. Keim, D. A.: Information visualization and visual data mining. IEEE Transactions on Visualization and Computer Graphics. **8**, no.1, 1–8 (2002)
11. http://otal.umd.edu/Olive/
12. http://www.visokio.com
13. Sun, T.: An icon-based data image construction method for production data visualization. Production Planning & Control. **14**, no.3, 290–303 (2003)
14. Tufte, E. R.: The Visual Display of Quantitative Information. Graphics Press, Cheshire, CT. (1983)
15. Ward, M. O.: A taxonomy of glyph placement strategies for multidimensional data visualization. Information Visualization. **1**, 194–210 (2002)
16. Xiong, B., Donath, J.: PeopleGarden: Creating data portraits for users. Proceedings UIST '99 Conference, ACM 37–44 (1999)

Chapter 19
Data Mining For Robust Flight Scheduling

Ira Assent, Ralph Krieger, Petra Welter, Jörg Herbers, and Thomas Seidl

Abstract In scheduling of airport operations the unreliability of flight arrivals is a serious challenge. Robustness with respect to flight delay is incorporated into recent scheduling techniques. To refine proactive scheduling, we propose classification of flights into delay categories. Our method is based on archived data at major airports in current flight information systems. Classification in this scenario is hindered by the large number of attributes, that might occlude the dominant patterns of flight delays. As not all of these attributes are equally relevant for different patterns, global dimensionality reduction methods are not appropriate. We therefore present a technique which identifies locally relevant attributes for the classification into flight delay categories. We give an algorithm that efficiently identifies relevant attributes. Our experimental evaluation demonstrates that our technique is capable of detection relevant patterns useful for flight delay classification.

19.1 Introduction

In airport operations, unreliability of flight schedules is a major concern. Airlines try to build flight schedules that incorporate buffer times in order to minimize disruptions of aircraft rotations as well as passenger and crew connections [22]. However, flight delays are still significant as can be studied in the reports of the

---

Ira Assent, Ralph Krieger, Thomas Seidl
Data Management and Exploration Group, RWTH Aachen University, Germany, phone: +492418021910, e-mail: {assent,krieger,seidl}@cs.rwth-aachen.de

Petra Welter
Dept. of Medical Informatics, RWTH Aachen University, Germany, e-mail: pwelter@mi.rwth-aachen.de

Jörg Herbers
INFORM GmbH, Pascalstraße 23, Aachen, Germany, e-mail: joerg.herbers@inform-ac.com

Bureau of Transportation Statistics in the U.S. [10] and the Central Office for Delay Analysis of Eurocontrol [16]. As a downstream effect, delays have a considerable impact on resource scheduling for airports and ground handlers. Modern approaches to ground staff scheduling and flight-gate assignment therefore aim at incorporating robustness with regard to flight delays, see e.g. [9, 14]. Classification of flights into delay categories is essential in refining proactive scheduling methods in order to minimize expected disruptions, e.g. by scheduling larger buffer times after heavily delayed flights.

Extensive flight data is recorded by flight information systems at all major airports. Using such databases, we aim at classifying flights to provide crucial information for actionable and dependable scheduling systems. For classification, numerous techniques in data mining have been proposed and employed successfully in various application domains. It has been shown that each type of classifier has its merit; there is no inherent superiority of any classifier [15].

In high-dimensional or noisy data like the automatically recorded flight data, however, classification accuracy may drop below acceptable levels. Locally irrelevant attributes often occlude class-relevant information. Recently, there has been some work on adapting to locally relevant attributes [12]. To detect locally relevant attributes, we use a subspace classification approach. In subspace clustering, the aim is detecting meaningful clusters in subspaces of the attributes [25]. This has proven to work well in high-dimensional domains. However, it does not utilize class labels and thus does not provide appropriate groupings for classification according to these labels. We overcome this shortcoming by incorporating class information into subspace search. Our contributions include:

- analysis and mining in the flight delay domain
- subspace classifier for flight delays and related application domains
- a novel, efficient and effective algorithm for subspace classifier training

Our experiments demonstrate that we are able to identify relevant attributes to improve classification accuracy for data which follows local patterns, providing more information for robust scheduling. The flight delay classification problem drove the development of this model. Its applicability to real world classification purposes, however, goes beyond this scenario.

19.2 Flight Scheduling in the Presence of Delays

We aim at supporting robust scheduling of flights in the presence of delays. Classifying incoming flights reliably as "ahead of time", "on time" and "delayed" allows users to update airport schedules accordingly. At the airport, information on flights is routinely recorded. Figure 19.1 illustrates the type of attributes for which information is stored, such as position of the aircraft, its gate, its airline, its type, etc. Note that we alienated the data as we are not allowed to disclose the original data. The type of attributes, however, reflects the information recorded for the real data.

The class label information was provided to us from scheduling experts as a basis for our technique.

Based on an in-depth analysis of the data and discussions with experts, we were able to identify the following requirements for our classification of flights approach:

Dealing with many attributes of locally varying relevance

We target at grouping flights with similar characteristics and identifying structure on the attribute level. In the flight domain, several aspects support the locality of flight delay structures. As an example, passenger figures may only influence departure delays when the aircraft is parked at a remote stand, i.e. when bus transportation is required. We have validated the hypothesis that relevance is not globally uniform but differs from class to class and from instance to instance for the flight delay data by training several types of classifiers. When using only relevant attributes – found using standard statistical tests – classification accuracy drops surprisingly. This suggests that globally irrelevant attributes are nonetheless locally relevant for individual patterns.

Providing explanatory components for schedulers

We are interested in techniques that are transparent with regard to the reproduction of classification results, allowing for interventions by experts.

Robustness to noise, variance in delay patterns

At some times of the day, flight delay patterns may be superposed by other factors like runway congestion. Weather conditions and other influences are not recorded in the data and cannot be used for proactive scheduling methods. These factors therefore cause significant noise.

Mining patterns for dynamic scheduling

To ensure that schedulers may adapt their resource planning dynamically, we propose a novel efficient algorithm for subspace classifier training. We give two pruning

No. Nominal	Pos Nominal	Gat Nominal	Airline Nominal	AircraftType Nominal	Rwy Nominal	Terminal Nominal	HandlingType Nominal	Info Nominal	MasPos Nominal	NumSeats Nominal	DayOfWeek Nominal	Hour Nominal	Class Nominal
1	PI67	A4IK77	Air1	142	3	1	J	U	60	300	4	14	delayed
2	PK14	H4AI51	Air2	100	6	U	J	U	72	100	3	10	on_time
3	PI67	H4AI51	Air1	100	6	U	J	U	61	100	4	13	on_time
4	PA07	T1I0ZZ	Air4	142	6	3	J	U	61	300	4	19	ahead_of_time
5	PL35	H4AI51	Air3	100	6	U	J	U	55	100	3	20	delayed
6	PT90	H4AI51	Air3	313	6	2	E	U	*	200	3	21	on_time
7	PK14	H4AI51	Air2	100	6	U	M	U	18	100	4	10	ahead_of_time
8	PT90	H4AI51	Air1	100	3	U	P	U	*	200	6	16	on_time
9	PW76	A4IK77	Air2	737	3	1	J	U	9	350	3	7	delayed
10	PT90	H4AI51	Air3	737	3	U	J	16	U	450	3	21	on_time
11	PI67	H4AI51	Air1	319	6	U	J	U	PT90	350	1	12	delayed
12	PT90	H4AI51	Air4	319	6	2	J	U	55	350	2	20	delayed
13	PT90	H4AI51	Air4	319	6	2	J	71	U	400	6	16	on_time
14	PB22	H4AI51	Air4	319	6	U	M	U	18	400	6	20	ahead_of_time
15	PT90	H4AI51	Air1	319	6	4	J	U	71	400	4	11	on_time
16	PC87	H4AI51	Air1	319	6	4	J	U	50	400	6	20	on_time
17	PT90	H4AI51	Air1	319	3	4	J	U	31	350	6	11	on_time
18	PK14	A4IK77	Air2	319	6	4	J	PT90	U	350	1	10	on_time
19	PK14	H4AI51	Air2	CR7	6	U	J	U	54	200	6	10	delayed
20	PT90	H4AI51	Air2	CR7	6	U	J	U	51	200	6	21	on_time
21	PW76	H4AI51	Air1	CR7	6	2	J	U	51	200	2	11	on_time
22	PT90	H4AI51	Air3	CR7	3	2	J	U	51	200	3	16	ahead_of_time
23	PC71	H4AI51	Air2	CR7	3	U	J	U	51	200	3	20	delayed

Fig. 19.1 Flight data

criteria that are exploited in an interleaved process to greatly reduce the computational cost of identifying locally relevant attributes.

19.3 Related Work

Classification is a field of extensive research. Several distinct branches of classification techniques have been developed. Neural networks learn a discriminant function between individual classes [11, 23, 32]. Bayes classifiers estimate the data distribution globally based on training data [7, 15]. Support vector machines (SVMs) compute a separating hyperplane in a higher dimensional space [13, 27, 34]. All of these approaches do not provide an explanatory component for the patterns learned. In our application, however, users wish to validate the decision basis for classification of flight delays.

Decision trees provide explanatory components by visualizing the decision taken during classification [24, 28, 29]. The idea is to build a model on training data by successively partitioning the data along some splitting attribute which best separates the data according to the given class label. The resulting tree is then used to classify incoming data tuples by following the branches corresponding to this tuple until a leaf containing class information is found. In general, decision trees have been successfully applied to predict class labels in application domains where global patterns are present. However, when it comes to noise and local patterns, decision trees do not necessarily reflect class structures. This is due to the fact that decision trees are built level by level, i.e. the choice of splitting attributes is based on a greedy-style evaluation strategy. Moreover, even if decision trees were to evaluate multiple split levels before making a split decision, they would not be able to represent parallel patterns in subspaces of the attribute domain. See Figure 19.2 for an example. As we can see, a parallel subspace pattern (gray square area, values b/c in attribute X_1 and a/b in X_2) is split due to other seemingly more prevailing patterns (values a/b, c, and d in X_1). A hierarchical structure thus cannot properly reflect both patterns, as there is only a single attribute per level which is split.

Nearest neighbors do not build a model beforehand, but instead query data in a "lazy" fashion [26]. As local relevance is not clear beforehand, "nearest neighbors" are chosen based on a global distance function. In [12], local weights are introduced, but as a starting point for iterative weighting, a global distance function is used. Thus, the local distribution used for weighting is based on all attributes which may not necessarily contribute to a query's class membership or even conceal it. Moreover, in high-dimensional spaces, i.e. faced with many attributes, distances become more and more similar due to the so-called "curse of dimensionality" [8]. Consequently, nearest neighbors lose their meaning and classification power. One solution approach generates an appropriate subset of attributes deemed meaningful for each incoming query data set [19]. Given a fixed number of target attributes, a generic algorithm followed by a greedy approach, evaluates individual attributes for each query. This requires a-priori knowledge on the number of attributes relevant as

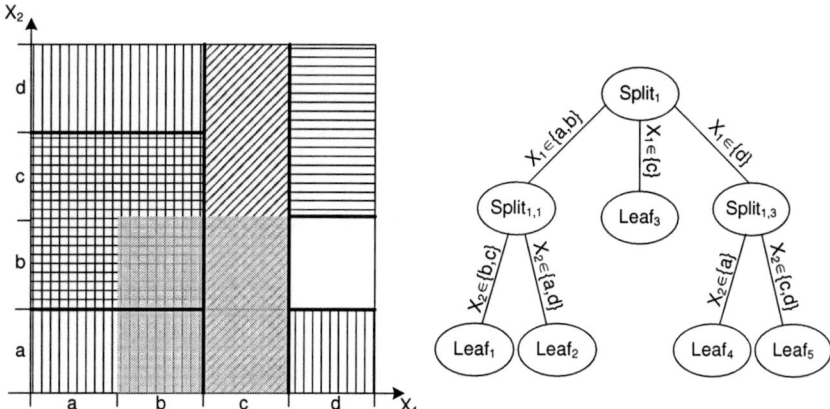

Fig. 19.2 Parallel patterns splitted by decision tree

well as a quality criterion for the choice of attributes for a given query, which is not available for the flights in general.

In [1], the authors develop a specialized approach for flight delay mining. Their premise is availability of weather information in the data which is not the case for our project. Moreover, periodicity of patterns as well as global relevance of attributes is assumed. As discussed before, in our flight data, the attributes show locally varying relevance.

For clustering, i.e. grouping of data with respect to mutual similarity, it is well known that traditional algorithms do not scale to high-dimensional spaces. They also suffer from the "curse of dimensionality", i.e. distances grow increasingly similar and meaningful clusters can no longer be detected [8]. Dimensionality reduction techniques like PCA (principle components analysis) aim at discarding irrelevant dimensions [20]. In many practical applications, however, no globally irrelevant dimensions exist, but only locally irrelevant dimensions for each cluster are observed. This observation has led to the development of different subspace clustering algorithms [2, 4, 5, 21, 30, 33]. Subspace clustering searches for clusters in low dimensional subspaces of the original high dimensional space. It has been shown to successfully identify locally relevant projections for clusters even in very high dimensional data. Subspace clustering, however, does not, by its very definition, take class labels into account, but aims at identifying patterns in an "unsupervised" manner.

In classification, i.e. in "supervised" learning tasks, the class labels are important. In flight delay classification, it is important to identify those patterns that provide information on the delay class. Our approach therefore takes class labels into account to identify those subspace clusters that contain information on flight delays.

19.4 Classification of Flights

For robust scheduling of flights, our aim is reliable classification of flights as "ahead of time", "on time", and "delayed". To account for locally varying attribute relevance, our first step is detection of the relevant subspaces. Subspace detection is followed by the actual classifying subspace clusters detection and the final assignment of class labels for incoming flights. A detailed discussion of this model can be found in [6].

19.4.1 Subspaces for Locally Varying Relevance

Interesting subspaces for flight delays exhibit a clustering structure in their attributes as well as coherent class label information. Such a structure is reflected by homogeneity which can be measured using conditional entropy $H(X|Y) = -\sum_{y \in Y} p(y) \sum_{x \in X} p(x|y) \cdot \log_2 p(x|y)$ [31].

Conditional attribute entropy $H(X_i|C)$ measures the uncertainty in an attribute given a class label. A low attribute entropy means that the attribute has a high cluster tendency w.r.t. a class label and is not blurred by noise. Conditional class entropy $H(C|X_i)$ measures the uncertainty of the class given an attribute, i.e. the confidence of class label prediction based on the attribute.

We define interestingness of a subspace $S = \{X_1, \ldots, X_m\}$ as the convex combination of normalized attribute entropy $H_N(S|C)$ and class entropy $H_N(C|S)$ (details on normalization can be found in [6]).

Definition 19.1. Subspace Interestingness. Given attributes $S = \{X_1, \ldots, X_m\}$, a class label C, and a weighting factor $0 \leq w \leq 1$, a subspace is interesting with respect to thresholds β, λ iff:

$$w \cdot H_N(S|C) + (1-w) \cdot H_N(C|S) \leq \beta \ \wedge \ H_N(S|C) \leq \lambda \ \wedge \ H_N(C|S) \leq \lambda$$

Thus, a subspace is interesting for subspace classification if it shows low normalized class and attribute entropy. w allows assigning different weights to these two aspects, while λ ensures that both individually fulfill minimum entropy requirements.

19.4.2 Integrating Subspace Information for Robust Flight Classification

Classifying subspace clusters are those clusters that are homogeneous with respect to flight delay class and that show frequent attribute value combinations [6].

Definition 19.2. Classifying Subspace Cluster. A set $SC = \{(X_1, v_1), \ldots, (X_m, v_m)\}$ of values $v_1, \ldots, v_m$ in a subspace $S = \{X_1, \ldots, X_m\}$ is a classifying subspace cluster with respect to a minimum frequency thresholds ϕ_1, ϕ_2, and maximum entropy γ iff:

$$H_N(C|SC) \leq \gamma \wedge AbsFreq(SC) \geq \phi_1 \wedge NormFreq(SC) \geq \phi_2$$

Classifying subspace clusters have low normalized class entropy, as well as high frequency in terms of attribute values. To ensure non-trivial subspace clusters, both absolute frequency of values and normalized frequency thresholds have to be exceeded. Details on normalizing thresholds for subspace clustering can be found in [5,6].

Classification of a given flight $f = (f_1, \ldots, f_d)$ is based on the class label distribution of relevant classifying subspace clusters. Let $CSC(f) = \{SC_i | \forall (X_k, v_k) \in SC_i : v_k = f_k\}$ denote the set of all classifying subspace clusters containing flight f. Simply assigning the class label based on the complete set $CSC(f)$ of classifying subspace clusters would be biased with respect to very large and redundant subspace clusters, where redundancy means similar clusters in slightly varying subspaces. We therefore propose selecting non-redundant locally relevant attributes for classification of a flight f from the set $CSC(f)$. The relevant attribute decision set DS_k is built iteratively by choosing those classifying subspace clusters $SC_i \in CSC(f)$ which have the highest information gain w.r.t the flight delay.

Definition 19.3. Classification. For a given a dataset D and parameter k, a flight $f = (f_1, \ldots, f_d)$ is classified to the majority class label of decision set DS_k. Based on the set of all classifying subspace clusters $CSC(f)$ for the flight f, DS_k is iteratively constructed from $DS_0 = \emptyset$ by selecting the subspace cluster $SC_j \in CSC(f)$ which maximizes the information gain about the class label:

$$DS_j = DS_{j-1} \cup SC_j, \; SC_j = \left\{ \underset{SC_i \in CSC(f)}{\mathrm{argmax}} \; \{H(C|DS_{j-1}) - H(C|DS_{j-1} \cup SC_i)\} \right\}$$

under the constraints that the decision space contains at least ϕ_1 objects: $|\{f \in D, \forall (X_k, v_k) \in SC_i : f_k = v_k\}| \geq \phi_1$ and that the information gain is positive: $H(C|DS_{j-1}) - H(C|DS_{j-1} \cup SC_i) > 0$

Hence, the decision set of a flight f is created by choosing those k subspace clusters containing f that provide most information on the class label, as long as more than a minimum number of flights are in the decision space. f is then classified according to the majority in the decision set DS_k. The flights and attributes in the decision set are helpful for users wishing to understand the information that led to classification of flight delays.

19.5 Algorithmic Concept

Our flight classification algorithm is based on three major steps: subspace search, clustering in these subspaces, and the actual class label assignment. Searching all possible subspaces is far too costly, thus we suggest a lossless pruning strategy based on monotonicity properties.

19.5.1 Monotonicity Properties of Relevant Attribute Subspaces

Attribute entropy decreases monotonically with growing number of attributes. We denote as *downward monotony* of attribute entropy that for a set of m attributes, subspace $S = \{X_1, .., X_m\}$, $e \in \mathbb{R}^+$ and $T \subseteq S$,

$$H(S|C) < e \quad \Rightarrow \quad H(T|C) < e$$

This downward monotony follows immediately from $H(X_i, X_j) \geq H(X_i)$ [17].

Thus, any subspace S whose lower dimensional subspace projections T do not exceed a threshold e, does not exceed this threshold e either. This downward monotony can be used for lossless pruning in a **bottom-up** apriori style algorithm. Apriori, originally from association rule mining, means joining two interesting subspaces with $m-1$ common attributes of size m to create a candidate subspace of size $m+1$ (cf. Fig. 19.4). Only the resulting candidates have to be analyzed further, the remainder may be safely discarded [2, 3, 21].

For class entropy, the converse holds: It grows monotonically with the number of attributes. We denote as *upward Monotony* of the class entropy that for a set of m attributes, subspace $S = \{X_1, .., X_m\}$, $e \in \mathbb{R}^+$ and $T \subseteq S$,

$$H(C|T) < e \quad \Rightarrow \quad H(C|S) < e$$

This upward monotony follows immediately from $H(X|X_i, X_j) \leq H(X|X_i)$ [17].

As class entropy grows with the number of attributes, **bottom-up** apriori algorithms cannot be applied. Applying apriori **top-down** in a naive manner, all subspaces with $m-1$ attributes would have to be generated before pruning of lower-dimensional projections with $m-1$ attributes is possible. To avoid this, we suggest a more sophisticated **top-down** approach and prove its losslessness in the next subsection (cf. Fig. 19.4).

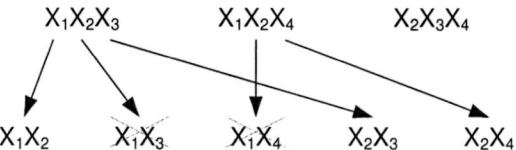

Fig. 19.3 Example top down generation

19.5.2 Top-down Class Entropy Algorithm: Lossless Pruning Theorem

We first illustrate our top down algorithm in an example before formally proving its losslessness. The main idea is to generate all subspace candidates with respect to class entropy in a top down manner without any duplicate candidates from those subspaces already found to satisfy the class entropy criterion (termed "*class homogeneous subspaces*").

In Figure 19.3, assume that we have four attributes $X_1,\ldots,X_4$ and that in the previous step, we have found the class homogeneous subspaces $X_1X_2X_3$, $X_1X_2X_4$, and $X_2X_3X_4$. In order to generate candidates, we iterate over these subspaces in lexicographic order.

From subspaces of dimensionality m, we generate in a top-down fashion only those subspace candidates of dimensionality $m-1$ which contain all attributes in lexicographic order up to a certain point. After this, just as with apriori, we check whether all super-subspaces containing the newly generated candidates exceed the threshold. Otherwise the newly generated subspace is removed from the candidate set.

The first three-dimensional subspace $X_1X_2X_3$ generates the two-dimensional subspaces X_1X_2 (drop X_3), X_1X_3 (drop X_2), X_2X_3 (drop X_1). Next, $X_1X_2X_4$ generates X_1X_4 and X_2X_4. X_1X_2 is not generated again by $X_1X_2X_4$, because dropping X_4 is not possible, as it is preceded by X_3 which is not contained in this subspace. The last three-dimensional subspace $X_2X_3X_4$ does not generate any two-dimensional subspace since the leading X_1 is not contained; its subsets X_2X_3 and X_2X_4 have been generated by other three- dimensional subspaces.

After candidate generation, we check whether the newly generated two-dimensional subspaces are really candidates by checking their respective supersets. For example, for X_1X_2, its supersets $X_1X_2X_3$ and $X_1X_2X_4$ exist. For X_1X_3, its supersets $X_1X_2X_3$ exists, but $X_1X_3X_4$ does not, so it is removed from further consideration. Likewise, X_1X_4 is removed as $X_1X_3X_4$ is missing, but X_2X_3 and X_2X_4 are kept.

In general, the set $Gen(m)$ of all generated candidate subspaces of dimensionality m, is created from all class homogeneous subspaces S of dimensionality $m+1$, denoted as $CHS(m+1)$. Each super-subspace $S' \in CHS(m+1)$ generates candidates by dropping one attribute X_k. Only those X_k can be dropped where all "smaller" attributes (w.r.t. to lexicographic ordering) are consecutively contained in S'.

The set $Gen(m)$ of generated candidate subspaces of dimensionality m is defined as $Gen(m) = \{CS, CS \in Cand(S'), S' \in CHS(m+1)\}$ with respect to $CHS(m+1)$, the set of all class homogeneous subspaces of dimensionality $m+1$ and the set of all candidate subspaces that can be generated by dropping the first subspace in lexicographic order: $Cand(S') = \{CS, CS = S' \setminus \{X_k\}, \forall k' \leq k : X_{k'} \in S'\}$.

The correctness of generating only these subspaces is stated in the following theorem:

Theorem 19.1. *Lossless top-down pruning*
Let $CHS_{Cand}(m) = \{CS, CS \cup \{X_i\} \in CHS(m+1) \; \forall X_i \notin CS\}$ be the set of all class homogeneous subspaces candidates of dimensionality m, then:

$$CS \in CHS_{Cand}(m) \Rightarrow CS \in Gen(m)$$

This theorem states that any class homogeneous subspace CS is contained in the set of generated candidate subspaces $Gen(m)$, i.e. our algorithm which detects these $Gen(m)$ is lossless.

Proof. The proof consists of two parts: first we show that all potential homogeneous candidate subspaces are generated, then we prove that no duplicates are generated.

To see that all candidates are generated, assume to the contrary that there is an m-dimensional candidate subspace $CS \in CHS_{Cand}(m)$ but $CS \notin Gen(m)$. Now let k be the smallest index such that $X_k \notin CS$ (note that at least one such k exist, since we generated CS in a top down fashion, and thus it cannot contain all attributes). Then we have that $CS \cup \{X_k\} \in CHS(m+1)$, per definition of $CHS_{Cand}(m)$. Then, $CS \in Cand(CS \cup \{X_k\})$ since for all $k' \leq k : X_{k'} \in CS \cup \{X_k\}$ and thus $CS \in Gen(m)$. This is a contradiction to our assumption. Thus, the theorem holds.

We now show that no duplicates are generated. Assume that there is a candidate subspace CS which is generated twice: $CS \in Cand(S_1)$ and $CS \in Cand(S_2)$, where $S_1 \neq S_2$. From the definition of $Cand$ directly follows that the dimensionality of S_1 is equal to the dimensionality of S_2. Further on if $S_1 \neq S_2$ there exist i, j such that $S_1 \setminus \{X_i\} = CS$ and $S_2 \setminus \{X_j\} = CS$. If $i = j$, we have $S_1 = S_2$ which contradicts our assumption. If $i \neq j$, let k be the smallest index such that $X_k \notin CS$. Then, if $X_k \notin S_1$, there is a $k' < k$ such that $X_{k'} \notin S_1$, thus $CS \notin Cand(S_1)$. This is in contrast to our assumption. If $X_k \in S_1$, then the attribute dropped is $X_k = X_i$ ($S_1 \setminus \{X_k\} = CS$). The same holds for S_2: $S_2 \setminus \{X_k\} = CS$, $X_k = X_j$. Then, $X_i = X_j$ which contradicts our assumption that $S_1 \neq S_2$. Thus, no duplicates are generated.

19.5.3 Algorithm: Subspaces, Clusters, Subspace Classification

The upward and downward closure is thus used to reduce the number of classifying subspace candidates. Each closure can be illustrated as a boundary in the lattice of the subspaces. Figure 19.4 illustrates this fact for a lattice of four attributes. The solid line is the boundary for the attribute entropy and the dashed line illustrates

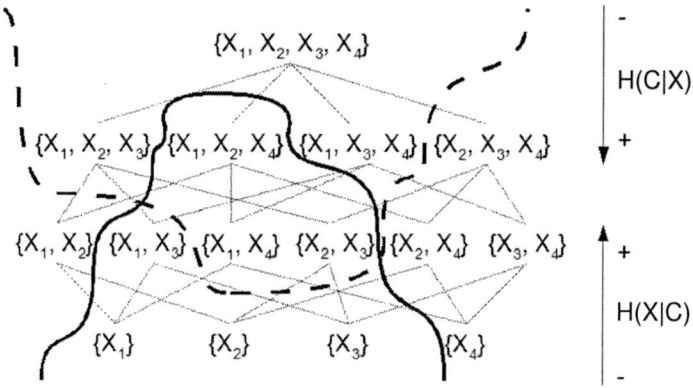

Fig. 19.4 Lattice of Subspaces and their projections used for up- and downward pruning

the boundary for the class entropy. Each subspace below the attribute boundary and above the class boundary is homogeneous with respect to the entropy considered. The subspaces between both boundaries are interesting subspace candidates, whose combined entropy has to be computed in the next step.

This leads to our proposed algorithm which alternatively determines smaller (downward) and larger (upward) candidate subspaces. By using a bottom up and top down approach simultaneously, both monotonicity properties can be exploited at the same time.

Figure 19.5 presents the pseudo code for identifying all subspaces having a combined entropy less than threshold β. The combined entropy used by the algorithm is specified by the weight (w) and threshold (β). The algorithm starts with a candidate set containing all one-dimensional subspaces for the bottom up method ($C_{Bot}(1)$) and the candidate-subspace containing all dimensions for the top down method ($C_{Top}(N)$). N denotes the dimensionality of the complete data-space. The sets of subspaces $S_{Bot}(0)$ and $S_{Top}(N+1)$ initialized in Line 3 are used for the first iteration of the while-loop when no information about the last step is known. Main part of the method is a loop over all dimensionalities. Simultaneously the class and attribute entropy are computed in a bottom up and top down manner. The subroutines called in Line 5 and 6 identify the sets of subspaces of dimensionality i and $N-i$, respectively, with entropy below β and γ. Based on these sets, new candidate sets are generated in Line 8 and 9. As discussed before, the bottom up candidates are generated following the apriori algorithm [3] while top down candidates are generated as described in Subsection 19.5.2.

Once the interesting subspaces have been detected, they are clustered according to the definition of classifying subspace clusters (Def. 19.2). Incoming flights are then assigned a class label based on the relevant attribute value combinations extracted from the set of classifying subspace clusters (Def. 19.3).

Algorithm 1: FlightDelay(w, β, N)

1 $C_{Bot}[1] = \{(\{X_1\}, 0), \ldots, (\{X_N\}, 0)\}$;   /* initial bottom up cand. set */
2 $C_{Top}[N] = \{(\{X_1 \ldots X_N\}, 0)\}$;   /* initial top up cand. set */
3 $S_{Bot}[0] = C_{Bot}[1]; S_{Top}[N+1] = C_{Top}[N]$;   /* assume cand. as results */
4 **for** $i = 1$ **to** N **do**   /* for each dimensionality */
5     $S_{Bot}[i] \leftarrow Subspace_{Attribute}(C_{Bot}[i], S_{Bot}[i-1], w, \beta)$
6     $S_{Top}[N-(i+1)] \leftarrow Subspace_{Class}(C_{Top}[N-(i+1)], S_{Top}[N-i], 1-w, \beta)$
7     $C_{Bot}[i+1] \leftarrow GenerateCand_{BottomUp}(S_{Bot}[i])$
8     $C_{Top}[N-i] \leftarrow GenerateCand_{TopDown}(S_{Bot}[N+1-i])$
9 **end**
10 $S = \bigcup_{k=1}^{N} (S_{Bot}[k] \cup S_{Top}[k])$;   /* join mined subspaces */
11 $result = FilterFalsePositives(S)$;   /* filter interesting subspaces */
12 **return** $result$;   /* return result */

Fig. 19.5 Algorithm for the flight delay subspace search

19.6 Evaluation of Flight Delay Classification in Practice

The flight data contains historic data from a large European airport. For a three-month period, we trained the classifier on arrivals of two consecutive months and tested on the following month. Outliers with delays outside [-60, 120] minutes have been eliminated. In total, 11.072 flights have been used for training and 5.720 flights for testing. For each flight, many different attributes are recorded, including e.g. the airline, flight number, aircraft type, routing, and the scheduled arrival time within the day. The class labels are "ahead of time", "on time" and "delayed".

As mentioned before, preliminary experiments on the flight data indicate that no global relevance of attributes exist. Moreover, the data is inherently noisy, and important influences like weather conditions are not collected from scheduling. For realistic testing as in practical application, classifiers can only draw from existing at-

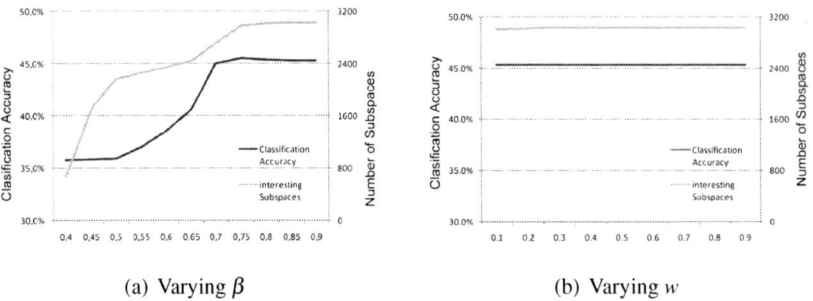

(a) Varying β (b) Varying w

Fig. 19.6 Parameter evaluation

19 Data Mining For Robust Flight Scheduling

	Flight data	Synthetic data	Glass	Iris
Proposed method	**45.4%**	**65.9%**	70.9%	**96.27%**
C4.5	43.9%	58.0%	66.8%	95.94%
K-NN	42.4%	54.3%	**71.1%**	93.91%
Naive Bayes	42.8%	64.1%	46.7%	95.27%

Fig. 19.7 Classification accuracy on four data sets

tributes. Missing or not collected parameters are not available for training or testing neither in our experiments nor during the actual scheduling process.

We have conducted prior experiments to evaluate the effect of ϕ and γ for minimum frequency and maximum entropy thresholds, respectively. For each data set we used a cross validation to chose ϕ_1 (absolute frequency), ϕ_2 (relative frequency) and γ. For λ we have chosen 0.9. This value corresponds to a rather relaxed setting as we only want to remove completely inhomogeneous subspaces from consideration. To restrict the search space β can be set to a low value.

First, we set up reasonable parameters for the threshold β of the interestingness and the weight w of the class and attribute entropy, respectively. Figure 6(a) illustrates varying β from 0.45 to 0.95, measuring classification accuracy and the number of classifying subspaces. The weight w for interestingness was set to 0.5. As expected, the number of classifying subspaces decreases when lowering the threshold β. At the same time, the classification accuracy does not change substantially until β takes a value of about 0.7. Varying parameter w yields the results depicted in Figure 6(b). As we can see, our approach is robust with respect to this parameter. This robustness is due to the ensuing subspace clustering phase. As classification accuracy does not change this confirms that our classifying subspace cluster definition selects the relevant patterns. Setting $w = 0.5$ gives equivalent weight to the class and attribute entropy and hence is a good choice for pruning subspaces.

Next, we evaluate classification accuracy thoroughly by comparing our flight delay classifier with competing classifiers that are applicable on the nominal attributes: the k-NN classifier with Manhattan distance, the C4.5 decision tree that also uses a class and attribute entropy model [29], and a Naive Bayes classifier, a probabilistic classifier that assumes independence of attributes. Parameter settings use the best values from the preceding experiments.

Besides the flight data, we use three additional data sets for a general evaluation of the performance of these classifiers. Two further real world data sets from the UCI KDD repository, Glass and Iris, are used [18], as well as synthetic data. Synthetic data is used to show the correctness of our approach. Local patterns are hidden in a data set of 7.000 objects and eight attributes. As background noise, each attribute of the synthetic data set is uniformly distributed over ten values. On top of this, 16 different local patterns (subspace clusters) with different dimensionalities and different numbers of objects are hidden in the data set. Each local pattern contains two or three class labels among which one class label is dominating. We randomly picked 7.000 objects for training and 1.000 objects for testing.

Figure 19.7 illustrates the classification accuracy using four different data sets. In the noisy synthetic data set, our approach outperforms other classifiers. The large degree of noise and the varying class label distribution within the subspace clusters make this a challenging task. From the real world experiment on the flight data, depicted in Figure 19.7, we see that the situation is even more complex. Still, our method performs better than its competitors. This result supports our analysis that locally relevant information for classification exists that should be used for model building. Experts from flight scheduling confirm that additional information on further parameters, e.g. weather conditions, is likely to boost classification. This information is inexistent in the current scheduling data that is collected routinely. We exploit all the information available, especially locally relevant attribute and value combinations, for the best classification in this noisy scenario. Finally we evaluated the performance on Glass and Iris. The results indicate that even in settings containing no or little noise our classifier performs well.

In summary, our flight classifier is capable of detecting locally relevant attributes in the noisy data, thus providing the desired information for dynamic scheduling. The relevant attributes serve as an explanatory component for experts working on airport schedules.

19.7 Conclusion

For our project in scheduling at airports, we developed a specialized approach that classifies incoming flights as "ahead of time", "on time", and "delayed" to aid dynamic scheduling in the presence of delays. Our classifier is capable of dealing with the high dimensionality of the data as well as with locally varying relevance of the attributes. Our experiments show that our classifier successfully exploits subspace structures even in highly noisy data.

This shows that even when important data (e.g. weather) is missing, proactive scheduling methods can successfully exploit data that is known at planning time. Our flight classification method can therefore provide an important basis for the feedback of operational data into the planning phase. Additionally to providing scheduling methods with means for advanced robustness measures, it can be valuable for airport resource managers in explaining delay structures and providing decision support.

References

1. M. Abdel-Aty, C. Lee, Y. Bai, X. Li, and M. Michalak. Detecting periodic patterns of arrival delay. *Journal of Air Transport Management*, pages 355–361, 2007.
2. R. Agrawal, J. Gehrke, D. Gunopulos, and P. Raghavan. Automatic subspace clustering of high dimensional data for data mining applications. In *Proceedings of the ACM International Conference on Management of Data (SIGMOD)*, pages 94–105, 1998.

3. R. Agrawal and R. Srikant. Fast algorithms for mining association rules. In *Proceedings of the International Conference on Very Large Data Bases (VLDB)*, pages 487–499, 1994.
4. I. Assent, R. Krieger, B. Glavic, and T. Seidl. Clustering multidimensional sequences in spatial and temporal databases. *International Journal on Knowledge and Information Systems (KAIS)*, 2008.
5. I. Assent, R. Krieger, E. Müller, and T. Seidl. DUSC: Dimensionality unbiased subspace clustering. In *Proceedings of the IEEE International Conference on Data Mining (ICDM)*, pages 409–414, 2007.
6. I. Assent, R. Krieger, P. Welter, J. Herbers, and T. Seidl. Subclass: Classification of multidimensional noisy data using subspace clusters. In *Proceedings of the Pacific-Asia Conference on Knowledge Discovery and Data Mining (PAKDD), Osaka, Japan.* Springer, 2008.
7. T. Bayes. An essay towards solving a problem in the doctrine of chances. *Philosophical Transactions of the Royal Society*, 53:370–418, 1763.
8. K. Beyer, J. Goldstein, R. Ramakrishnan, and U. Shaft. When is nearest neighbors meaningful. In *Proceedings of the 7th International Conference on Database Theory (ICDT)*, pages 217–235, 1999.
9. A. Bolat. Procedures for providing robust gate assignments for arriving aircrafts. *European Journal of Operational Research*, 120:63–80, 2000.
10. Bureau of Transportation Statistics. Airline on-time performance data. Available from http://www.transtats.bts.gov.
11. Y. Cao and J. Wu. Projective art for clustering data sets in high dimensional spaces. *Neural Networks*, 15(1):105–120, 2002.
12. C. Domeniconi, J. Peng, and D. Gunopulos. Locally adaptive metric nearest-neighbor classification. *IEEE Transactions on Pattern Analysis and Machine Intelligence (PAMI)*, 24(9):1281–1285, 2002.
13. J. Dong, A. Krzyak, and C. Suen. Fast SVM Training Algorithm with Decomposition on Very Large Data Sets. *IEEE Transactions Pattern Analysis and Machine Intelligence (PAMI)*, pages 603–618, 2005.
14. U. Dorndorf, F. Jaehn, and E. Pesch. Modelling robust flight-gate scheduling as a clique partitioning problem. *Transportation Science*, 2008.
15. R. Duda, P. Hart, and D. Stork. *Pattern Classification (2nd Edition)*. Wiley, 2000.
16. Eurocontrol Central Office for Delay Analysis. Delays to air transport in europe. Available from http://www.eurocontrol.int/eCoda.
17. R. Gray. *Entropy and Information Theory*. Springer, 1990.
18. S. Hettich and S. Bay. The UCI KDD archive [http://kdd.ics.uci.edu]. Irvine, CA: University of California, Department of Information and Computer Science, 1999.
19. A. Hinneburg, C. Aggarwal, and D. Keim. What is the nearest neighbor in high dimensional spaces? In *Proceedings of the International Conference on Very Large Data Bases (VLDB)*, pages 506–515, September 2000.
20. I. Joliffe. *Principal Component Analysis*. Springer, New York, 1986.
21. K. Kailing, H.-P. Kriegel, and P. Kröger. Density-connected subspace clustering for high-dimensional data. In *Proceedings of the IEEE International Conference on Data Mining (ICDM)*, pages 246–257, 2004.
22. S. Lan, J.-P. Clarke, and C. Barnhart. Planning for robust airline operations: Optimizing aircraft routings and flight departure times to minimize passenger disruptions. *Transportation Science, 40(1):15-28*, 2006.
23. W. McCulloch and W. Pitts. A logical calculus of the ideas immanent in nervous activity. *Bulletin of Mathematical Biophysics*, 5:115–137, 1943.
24. S. Murthy. Automatic construction of decision trees from data: A multi-disciplinary survey. *Data Mining and Knowledge Discovery*, 2(4):345–389, 1998.
25. L. Parsons, E. Haque, and H. Liu. Subspace clustering for high dimensional data: a review. *SIGKDD Explorations Newsletter*, 6(1):90–105, 2004.
26. E. Patrick and F. Fischer. A generalized k-nearest neighbor rule. *Information and Control*, 16(2):128–152, 1970.

27. J. Platt. Fast training of support vector machines using sequential minimal optimization. In Schoelkopf, Burges, and Smola, editors, *Advances in Kernel Methods*. MIT Press, 1998.
28. J. Quinlan. Induction of decision trees. *Machine Learning*, 1:81–106, 1986.
29. J. Quinlan. *C4.5: Programs for Machine Learning*. Morgan Kaufmann, 1992.
30. K. Sequeira and M. Zaki. SCHISM: A new approach for interesting subspace mining. In *Proceedings of the IEEE International Conference on Data Mining (ICDM)*, pages 186–193, 2004.
31. C. Shannon and W. Weaver. *The Mathematical Theory of Communication*. University of Illinois Press, Urbana, Illinois, 1949.
32. L. Silva, J. M. de Sa, and L. Alexandre. Neural network classification using Shannon? Entropy. In *Proceedings of the European Symposium on Artificial Neural Networks (ESANN)*, 2005.
33. M. Zaki, M. Peters, I. Assent, and T. Seidl. Clicks: An effective algorithm for mining subspace clusters in categorical datasets. *Data & Knowledge Engineering (DKE)*, 57, 2007.
34. H. Zhang, A. Berg, M. Maire, and J. Malik. SVM-KNN: Discriminative nearest neighbor classification for visual category recognition. *Proceedings of the IEEE Conference on Computer Vision and Pattern Recognition (CVPR)*, 2, 2006.

Chapter 20
Data Mining for Algorithmic Asset Management

Giovanni Montana and Francesco Parrella

Abstract Statistical arbitrage refers to a class of algorithmic trading systems implementing data mining strategies. In this chapter we describe a computational framework for statistical arbitrage based on support vector regression. The algorithm learns the fair price of the security under management by minimining a regularized ε-insensitive loss function in an on-line fashion, using the most recent market information acquired by means of streaming financial data. The difficult issue of adaptive learning in non-stationary environments is addressed by adopting an ensemble learning approach, where a meta-algorithm strategically combines the opinion of a pool of experts. Experimental results based on nearly seven years of historical data for the iShare S&P 500 ETF demonstrate that satisfactory risk-adjusted returns can be achieved by the data mining system even after transaction costs.

20.1 Introduction

In recent years there has been increasing interest for active approaches to investing that rely exclusively on mining financial data, such as *market-neutral* strategies [11] . This is a general class of investments that seeks to neutralize certain market risks by detecting market inefficiencies and taking offsetting long and short positions, with the ultimate goal of achieving positive returns independently of market conditions. A specific instance of market-neutral strategies that heavily relies on temporal data mining is referred to as *statistical arbitrage* [11, 14]. Algorithmic asset management systems embracing this principle are developed to make *spread trades*, namely trades that derive returns from the estimated relationship between two statistically related securities.

Giovanni Montana, Francesco Parrella
Imperial College London, Department of Mathematics, 180 Queen's Gate, London SW7 2AZ, UK,
e-mail: {g.montana,f.parrella}@imperial.ac.uk

An example of statistical arbitrage strategies is given by *pairs trading* [6]. The rationale behind this strategy is an intuitive one: if the difference between two statistically depending securities tends to fluctuate around a long-term equilibrium, then temporary deviations from this equilibrium may be exploited by going long on the security that is currently under-valued, and shorting the security that is over-valued (relatively to the paired asset) in a given proportion. By allowing short selling, these strategies try to benefit from decreases, not just increases, in the prices. Profits are made when the assumed equilibrium is restored.

The system we describe in this chapter can be seen as a generalization of pairs trading . In our setup, only one of the two dependent assets giving raise to the spread is a tradable security under management. The paired asset is instead an artificial one, generated as a result of a data mining process that extracts patterns from a large population of data streams, and utilizes these patterns to build up the synthetic stream in real time. The extracted patterns will be interpreted as being representative of the current market conditions, whereas the synthetic asset will represent the *fair* price of the target security being traded by the system. The underlying concept that we try to exploit is the existence of time-varying cross-sectional dependencies among securities. Several data mining techniques are being developed lately to capture dependencies among data streams in a time-aware fashion, both in terms of latent factors [12] and clusters [1]. Recent developments include novel database architectures and paradigms such as CEP (Complex Event Processing) that discern patterns in streaming data, from simple correlations to more elaborated queries.

In financial applications, data streams arrive into the system one data point at a time, and quick decisions need to be made. A prerequisite for a trading system to operate efficiently is to learn the novel information content obtained from the most recent data in an *incremental* way, slowly forgetting the previously acquired knowledge and, ideally, without having to access all the data that has been previously stored. To meet these requirements, our system builds upon incremental algorithms that efficiently process data points as they arrive. In particular, we deploy a modified version of on-line support vector regression [8] as a powerful function approximation device that can discover non-negligible divergences between the paired assets in real time. Streaming financial data are also characterized by the fact that the underlying data generating mechanism is constantly evolving (i.e. it is non-stationary), a notion otherwise referred to as *concept drifting* [12] . Due to this difficulty, particularly in the high-frequency trading spectrum, a trading system's ability to capture profitable inefficiencies has an ever-decreasing half life: where once a system might have remained viable for long periods, it is now increasingly common for a trading system's performance to decay in a matter of days or even hours. Our attempt to deal with this challenge in an autonomous way is based on an *ensemble learning* approach , where a pool of trading algorithms or *experts* are evolved in parallel, and then strategically combined by a master algorithm. The expectation is that combining expert opinion can lead to fewer trading mistakes in all market conditions.

20.2 Backbone of the Asset Management System

In this section we outline the rationale behind the statistical arbitrage system that forms the theme of this chapter, and provide a description of its main components. Our system imports $n+1$ cross-sectional financial data streams at discrete time points $t = 1, 2, \ldots$. In the sequel, we will assume that consecutive time intervals are all equal to 24 hours, and that a trading decision is made on a daily basis. Specifically, after importing and processing the data streams at each time t, a decision to either buy or short sell a number of shares of a *target* security Y is made, and an order is executed. Different sampling frequencies (e.g. irregularly spaced intervals) and trading frequencies could also be incorporated with only minor modifications.

The imported data streams represent the prices of $n+1$ assets. We denote by y_t the price of the security Y being traded by the system, whereas the remaining n streams, collected in a vector $s_t = (s_{t1}, \ldots, s_{tn})^T$, refer to a large collection of financial assets and economic indicators, such as other security prices and indices, which possess some explanatory power in relation to Y. These streams will be used to estimate the *fair* price of the target asset Y at each observational time point t, in a way that will be specified below. We postulate that the price of Y at each time t can be decomposed into two components, that is $y_t = z_t + m_t$, where z_t represents the current *fair* price of Y, and the additive term m_t represents a potential *misprising*. No further assumptions are made regarding the data generating process. Clearly, if the markets were always perfectly efficient, we would have that $y_t = z_t$ at all times. However, when $|m_t| > 0$, an arbitrage opportunity arises. For instance, a negative m_t indicates that Y is temporarily under-valued. In this case, it is sensible to expect that the market will promptly react to this temporary inefficiency with the effect of moving the target price up. Under this scenario, an investor would then buy a number of shares hoping that, by time $t+1$, a profit proportional to $y_{t+1} - y_t$ will be made. Our system is designed to identify and exploit possible statistical arbitrage opportunities of this sort in an automated fashion. This trading strategy can be formalized by means of a binary decision rule $d_t \in \{0,1\}$ where $d_t = 0$ encodes a sell signal, and $d_t = 1$ a buy signal. Accordingly, we write

$$d_t(m_t) = \begin{cases} 0 & m_t > 0 \\ 1 & m_t < 0 \end{cases} \quad (20.1)$$

where we have made explicit the dependence on the current misprising $m_t = y_t - z_t$. If we denote the change in price observed on the day following the trading decision as $r_{t+1} = y_{t+1} - y_t$, we can also introduce a $0-1$ loss function $L_{t+1}(d_t, r_{t+1}) = |d_t - 1_{(r_{t+1} > 0)}|$, where the indicator variable $1_{(r_{t+1} > 0)}$ equals one if $r_{t+1} > 0$ and zero otherwise. For instance, if the system generates a sell signal at time t, but the security's price increases over the next time interval, the system incurs a unit loss.

Obviously, the fair price z_t is never directly observable, and therefore the misprising m_t is also unknown. The system we propose extracts knowledge from the large collection of data streams, and incrementally imputes the fair price z_t on the basis of the newly extracted knowledge, in an efficient way. Although we expect some streams to have high explanatory power, most streams will carry little signal and will mostly contribute to generate noise. Furthermore, when n is large, we expect several streams to be highly correlated over time, and highly dependent streams will provide redundant information. To cope with both of these issues, the system extracts knowledge in the form of a feature vector x_t, dynamically derived from s_t, that captures as much information as possible at each time step. We require for the components of the feature vector x_t to be in number less than n, and to be uncorrelated with each other. Effectively, during this step the system extracts informative patterns while performing dimensionality reduction.

As soon as the feature vector x_t is extracted, the pattern enters as input of a non-parametric regression model that provides an estimate of the fair price of Y at the current time t. The estimate of z_t is denoted by $\hat{z}_t = f_t(x_t; \phi)$, where $f_t(\cdot; \phi)$ is a time-varying function depending upon the specification of a hyperparameter vector ϕ. With the current $\hat{z}_t$ at hand, an estimated mispricing $\hat{m}_t$ is computed and used to determine the trading rule (20.1). The major difficulty in setting up this learning step lies in the fact that the true fair price z_t is never made available to us, and therefore it cannot be learnt directly. To cope with this problem, we use the observed price y_t as a surrogate for the fair price and note that proper choices of ϕ can generate sensible estimates $\hat{z}_t$, and therefore realistic mispricing $\hat{m}_t$.

We have thus identified a number of practical issues that will have to be addressed next: (a) how to recursively extract and update the feature vector x_t from the the streaming data, (b) how to specify and recursively update the pricing function $f_t(\cdot; \phi)$, and finally (c) how to select the hyperparameter vector ϕ.

20.3 Expert-based Incremental Learning

In order to extract knowledge from the streaming data and capture important features of the underlying market in real-time, the system recursively performs a principal component analysis, and extracts those components that explain a large percentage of variability in the n streams. Upon arrival, each stream is first normalized so that all streams have equal means and standard deviations. Let us call $C_t = E(s_t s_t^T)$ the unknown population covariance matrix of the n streams. The algorithm proposed by [16] provides an efficient procedure to incrementally update the eigenvectors of C_t when new data points arrive, in a way that does not require the explicit computation of the covariance matrix. First, note that an eigenvector g_t of C_t satisfies the characteristic equation $\lambda_t g_t = C_t g_t$, where λ_t is the corresponding eigenvalue. Let us call $\widehat{h}_t$ the current estimate of $C_t g_t$ using all the data up to the current time t. This is given by $\widehat{h}_t = \frac{1}{t} \sum_{i=1}^{t} s_i s_i^T g_i$,

which is the incremental average of $s_i s_i^T g_i$, where $s_i s_i^T$ accounts for the contribution to the estimate of C_i at point i. Observing that $g_t = h_t / ||h_t||$, an obvious choice is to estimate g_t as $\widehat{h}_{t-1} / ||\widehat{h}_{t-1}||$. After some manipulations, a recursive expression for $\widehat{h}_t$ can be found as

$$\widehat{h}_t = \frac{t-1}{t}\widehat{h}_{t-1} + \frac{1}{t}s_t s_t^T \frac{\widehat{h}_{t-1}}{||\widehat{h}_{t-1}||} \tag{20.2}$$

Once the first k eigenvectors are extracted, recursively, the data streams are projected onto these directions in order to obtain the required feature vector x_t. We are thus given a sequence of paired observations $(y_1, x_1), \ldots, (y_t, x_t)$ where each x_t is a k-dimensional feature vector representing the latest market information and y_t is the price of the security being traded.

Our objective is to generate an estimate of the target security's fair price using the data points observed so far. In previous work [9, 10], we assumed that the fair price depends linearly in x_t and that the linear coefficients are allowed to evolve smoothly over time. Specifically, we assumed that the fair price can be learned by recursively minimizing the following loss function

$$\sum_{i=1}^{t-1}(y_i - w_i^T x_i) + C(w_{i+1} - w_i)^T(w_{i+1} - w_i) \tag{20.3}$$

that is, a penalized version of ordinary least squares. Temporal changes in the time-varying linear regression weights w_t result in an additional loss due to the penalty term in (20.3). The severity of this penalty depends upon the magnitude on the regularization parameter C, which is a non-negative scalar: at one extreme, when C gets very large, (20.3) reduces to the ordinary least squares loss function with time-invariant weights; at the other extreme, as C is small, abrupt temporal changes in the estimated weights are permitted. Recursive estimation equations and a connection to the Kalman filter can be found in [10], which also describes a related algorithmic asset management system for trading futures contracts. In this chapter we depart from previous work in two main directions. First, the rather strong linearity assumption is released so as to add more flexibility in modelling the relationship between the extracted market patterns and the security's price. Second, we adopt a different and more robust loss function. According to our new specification, estimated prices $f_t(x_t)$ that are within $\pm \varepsilon$ of the observed price y_t are always considered *fair* prices, for a given user-defined positive scalar ε related to the noise level in the data. At the same time, we would also like $f_t(x_t)$ to be as flat as possible. A standard way to ensure this requirement is to impose an additional penalization parameter controlling the norm of the weights, $||w||^2 = w^T w$. For simplicity of exposition, let us suppose again that the function to be learned is linear and can be expressed as $f_t(x_t) = w^T x_t + b$, where b is a scalar representing the bias. Introducing slack variables ξ_t, ξ_t^* quantifying estimation errors greater than ε, the learning task can be casted into the following minimization problem,

$$\min_{w_t, b_t} \frac{1}{2} w_t^T w_t + C \sum_{i=1}^{t}(\xi_i + \xi_i^*) \tag{20.4}$$

$$\text{s.t.} \begin{cases} -y_i + (w_i^T x_i + b_i) + \varepsilon + \xi_i \geq 0 \\ y_i - (w_i^T x_i + b_i) + \varepsilon + \xi_i^* \geq 0 \\ \xi_i, \xi_i^* \geq 0, \ i = 1, \ldots, t \end{cases} \quad (20.5)$$

that is, the support vector regression framework originally introduced by Vapnik [15]. In this optimization problem, the constant C is a regularization parameter determining the trade-off between the flatness of the function and the tolerated additional estimation error. A linear loss of $|\xi_t| - \varepsilon$ is imposed any time the error $|\xi_t|$ is greater than ε, whereas a zero loss is used otherwise. Another advantage of having an ε-insensitive loss function is that it will ensure sparseness of the solution, i.e. the solution will be represented by means of a small subset of sample points. This aspect introduces non negligible computational speed-ups, which are particularly beneficial in time-aware trading applications. As pointed out before, our objective is learn from the data in an incremental way. Following well established results (see, for instance, [5]), the constrained optimization problem defined by Eqs. (20.4) and (20.5) can be solved using a Lagrange function,

$$\begin{aligned} L = \frac{1}{2} w_t^T w_t + C \sum_{i=1}^{t} (\xi_i + \xi_i^*) - \sum_{i=1}^{t} (\eta_i \xi_i + \eta_i^* \xi_i^*) \\ - \sum_{i=1}^{t} \alpha_i (\varepsilon + \xi_i - y_t + w_t^T x_t + b_t) - \sum_{i=1}^{t} \alpha_i^* (\varepsilon + \xi_i^* + y_t - w_t^T x_t - b_t) \end{aligned} \quad (20.6)$$

where $\alpha_i, \alpha_i^*, \eta_i$ and η_i^* are the Lagrange multipliers, and have to satisfy positivity constraints, for all $i = 1, \ldots, t$. The partial derivatives of (20.6) with respect to w, b, ξ and ξ^* are required to vanish for optimality. By doing so, each η_t can be expressed as $C - \alpha_t$ and therefore can be removed (analogously for η_t^*). Moreover, we can write the weight vector as $w_t = \sum_{i=1}^{t} (\alpha_i - \alpha_i^*) x_i$, and the approximating function can be expressed as a support vector expansion, that is

$$f_t(x_t) = \sum_{i=1}^{t} \theta_i x_i^T x_i + b_i \quad (20.7)$$

where each coefficient θ_i has been defined as the difference $\alpha_i - \alpha_i^*$. The dual optimization problem leads to another Lagrangian function, and its solution is provided by the Karush-Kuhn-Tucker (KKT) conditions, whose derivation in this context can be found in [13]. After defying the *margin function* $h_i(x_i)$ as the difference $f_i(x_i) - y_i$ for all time points $i = 1, \ldots, t$, the KKT conditions can be expressed in terms of $\theta_i, h_i(x_i), \varepsilon$ and C. In turn, each data point (x_i, y_i) can be classified as belonging to each one of the following three auxiliary sets,

$$\mathscr{S} = \{i \mid (\theta_i \in [0,+C] \wedge h_i(x_i) = -\varepsilon) \vee (\theta_i \in [-C,0] \wedge h_i(x_i) = +\varepsilon)\}$$
$$\mathscr{E} = \{i \mid (\theta_i = -C \wedge h_i(x_i) \geq +\varepsilon) \vee (\theta_i = +C \wedge h_i(x_i) \leq -\varepsilon)\} \quad (20.8)$$
$$\mathscr{R} = \{i \mid \theta_i = 0 \wedge |h_i(x_i)| \leq \varepsilon\}$$

and an incremental learning algorithm can be constructed by appropriately allocating new data points to these sets [8]. Our learning algorithm is based on this idea, although our definition (20.8) is different. In [13] we argue that a sequential learning algorithm adopting the original definitions proposed by [8] will not always satisfy the KKT conditions, and we provide a detailed derivation of the algorithm for both incremental learning and forgetting of old data points[1].

In summary, three parameters affect the estimation of the fair price using support vector regression. First, the C parameter featuring in Eq. (20.4) that regulates the trade-off between model complexity and training error. Second, the parameter ε controlling the width of the ε-insensitive tube used to fit the training data. Finally, the σ value required by the kernel. We collect these three user-defined coefficients in the hyperparameter vector ϕ. Continuous or adaptive tuning of ϕ would be particularly important for on-line learning in non-stationary environments, where previously selected parameters may turn out to be sub-optimal in later periods. Some variations of SVR have been proposed in the literature (e.g. in [3]) in order to deal with these difficulties. However, most algorithms proposed for financial forecasting with SVR operate in an off-line fashion and try to tune the hyperparameters using either exhaustive grid searches or other search strategies (for instance, evolutionary algorithms), which are very computationally demanding.

Rather than trying to optimize ϕ, we take an ensemble learning approach: an entire population of p SVR *experts* is continuously evolved, in parallel, with each expert being characterized by its own parameter vector $\phi^{(e)}$, with $e = 1, \ldots, p$. Each expert, based on its own opinion regarding the current fair value of the target asset (i.e. an estimate $z_t^{(e)}$) generates a binary trading signal of form (20.1), which we now denote by $d_t^{(e)}$. A meta-algorithm is then responsible for combining the p trading signals generated by the experts. Thus formulated, the algorithmic trading problem is related to the task of predicting binary sequences from expert advice which has been extensively studied in the machine learning literature and is related to sequential portfolio selection decisions [4]. Our goal is for the trading algorithm to perform nearly as well as the *best* expert in the pool so far: that is, to guarantee that at any time our meta-algorithm does not perform much worse than whichever expert has made the fewest mistakes to date. The implicit assumption is that, out of the many SVR experts, some of them are able to capture temporary market anomalies and therefore make good predictions.

The specific expert combination scheme that we have decided to adopt here is the *Weighted Majority Voting* (WMV) algorithm introduced in [7]. The WMV algorithm maintains a list of non-negative weights $\omega_1, \ldots, \omega_p$, one for each expert, and predicts based on a weighted majority vote of the expert opinions. Initially, all weights are set to one. The meta-algorithm forms its prediction by comparing the total weight

---
[1] C++ code of our implementation is available upon request.

of the experts in the pool that predict 0 (short sell) to the total weight q_1 of the algorithms predicting 1 (buy). These two proportions are computed, respectively, as $q_0 = \sum_{e:d_t^{(e)}=0} \omega_e$ and $q_1 = \sum_{e:d_t^{(e)}=1} \omega_e$. The final trading decision taken by the WMV algorithm is

$$d_t^{(*)} = \begin{cases} 0 & \text{if } q_o > q_1 \\ 1 & \text{otherwise} \end{cases} \qquad (20.9)$$

Each day the meta algorithm is told whether or not its last trade was successfull, and a $0-1$ penalty is applied, as described in Section 20.2. Each time the WMV incurs a loss, the weights of all those experts in the pool that agreed with the master algorithm are each multiplied by a fixed scalar coefficient β selected by the user, with $0 < \beta < 1$. That is, when an expert e makes as mistake, its weight is downgraded to $\beta \omega_e$. For a chosen β, WMW gradually decreases the influence of experts that make a large number of mistakes and gives the experts that make few mistakes high relative weights.

20.4 An Application to the iShare Index Fund

Our empirical analysis is based on historical data of an exchange-traded fund (ETF) . ETFs are relatively new financial instruments that have exploded in popularity over the last few years. ETFs are securities that combine elements of both index funds and stocks: like index funds, they are pools of securities that track specific market indexes at a very low cost; like stocks, they are traded on major stock exchanges and can be bought and sold anytime during normal trading hours. Our target security is the iShare S&P 500 Index Fund, one of the most liquid ETFs. The historical time series data cover a period of about seven years, from 19/05/2000 to 28/06/2007, for a total of 1856 daily observations. This fund tracks very closely the S&P 500 Price Index and therefore generates returns that are highly correlated with the underlying market conditions. Given the nature of our target security, the explanatory data streams are taken to be a subset of all

constituents of the underlying S&P 500 Price Index comprising $n = 455$ stocks, namely all those stocks whose historical data was available over the entire period chosen for our analysis. The results we present here are generated out-of-sample by emulating the behavior of a real-time trading system. At each time point, the system first projects the lastly arrived data points onto a space of reduced dimension. In order to implement this step, we have set $k = 1$ so that only the first eigenvector is extracted. Our choice is backed up by empirical evidence, commonly reported in the financial literature, that the first principal component of a group of securities captures the *market factor* (see, for instance, [2]). Optimal values of $k > 1$ could be inferred from the streaming data in an incremental way, but we do not discuss this direction any further here.

Table 20.1 Statistical and financial indicators summarizing the performance of the 2560 experts over the entire data set. We use the following notation: SR=Sharpe Ratio, WT=Winning Trades, LT=Losing Trades, MG=Mean Gain, ML=Mean Loss, and MDD=Maximum Drawdown. PnL, WT, LT, MG, ML and MDD are reported as percentages.

Summary	Gross SR	Net SR	Gross PnL	Net PnL	Volatility	WT	LT	MG	ML	MDD
Best	1.13	1.10	17.90	17.40	15.90	50.16	45.49	0.77	0.70	0.20
Worst	-0.36	-0.39	-5.77	-6.27	15.90	47.67	47.98	0.72	0.76	0.55
Average	0.54	0.51	8.50	8.00	15.83	48.92	46.21	0.75	0.72	0.34
Std	0.36	0.36	5.70	5.70	0.20	1.05	1.01	0.02	0.02	0.19

With the chosen grid of values for each one of the three key parameters (ε varies between 10^{-1} and 10^{-8}, while both C and σ vary between 0.0001 and 1000), the pool comprises 2560 experts. The performance of these individual experts is summarized in Table 20.1, which also reports on a number of financial indicators (see the caption for details). In particular, the *Sharpe Ratio* provides a measure of risk-adjusted return, and is computed as the ratio between the average return produced by an expert over the entire period, divided by its standard deviation. For instance, the best expert over the entire period achieves a promising 1.13 ratio, while the worst expert yields negative risk-adjusted returns. The *maximum drawdown* represents the total percentage loss experienced by an expert before it starts winning again. From this table, it clearly emerges that choosing the right parameter combination, or expert, is crucial for this application, and relying on a single expert is a risky choice.

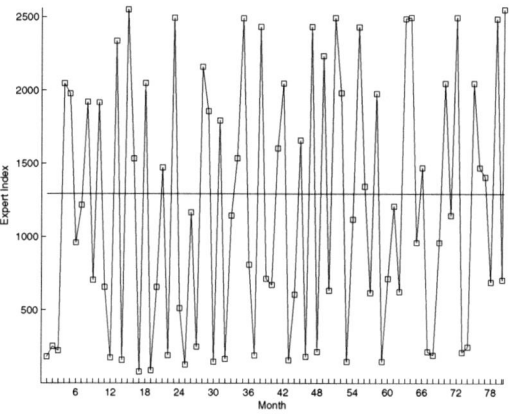

Fig. 20.1 Time-dependency of the best expert: each square represents the expert that produced the highest Sharpe ratio during the last trading month (22 days). The horizontal line indicates the best expert overall. Historical window sizes of different lengths produced very similar patterns.

However, even if an optimal parameter combination could be quickly identified, it would soon become sub-optimal. As anticipated, the best performing expert in the pool dynamically and quite rapidly varies across time. This important aspect can be appreciated by looking at the pattern reported in Figure 20.1, which identifies the best expert over time by considering the Sharpe Ratio generated in the last trading month. From these results, it clearly emerges that the overall performance of the

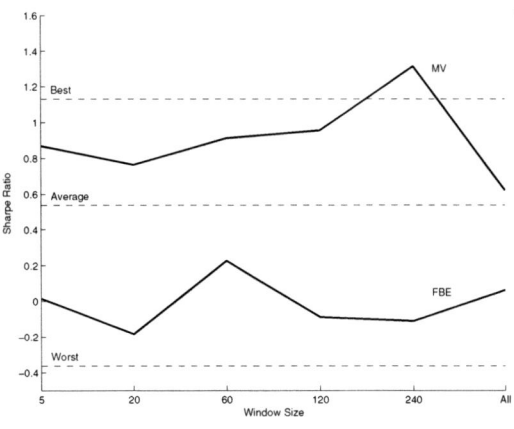

Fig. 20.2 Sharpe Ratio produced by two competing strategies, *Follow the Best Expert* (FBE) and *Majority Voting* (MV), as a function of window size.

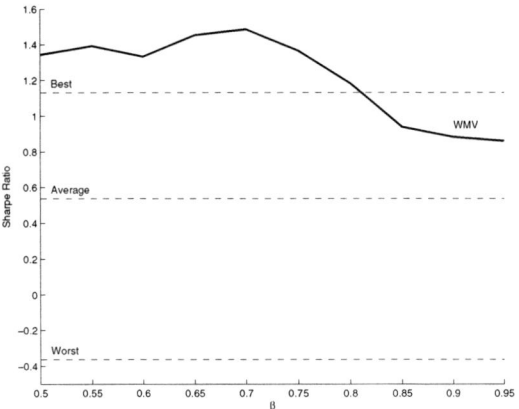

Fig. 20.3 Sharpe Ratio produced by *Weighted Majority Voting* (WMV) as a function of the β parameter. See Table 20.2 for more summary statistics.

Fig. 20.4 Comparison of profit and losses generated by *Buy-and-Hold* (B&H) versus *Weighted Majority Voting* (WMV), after costs (see the text for details).

system may be improved by dynamically selecting or combining experts. For comparison, we also present results produced by two alternative strategies. The first one, which we call *Follow the Best Expert* (FBE), consists in following the trading decision of the best performing expert seen to far, where again the optimality criterion used to elect the best expert is the Sharpe Ratio. That is, on each day, the best expert is the one that generated the highest Share Ratio over the last m trading days, for a given value of m. The second algorithm is *Majority Voting* (MV). Analogously to WMV, this meta algorithm combines the (unweighted) opinion of all the experts in the pool and takes a majority vote. In our implementation, a majority vote is reached if the number of experts deliberating for either one of the trading signals represents a fraction of the total experts at least as large as q, where the optimal q value is learnt by the MV algorithm on each day using the last m trading days. Figure 20.2 reports on the Sharpe Ratio obtained by these two competing strategies, FBW and MV, as a function of the window size m. The overall performance of a simple minded strategy such a FBE falls well below the average expert performance, whereas MV always outperforms the average expert. For some specific values of the window size (around 240 days), MV even improves upon the best model in the pool.

The WMV algorithm only depends upon one parameter, the scalar β. Figure 20.3 shows that WMV always consistently outperforms the average expert regardless of the chosen β value. More surprisingly, for a wide range of β values, this algorithm also outperforms the best performing expert by a large margin (Figure 20.3). Clearly, the WMV strategy is able to strategically combine the expert opinion in a dynamic way. As our ultimate measure of profitability, we compare financial returns generated by WMV with returns generated by a simple *Buy-and-Hold* (B&H) investment strategy. Figure 20.4 compares the profits and losses obtained by our algorithmic trading system with B&H, and illustrates the typical market neutral behavior of the active trading system. Furthermore, we have attempted to include realistic estimates of transaction costs, and to characterize the statistical significance of these results. Only estimated and visible costs are considered here, such as *bid-ask spreads* and fixed commission fees. The bid-ask spread on a security represents the difference between the lowest available quote to sell the security under consideration (the ask or the offer) and the highest available quote to buy the same security (the bid). Historical *tick by tick* data gathered from a number of exchanges using the OpenTick provider have been used to estimate bid-ask spreads in terms of base points or bps^2. In 2005 we observed a mean bps of 2.46, which went down to 1.55 in 2006 and to 0.66 in 2007. On the basis of these findings, all the net results presented in Table 20.2 assume an indicative estimate of 2 bps and a fixed commission fee ($10).

Finally, one may tempted to question whether very high risk-adjusted returns, as those generated by WMV with our data, could have been produced only by chance. In order to address this question and gain an understanding of the statistical significance of our empirical results, we first approximate the Sharpe Ratio distribution (after costs) under the hypothesis of random trading decisions, i.e. when sell and buy signals are generated on each day with equal probabilities, using Monte Carlo

---

[2] A base point is defined as $10000 \frac{(a-b)}{m}$, where a is the ask, b is the bid, and m is their average.

simulation. Based upon 10,000 repetitions, this distribution has mean -0.012 and standard deviation 0.404. With reference to this distribution, we are then able to compute empirical p-values associated to the observed Sharpe Ratios, after costs; see Table 20.2. For instance, we note that a value as high as 1.45 or even higher ($\beta = 0.7$) would have been observed by chance only in 10 out of 10,000 cases. These findings support our belief that the SVR-based algorithmic trading system does capture informative signals and produces statistically meaningful results.

Table 20.2 Statistical and financial indicators summarizing the performance of Weighted Majority Voting (WMV) as function of β. See the caption of Figure 20.1 and Section 20.4 for more details.

β	Gross SR	Net SR	Gross PnL	Net PnL	Volatility	WT	LT	MG	ML	MDD	p-value
0.5	1.34	1.31	21.30	20.80	15.90	53.02	42.63	0.74	0.73	0.24	0.001
0.6	1.33	1.30	21.10	20.60	15.90	52.96	42.69	0.75	0.73	0.27	0.001
0.7	1.49	1.45	23.60	23.00	15.90	52.71	42.94	0.76	0.71	0.17	0.001
0.8	1.18	1.15	18.80	18.30	15.90	51.84	43.81	0.75	0.72	0.17	0.002
0.9	0.88	0.85	14.10	13.50	15.90	50.03	45.61	0.76	0.71	0.25	0.014

References

1. C.C. Aggarwal, J. Han, J. Wang, and Yu P.S. *Data Streams: Models and Algorithms*, chapter On Clustering Massive Data Streams: A Summarization Paradigm, pages 9–38. Springer, 2007.
2. C. Alexander and A. Dimitriu. Sources of over-performance in equity markets: mean reversion, common trends and herding. Technical report, ISMA Center, University of Reading, UK, 2005.
3. L. Cao and F. Tay. Support vector machine with adaptive parameters in financial time series forecasting. *IEEE Transactions on Neural Networks*, 14(6):1506–1518, 2003.
4. N. Cesa-Bianchi and G. Lugosi. *Prediction, learning, and games*. Cambridge University Press, 2006.
5. N. Cristianini and J. Shawe-Taylor. *An Introduction to Support Vector Machines*. Cambridge University Press, 2000.
6. R.J. Elliott, J. van der Hoek, and W.P. Malcolm. Pairs trading. *Quantitative Finance*, pages 271–276, 2005.
7. N. Littlestone and M.K. Warmuth. The weighted majority algorithm. *Information and Computation*, 108:212–226, 1994.
8. J. Ma, J. Theiler, and S. Perkins. Accurate on-line support vector regression. *Neural Computation*, 15:2003, 2003.
9. G. Montana, K. Triantafyllopoulos, and T. Tsagaris. Data stream mining for market-neutral algorithmic trading. In *Proceedings of the ACM Symposium on Applied Computing*, pages 966–970, 2008.
10. G. Montana, K. Triantafyllopoulos, and T. Tsagaris. Flexible least squares for temporal data mining and statistical arbitrage. *Expert Systems with Applications*, doi:10.1016/j.eswa.2008.01.062, 2008.
11. J. G. Nicholas. *Market-Neutral Investing: Long/Short Hedge Fund Strategies*. Bloomberg Professional Library, 2000.

12. S. Papadimitriou, J. Sun, and C. Faloutsos. *Data Streams: Models and Algorithms*, chapter Dimensionality reduction and forecasting on streams, pages 261–278. Springer, 2007.
13. F. Parrella and G. Montana. A note on incremental support vector regression. Technical report, Imperial College London, 2008.
14. A. Pole. *Statistical Arbitrage. Algorithmic Trading Insights and Techniques*. Wiley Finance, 2007.
15. V. Vapnik. *The Nature of Statistical Learning Theory*. Springer, 1995.
16. J. Weng, Y. Zhang, and W. S. Hwang. Candid covariance-free incremental principal component analysis. *IEEE Transactions on Pattern Analysis and Machine Intelligence*, 25(8):1034–1040, 2003.

Reviewer List

- Bradley Malin
- Maurizio Atzori
- HeungKyu Lee
- S. Gauch
- Clifton Phua
- T. Werth
- Andreas Holzinger
- Cetin Gorkem
- Nicolas Pasquier
- Luis Fernando DiHaro
- Sumana Sharma
- Arjun Dasgupta
- Francisco Ficarra
- Douglas Torres
- Ingrid Fischer
- Qing He
- Jaume Baixeries
- Gang Li
- Hui Xiong
- Jun Huan
- David Taniar
- Marcel van Rooyen
- Markus Zanker
- Ashrafi Mafruzzaman
- Guozhu Dong
- Kazuhiro Seki
- Yun Xiong
- Paul Kennedy
- Ling Qiu
- K. Selvakuberan
- Jimmy Huang
- Ira Assent
- Flora Tsai
- Robert Farrell
- Michael Hahsler
- Elias Roma Neto
- Yen-Ting Kuo
- Daniel Tao
- Nan Jiang
- Themis Palpanas
- Yuefeng Li
- Xiaohui Yu
- Vania Bogorny
- Annalisa Appice
- Huifang Ma
- Jaakko Hollmen
- Kurt Hornik
- Qingfeng Chen
- Diego Reforgiato
- Lipo Wang
- Duygu Ucar
- Minjie Zhang
- Vanhoof Koen
- Jiuyong Li
- Maja Hadzic
- Ruggero G. Pensa
- Katti Faceli
- Nitin Jindal
- Jian Pei
- Chao Luo
- Bo Liu
- Xingquan Zhu
- Dino Pedreschi
- Balaji Padmanabhan

Index

D^3M, 5
F_1 Measure, 74
N-same-dimensions, 117
SquareTiles software, 259, 260

Accuracy, 244
Action-Relation Modelling System, 25
actionability of a pattern, 7
actionable knowledge, 12
actionable knowledge discovery, 4, 6, 7
actionable pattern set, 6
actionable patterns, 6
actionable plan, 12
actionable results, 54
acute lymphoblastic leukaemia, 164
adaptivity, 100
adversary, 97
airport, 267, 268, 278, 280
AKD, 4
AKD-based problem-solving system, 4
algorithm MaPle, 37
algorithm MaPle+, 44
algorithmic asset management, 283
algorithms, 226, 246
analysis of variance, 92
anomaly detection algorithms, 102
anonymity, 107
anti–monotone principle, 212
application domain, 226
apriori, 274, 275, 277
  bottom-up, 274
  top-down, 274–276
ARSA model, 191
association mining, 84
association rule, 85
association rules, 83, 89, 106
AT model, 174

Author-Topic model, 174
automatic planning, 29
autoregressive model, 184

benchmarking analysis, 262, 265
biclustering, 112
bioinformatics, 114
biological sequences, 113
blog, 169
blog data mining, 170
blogosphere, 170
blogs, 183
business decision-making, 4
business intelligence, 4
business interestingness, 7
business objective, 54
Business success criteria, 55

C4.5, 246
CBERS, 247, 250
CBERS-2, 247
CCD Cameras, 247
CCM, 200
cDNA microarray, 160
chi-square test, 84
classification, 241, 267–274, 279, 280
  subspace, 268, 272, 273, 276, 277, 279
clustering, 92, 93, 112, 271, 272
  subspace, 268, 271, 273, 279
code compaction, 209, 211
combined association rules, 89, 90
Completeness, 244
concept drifting, 284
concept space, 202
conceptual semantic space, 199
confidentiality, 102
constrained optimization, 288

context-aware data mining, 227
context-aware trajectory data mining, 238
contextual aliases table, 149
contextual attributes vector, 149
cryptography, 107
CSIM, 201
customer attrition, 20
cyber attacks, 169

Data analysis, 248
data flow graph, 210
data groups, 245
data intelligence, 5
data mining, 3, 114, 128, 228, 241, 243, 245, 249, 250
data mining algorithms, 128
data mining application, 3
data mining for planning, 20
data mining framework, 227
data mining objective, 56
data quality, 243, 244, 248
data semantics, 238
data streams, 284
data-centered pattern mining framework, 5
data-mining generated state space, 14
DBSCAN, 92, 93
decision tree, 83, 84, 279
decision trees, 246
Demographic Census, 243
demography, 245
derived attributes, 58
digital image classification, 242
Digital image processing, 241
digital image processing, 241–243, 248
dimensionality reduction, 174
disease causing factors, 128
Domain Driven Data Mining, 3
domain driven data mining, 5
domain intelligence, 5
domain knowledge, 112
domain-centered actionable knowledge discovery, 53
domain-drive data mining, 53
Domain-driven, 117
domain-driven data mining, 232

education, 245
embedding, 212
ensamble learning, 284
entropy, 272, 277
 attribute, 272, 274, 276, 277, 279
 class, 272–275, 277
 combined, 277
 conditional, 272
 maximum, 273, 279
entropy detection, 105
Event template, 205
exchange-traded fund, 290
experts, 291
extracts actions from decision trees, 12

feature selection, 188
feature selection for microarray data, 161
flight, 267–269, 271–274, 277–280
 delay, 267–273, 278
fragment, 212
frequency, 212
frequent pattern analysis, 128
frequent subtree mining, 130

garbage collecting, 245, 249
gene expression, 112
gene feature ranking, 161
genomic, 111
geodesic, 174
geometric patterns, 226
GHUNT, 199
Gibbs sampling, 174

hidden pattern mining process, 4
high dimensionality, 113
High School Relationship Management (HSRM), 260, 263
high utility plans, 20
HITS, 204
household, 245
householder, 245
HowNet, 202
human actor, 53
human intelligence, 5, 53
human participation, 53
hypothesis test, 94

IMB3-Miner, 128
IMB3-Miner algorithm, 128
impact-targeted activity patterns, 86
incremental learning, 289
Information visualization, 255
intelligence analysis, 171
intelligence metasynthesis, 5
intelligent event organization and retrieval system, 204
interviewing profiles, 71
Isomap, 174

J48, 246

k-means, 92, 93

Index   301

kindOf, 66
KL distance, 174
knowledge discovery, 3, 226
knowledge hiding, 108
Kullback Leibler distance, 174

land usages categories, 248
Land use, 241
land use, 242
land use mapping, 241
Latent Dirichlet Allocation, 173
Latent Semantic Analysis, 173
LDA, 173
learn relational action models, 12
Library of Congress Subject Headings (LCSH), 66
link detection, 104
local instance repository, 67
LSA, 173

Machine Learning, 246, 249
manifold, 174
market neutral strategies, 283
Master Aliases Table, 144
mathematical model, 257
maximal pattern-based clustering, 35
maximal pCluster, 35
MDS, 174
mental health, 127
mental health domain, 128
mental health information, 128
mental illness, 128
microarray, 159
microarray data quality issues, 160
mining δ-pClusters, 34
mining DAGs, 211
mining graphs, 211
monotonicity, 274, 277
   downward, 274
   upward, 274
Monte Carlo, 294
MPlan algorithm, 18
Multi-document summarization, 206
multi-hierarchy text classification, 199
multidimensional, 244
Multidimensional Scaling, 174

naive bayes, 270, 279
nearest neighbor, 270, 279
network intelligence, 5
non-interviewing profiles, 71
non-obvious data, 103

Omniscope, 255

ontology, 66
ontology mining, 64
ontology mining model, 65
opinion mining, 185

PageRank, 203
pairs trading, 284
Pareto Charts, 255
partOf, 66
pattern-based cluster, 32, 34
pattern-based clustering, 34
pattern-centered data mining, 53
personalized ontology, 68
personalized search, 64
Plan Mining, 12
PLSA, 173, 189
post-processing data mining models, 12
postprocess association rules, 12
postprocess data mining models, 25
prediction, 184
preface, v
privacy, 102
Probabilistic Latent Semantic Analysis, 173
probabilistic model, 173
procedural abstraction, 210
pruning, 269, 274–277, 279
pScore, 34
pseudo-relevance feedback profiles, 71

quality data, 100

randomisation, 106
RCV1, 74
regression, 288
Relational Action Models, 25
relevance indices, 150
relevance indices table, 149
reliable data, 102
Remote Sensing, 242
Remote sensing, 241, 242
remote sensing, 243
resilience, 99

S-PLSA model, 184
Sabancı University, 254, 260
sales prediction, 185
satellite images, 242
scheduling, 267–269, 272, 278–280
secure multi-party computation, 106
security blogs, 170
security data mining, 97
security threats, 170
semantic focus, 70
semantic patterns, 226

semantic relationships, 65, 68
Semantic TCM Visualizer, 144
semantic trajectories, 229
semantics, 226
semi-structured data, 130
sentiment mining, 185
sentiments, 183
sequence pattern, 112
sequential pattern mining, 85
sequential patterns, 89, 235
Sharpe Rario, 291
similarity, 112
smart business, 4
social intelligence, 5
spatial data mining, 241
spatio-temporal clustering, 231
spatio-temporal data, 225
specificity, 70
square tiles visualization, 254, 255
stable data, 103
statistical arbitrage, 284
subject, 66
subject ontology, 68
subspace, 113
   interesting, 272, 274, 277
support, 212

tamper-resistance, 97
taxonomy of dirty data, 260
TCM Ontology Engineering System, 152
technical interestingness, 7
Term frequency, 145
Text mining, 144
the Yeast microarray data set, 47
time learner, 206

Timeliness, 244
trading strategy, 285
training set, 64
trajectories, 230
trajectory data mining, 238
trajectory patterns, 226
transcriptional regulatory, 112
TREC, 71, 72
tree mining, 128
tree mining algorithms, 130
tree structured data, 128
trustworthiness, 163
Turkey, 253

unforgeable data, 103
University Entrance Exam, 253
user background knowledge, 64
user information need, 65, 69
user profiles, 71
user rating, 187

visual data mining, 255
visualization, 174
volume detection, 104

water supply, 245, 249
Web content mining, 199
Web event mining, 204
Web Information Gathering System, 73
Web structure mining, 203
weighted majority voting, 289
weighted MAX-SAT solver, 26
Weka software, 246
world knowledge, 64

Printed in the United States
126361LV00002BA/22-27/P